Recent Trends in Microbiology, Mycology and Plant Pathology

The Editor

Prof. H.C. Lakshman, born in Huskur, small village of Anekal taluk, Bangalore district in Karnataka on 20th December, 1954. Dr. Lakshman obtained his B.Sc. in 1975 and M.Sc., in 1977, from Bangalore University, Bangalore, Karnataka, Ph.D. in Botany, during 1996, from Karnatak University, Dharwad, Karnataka, India. Presently he is working as a senior Professor at P.G. Department of Studies in Botany, Karnatak University, Dharwad, Karnataka, 2004 to date.

He served as a Lecturer in Botany, K.L.E's S. Nijalingappa Science College, Bangalore 1978-1983; Senior Grade Lecturer in Botany, K.L.E's S.K. Arts and H.S.K. Institute of Science, Vidyanagar, Hubli, Karnataka, 1983-1996 and as a Reader in Botany, 1996-2004 and Professor in Botany, 2004 to till date at P.G. Department of Studies in Botany, Karnatak University, Dharwad, Karnataka and acted as a Chairman, P.G. Department of Studies in Botany, K.U. Dharwad, 2009-2011. Dr. Lakshman successfully guided 28 students for Ph.D. and M.Phil. He has to his credit 15 Books, 325 research articles published in various national and international scientific research journals. He received many prestigious awards to quote few Dr. C.V. Raman Literature Award-2006, Eminent Scientist Award of the year 2011, NESA, New Delhi, Excellence in Research Award, Education ExpoTv, Noida. He is the Editorial Board member for Journal of Theoretical and Experimental Biology, Bulletin of Basic and Applied Plant Biology, Plant Sciences Feed, Bioscience Discovery, The Scientific Temper, International Journal of Biotechnology and Bioscience, International Journal of Environmental Sciences etc. He is the Fellow of Society of Environmental Sciences (F.S.E.Sc.) and Fellow of International Society of Ecological Communication (F.I.S.E.C.). Major research interest areas are Soil Microbiology, Mycology, Plant Pathology, Environmental Biology, Agricultural Sciences, Biofertilizers.

Recent Trends in Microbiology, Mycology and Plant Pathology

Editor
Dr. H.C. Lakshman
Professor and Formerly Chairman
P.G. Department of Studies in Botany
Karnatak University, Pavate Nagar, Dharwad – 580 003
Karnataka, India

2015

Daya Publishing House®
A Division of
Astral International Pvt. Ltd.
New Delhi – 110 002

Publisher's note:

Every possible effort has been made to ensure that the information contained in this book is accurate at the time of going to press, and the publisher and author cannot accept responsibility for any errors or omissions, however caused. No responsibility for loss or damage occasioned to any person acting, or refraining from action, as a result of the material in this publication can be accepted by the editor, the publisher or the author. The Publisher is not associated with any product or vendor mentioned in the book. The contents of this work are intended to further general scientific research, understanding and discussion only. Readers should consult with a specialist where appropriate.

Every effort has been made to trace the owners of copyright material used in this book, if any. The author and the publisher will be grateful for any omission brought to their notice for acknowledgement in the future editions of the book.

All Rights reserved under International Copyright Conventions. No part of this publication may be reproduced, stored in a retrieval system, or transmitted in any form or by any means, electronic, mechanical, photocopying, recording or otherwise without the prior written consent of the publisher and the copyright owner.

Cataloging in Publication Data—DK

Courtesy: D.K. Agencies (P) Ltd. <docinfo@dkagencies.com>

Recent trends in microbiology, mycology and plant pathology /
editor, Dr. H.C. Lakshman.

   pages   cm

Includes bibliographical references and index.
ISBN 9789351306542 (International Edition)

  1. Microbiology. 2. Mycology. 3. Plant diseases.   I. Lakshman, H. C., 1954-, editor.

DDC 579   23

Published by	:	**Daya Publishing House®** A Division of **Astral International Pvt. Ltd.** – ISO 9001:2008 Certified Company – 4760-61/23, Ansari Road, Darya Ganj New Delhi-110 002 Ph. 011-43549197, 23278134 E-mail: info@astralint.com Website: www.astralint.com
Laser Typesetting	:	**Classic Computer Services**, Delhi - 110 035
Printed at	:	**Thomson Press India Limited**

PRINTED IN INDIA

KUVEMPU **UNIVERSITY**

☎ (O) : (08282) 256254
Extn. : 325
(R) : (08182) 248059
Fax : (08282) 256 255, 256 262
Mobile : 94489 43864
E-mail : krishnappam4281@yahoo.com

DR. M. KRISHNAPPA, M. Sc., M. Phil., Ph. D.
PROFESSOR
ExRegistrar, Kuvempu University
ExDean-Faculty of Biological Sciences
JNANA SAHYADRI
SHANAKARAGHATTA

Dept. of Post Graduate Studies & Research in
APPLIED BOTANY
Jnana Sahyadri, Shankaraghatta-577451
Shimoga Dist., Karnataka, INDIA

Foreword

Micro-organisms are useful in human welfare and also the plant microbial interactions. Utilization of these micro-organisms in all the fields beneficially according to human welfare. These micro-organisms causes harmful effect to the crop plants and causes yield loss. Some group of the micro-organisms boost the development of the plants in the symbiotic association as endophytic fungi. Some of the endophytic fungi enhances the medicinal value of plants.

It is pleasure to write few words about the book on "Recent Trends in Microbiology, Mycology and Plant Pathology". Many new informations are available regarding the microbiology and plant pathology. This book provides all information in the chapters. The book is divided into three divisions, *i.e.,* microbiology, mycology and plant pathology.

In the microbiology division, all the chapters dealt about the techniques, detection, procedures, extraction of secondary metabolites and applications of different microbes in the agriculture and other industrial usage. In all the chapters the authors have presented their original findings. The said procedures techniques and other information would help the other researchers, students for their academic and for their research work.

In the mycology and plant pathology division all the chapters dealt on the application of fungi and management of the plant diseases. The disease control by using fungicides and by integrated pest management. This method is more appropriate and ecofriendly. It deals with diseases of viruses and mycoplasma and non infections diseases of crop plants and their management.

In this division the authors have studied the viral and mycoplasmal diseases their transmission and their management. There is information on the non-infectious diseases of crop plants, induction of disease resistance against viral disease.

The book is very useful to the graduate, post graduate, research scholars and to the staff members. It provides lot of information on disease management, defence mechanism, industrial applications and bioprospecting of microbes.

I congratulate the author Prof. H.C. Lakshman for compiling all the chapters and brought in to a book format. I wish this book should useful to the academic fraternity. I am thankful to the Prof. H.C. Lakshman for providing this opportunity to write comments on this book.

Place: Shankaraghatta

Prof. M. Krishnappa

Preface

Microbiology, Mycology and Plant Pathology is in essence the deciphering and use of biological knowledge. It has multidisciplinary access, since; it has the foundation in many disciplines including bio-geology, ecology, chemistry, biochemistry, and genetics. It may be viewed as a series of enabling techniques to technologies that involve the practical application of micro-organisms. Today, the micro-organisms are the basic tools of genetic engineering and biotechnology. This has been partly due to the use of micro-organisms in the production of organic chemicals, antibiotics, wine, beer, cheeses, bio-fertilizers, pharmaceuticals and supplements in boosting production and yield in agriculture, horticulture and forestry. They also control insect, pests causing diseases in crop plants.

Undoubtedly modern biological knowledge can only maximise its full potential to benefit mankind through achieving basic research findings and awareness in modern society. Participating scientists must learn to communicate in scientific records, books, and journals. By doing so, they will generate a greater level of confidence and trust between the scientific community and the society at large. The present book brought a significant research and review articles contributed by researchers and scientists especially on Microbiology, Mycology and Plant pathology. The microbiology consists of eleven chapters, each chapter dealing with different facets of microbial applications. Second part of this book dealing mainly on fungal metabolites, production of antioxidants, pectinase, pharmaceuticals.

The success in chemical control of plant diseases causing pathogens accompanied by environmental pollution and this leads to ecological imbalance with application of various pesticides resulted in their accumulation in soil, water and plants. Therefore, the scientific endeavour is directed towards the elaboration and wide application of biological methods and thus, the third part of this book includes nine chapters dealing with different diseases caused by fungi, bacteria, viruses and mycoplasma on leafy vegetables, horticultural crops. Similarly, each chapter gives the disease management techniques, biological control measures, and eco-friendly approaches. This book is aimed to give an integrated view of its complex in the subject for young students, researchers, scientists and teaching community as whole.

I am deeply indebted to Dr. M. Krishnappa, Ex-Registrar and Professor, Department of Studies and Research in Applied Botany, Kuvempu University, Shankarghatta, Shimoga, Karnataka, India, for encouraging me and writing foreword for this book. I am also thankful to Mr. Channabasava A., Research Scholar, P.G. Department of Studies in Botany, Karnatak University, Pavate Nagar, Dharwad for his help for completing this task. I extend my sincere thanks to Astral Publishing Company, New Delhi for timely printing and publishing this book.

Dr. H.C. Lakshman

Contents

Foreword   *v*

Preface   *vii*

Section I: MICROBIOLOGY

1. **Beneficial Microbial Life in Soil: The Basis for Sustainable
   Plant Life on Earth**   **3**
   H.C.Lakshman

2. **A Protocol for DNA Extraction from Rhizosphere of Field
   Growing Plant, Harboring Arbuscular Mycorrhizal Fungi**   **27**
   *Ashok Shukla, Deepak Vyas, Keerti Dehariya, Vandana Bharti Ahirwar,
   Onkar Salunkhe, Daulat Ram Gwalwanshi and Anuradha Jha*

3. **Mycorrhizae and Molecular Approaches to its Functioning**   **39**
   M.N. Sreenivasa

4. **Recent Techniques in Arbuscular Mycorrhizal
   Fungal Mass Production**   **47**
   P. Savitha, K.S. Jagadeesh, D.K. Dushyantha and P. Nirmalnath Jones

5. **Bioremediation: Basic Concepts, Approaches and Applications**   **61**
   M. Nagalakshmi Devamma, Kavya Deepthi M. and Shaik Thahir Basha

6. **Improved Seedling Emergence and Biomass Production of
   Crop Plants by *Trichoderma* spp.**   **83**
   J.N. Rajkonda and U.N. Bhale

7. Interactions between the Arbuscular Mycorrhizae,
    Glomus mosseae and *Trichoderma* and their Significance
    for Enhancing Plant Growth and Suppressing Wilt Disease
    of *Cajanus cajan* (L).   95
    Keerti Dehariya, Imtiyaz Ahmed Sheikh, Ashok shukla and Deepak Vyas

8. Role of Cyanobacteria in Phosphate Nutrition to Plants   105
    Ratna V. Airsang and H.C. Lakshman

9. Role of Micro-organism in Production of Single Cell Protein   115
    Kiran P. Kolkar and H.C. Lakshman

10. Micro-organisms as Pollution Indicators in Selected Lentic
    Habitats of Dharwad, Karnataka State, India   135
    Doris M. Singh

11. Short-Term Water Stress Induced Accumulation of Proline
    in *Triticum aestivum* L. Varieties Inoculated with AM
    Fungus *Rhizophagus fasciculatus*   143
    V.S. Bheemareddy, S.B. Gadi and H.C. Lakshman

Section II: MYCOLOGY

12. Antioxidants from Mushrooms may Antagonize Convulsions   153
    *Imtiyaz Ahmad Sheikh, Keerti Dehariya, Vinita Singh, Poonam Dehariya
    and Deepak Vyas*

13. Fungal Metabolites and their Importance   177
    A. Channabasava, H.C. Lakshman and T.C. Taranath

14. Fungal Pectinases and their Biotechnological Applications   197
    D.K. Dushyantha and K.S. Jagadeesh

15. Role of Fungi in the Contribution of Pharmaceuticals
    and Enzymes   213
    Jayshree M. Kurandawad and H.C. Lakshman

Section III: PLANT PATHOLOGY

16. Bioagents for the Management of Plant Diseases   245
    Shripad Kulkarni and S.P. Singh

17. Biological Control of Plant Pathogens   265
    *M. Nagalakshmi Devamma, Shaik Thahir Basha, M. Kavya Deepthi
    and H.C. Lakshman*

18. Eco-friendly Approaches in the Management of Post-harvest
    Fruit Diseases of Ivy Gourd (*Coccinia inidica* Wight and Arn.)   283
    V.S. Chatage and U.N. Bhale

19. Diseases on some Important Common Leafy Vegetables and
    their Control Management   297
    Shwetha C. Madgaonkar and H.C. Lakshman

20. Plant Diseases Caused by Viruses and Mycoplasma and
    their Control Measures   313
    Pushpa K. Kavatagi and H.C. Lakshman

21. Induction of Defense Responses in Tomato against Tolcv
    by Consortium of Plant Growth Promoting Rhizobacteria
    Formulated with Chitosan   325
    *Shefali Mishra, K.S. Jagadeesh, P.U. Krishnaraj, G. Jyothi, A.S. Byadagi
    and A.S. Vastrad*

22. Management of Diseases of Horticultural Crops
    through Organics   339
    Shripad Kulkarni and V.I. Benagi

23. Interaction between Host and Pathogens in Development
    of Diseases   349
    B.S. Agadi and H.C. Lakshman

24. Influence of Culture Media and Environmental Factors on
    Mycelial Growth and Conidial Yield of *Fusarium proliferatum*
    a Potential Pathogen of *Echinochloa crusgalli*
    (Major Weed in Rice)   359
    G. Jyothi, K.R.N. Reddy, K.R.K. Reddy, Shefali Mishra and A.R. Podile

25. Importance of Plant Disease Forecasting and
    Disease Management   373
    Romana M. Mirdhe and H.C. Lakshman

    Index   389

SECTION I
MICROBIOLOGY

2015, Recent Trends in Microbiology, Mycology and Plant Pathology   *Pages 3–25*
Editor: **Dr. H.C. Lakshman**
Published by: **DAYA PUBLISHING HOUSE, NEW DELHI**

Chapter 1

Beneficial Microbial Life in Soil: The Basis for Sustainable Plant Life on Earth

H.C.Lakshman

*Microbiology Laboratory, P.G. Department of Studies in Botany,
Karnatak University, Pavate Nagar, Dharwad – 580 003, Karnataka, India
E-mail: dr.hclakshman@gmail.com*

Introduction

Soil is the nature gift to mother earth. The soil mainly consists of sand, silt and clay particles with gaseous and mineral elements such as oxygen (O_2), silicon (Si), aluminum (Al), potassium (K), calcium (Ca), magnesium (Mg) etc and soil solution contains dissolved materials. Air in the soil contact with air above ground aerates the roots with oxygen and help to remove excess of carbon dioxide (CO_2) from respiring roots. Organic matter, the soul of soil is a major consideration in organic farming. Its presence or absence makes the soil living or dead. The India National Programme for Organic Production (NPOP) recognizes that "The fertility of soil is to be maintained and increased with the biological activity of the soil held intact". From the point of view of farming, soil can be considered having four important parts; solid minerals, water, air and organic matter (Foster, 1988). Soil contains the required mineral elements and nutrients for the growth and multiplication of several lower forms of plants and animals. These biological species could be broadly divided into microflora and microfauna. The number and kinds of micro-organisms present in soil depend on many environmental factors, such as pH, temperature, moisture, aeration and nutrients available. Soil is medium for growth, reproduction, respiration, nutrition and even decomposition after death for most micro-flora and fauna. They complete their life cycles and later provide the bio-mass for decomposition in the soils. Similarly, soil is also medium for physical anchorage, growth and proliferation of roots of all plants and thereby facilitates absorption of water and nutrients. The roots also provide large biomass for decomposition after completion of life cycle of plants. The whole environment in the soil provides production and recycling of biomass from lives of plant roots, animals and micro-organisms by closer integration of their life cycles.

Soil micro-flora plays a pivotal role in evaluation of soil conditions and in stimulating plant growth (Singh *et al.,* 2009). Micro-organisms are beneficial in increasing the soil fertility and plant growth as they are involved in several biochemical transformation and mineralization activities in soil. Type of cultivation and crop management practices found to have greater influence on the activity of soil micro-flora (Godfrey *et al.,* 2005). Continuous use of chemical fertilizers over a long period may cause imbalance in soil micro-flora and thereby indirectly affect biological properties of soil leading to soil degradation (Hawksworth and Colwell, 1992). The activity of soil microflora is comparatively more in surface than in subsurface horizons and decrease with depth due to decrease in organic matter (Roy, 2007). Soil micro-flora also plays fundamental roles in many ecosystem processes including decomposition and nutrient cycling and affects many important soil hydrological and chemical properties (Lynch, 1987a). Hence change in the soil microbial community may lead to changes in the structure and function of the overall ecosystem, and ultimately determine ecosystem sustainability (Richards, 1987). Soil management in Organic Farming centers on management of organic matter to sustain soil organic carbon, the key to production and productivity. In this process the basic and important methods is to increase the microbial activity in the soil by carefully managing and improving the quantity and quality of humus in it.

The soil organisms vary in number from a few per hectare to many millions per gram of soil. The density of population is determined by food supply, moisture, temperature, physical condition and the reaction of the soil. In neutral soils, bacteria dominate over types of microscopic life on the other hand; fungi predominate in acidic and organic matter such soil. Algae abound on the soil in constantly moist or shady situation. Under favourable conditions, the bacteria multiply enormously. In sandy desert soils and under water logged conditions they are very scarce.

Table 1.1: Distribution of Micro-organisms in Soil

Soils	Number/g of Soil in 15 cm Depth (x10,00,000)		
	Bacteria	Actinomycetes	Fungi
Deep black soil	1.29	0.10	0.03
Medium black soil	14.94	0.29	0.009
Alluvial soil	20.11	0.26	0.03
Latertic soil	1.06	0.47	0.12
Red sandy soil	1.34	0.62	0.008

Micro-organisms Associated with Rhizopshere and their Importance

The rhizosphere is considered to be that zone of the soil environment influenced by plant roots and it represents a highly complex ecosystem, which is influenced by a number of biotic and abiotic factors. It is widely believed that this zone will be variable in extent and be directly influenced by root physiology and by soil environment factors. Available evidence suggests that plants and rhizosphere organisms function in an interdependent fashion. Rhizosphere studies have shown that bacterial numbers are greatly increased in rhizosphere soils. Dilution plate count indicates that the ratio of bacterial numbers in rhizosphere compared to non-rhizosphere soil varies from 10:1 to 50:1 (Richards, 1987; Lynch, 1990). Comparison of rhizosphere and non-rhizosphere soil indicates that significant qualitative shift occurs in bacterial and fungal species detected (Farrar, 2003). Rhizosphere species composition is influenced by numerous factors, including plant species and genotype, plant nutrient status, presence and type of mycorrhiza, soil type, soil moisture, light supply and other factors. The continued maintenance of a normal rhizosphere is mediated by the release of a wide variety of organic carbon compounds. Available data, obtained from crops and tree seedlings suggest that 40-50 per cent of the net carbon fixed may be exuded or rapidly released to the rhizosphere (Rovira, 1965). More complex carbon compounds may enter the rhizosphere more slowly resulting from root aging. The rhizosphere contains diverse array of metabolic substrates such as exudates, secretions, plant mucilage, mucigel and lysates (Whipps, 2001). The diversity and complexity of released compounds is likely to be an important factor contributing to the high species diversity of rhizosphere micro-organisms.

The bacterial/cyanobacterial communities of the rhizosphere have strong influence on the growth and health of the plant as well as on their ability to adapt to changed environmental conditions. The enhanced degradation of pesticides in soil sections close to the root surface is related to the rhizosphere-induced co-metabolism of pesticides. The actively growing plant roots provide an excellent environment for intensive microbial activity, resulting in enhanced biodegradation of organic contaminants. Selective enrichment of micro-organisms is likely to have a significant impact on the rhizoremediation, rhizoextraction or rhizofiltration of recalcitrant organic contaminants in soils. Attempts would be given to provide nutrient and plant growth – promoting substances through micro-organisms for sustainable crop production.

Rhizosphere controls the transformation of nutrient ions and contaminants through changes in pH, redox potential, microbial population and mycorrhizal association. Changes in pH are brought about by the excretion of protons (H^+), hydroxyl (OH^-) or bicarbonate (HCO_3^-) ions due to cation/anion imbalance in the plant, the evolution of CO_2 by respiration, and the excretion of low-molecular-weight organic acids. Plants taking excess cation over anion (cation charge surplus) tend to balance the charge by releasing H^+, resulting in acidification of rhizosphere. Conversely, plants taking excess anion over cation (anion charge surplus) tend to balance the charge by releasing OH^- or HCO_3^- ions, resulting in alkalization. This paper deals the most beneficial micro-organisms commonly found in soil, they are: bacterial, actonomycetes, fungi, yeasts, cyanobacteria, algae, protozoa, nematodes and viruses.

Bacteria

Bacteria are the dominant group of micro-organisms in the soil. Because of their small size, the total biomass of bacteria is frequently less than that of fungi. Their population ranges from 100,000 to several hundred million per gram of soil, depending upon the physical, chemical and biological conditions of the soil. Bacteria live in soil as cocci, bacilli and spirilli. The rod shaped bacteria are common, whereas spirilli are rare.

The bacterial community is dominated mainly by 25 genera. The majority of the species belong to the genera *Arthrobacter, Pseudomonas, Agrobacterium, Flavobacterium* and *Bacillus*. Some small number of pathogens affecting humans, livestock and other animals, cultivated plants, and wild species of higher plants are also present. Among animal pathogens are *Clostridium, Bacillus, Listeria, Coxiella* and *Streptocococcus*. The pathogens of higher plants are of particular significance. These include species of *Agrobacterium, Corynebacterium, Erwinia, Pseudomonas* and *Xanthomonas*. It is generally assumed that bacteria require no growth factor in soil but there are many exceptions. A large number of local communities are unable to grow in the absence of one or several of B-vitamins or amino acids. These growth factors are not found in soil in the absence of biological activity.

Role of Bacteria in Biological Nitrogen Fixation

Prokaryotes are the only organisms that are able to utilize the nitrogen in the atmosphere and fix it in the utilizable form. The nitrogen fixers may be free-living organisms or symbionts with higher plants. They can carry out reactions that convert triple bonded nitrogen into organic compounds assimilable by plants. The symbiotic nitrogen fixation of leguminous plants can produce a nitrogen yield of 100-300 kg N/hectare/annum. The free-living organisms are estimated to contribute about 1-3 kg/N/hectare/annum. In addition, significant amounts of bound nitrogen may reach the soil via precipitation from the atmosphere. According to the degree of atmospheric pollution, this can amount to 3-30 kg N/ha/annum.

Inoculations of legumes with root colonizing bacteria and *Rhizobium* has been demonstrated to affect symbiotic nitrogen fixation by enhancing the colonization and nodulation of the legume roots by rhizobia through multitudinous desirable characteristics. Attributes of *Rhizobium*, host and associated rhizosphere microflora influence the outcome of the competition for nodulation sites (Tilak and Saxena, 1994). Besides indigenous rhizobia, other rhizospheric micro-organisms and AM fungi contribute to the competitive success of an inoculants strain (Saxena *et al.,* 1997). Information on the interaction of *Bradyrhizobium* sp. (vigna) and rhizospheric bacteria on mungbean has been reported (Gupta *et al.,* 1998 a,b,c).

Non-Symbiotic Nitrogen Fixation

In 1894-95 Winogradsky discovered the first bacteria which possess the property of fixing atmospheric nitrogen without symbiotic association. He named this bacterium as *Clostridium pasteurianum* and this is an anaerobic species. Aerotic non-symbiotic nitrogen fixing bacteria (Azotobacter and Beijerinckia) were discovered in the year 1901 by Beijerinck. Other than these aerobic and anaerobic bacteria the members of the blue green algae and many photosynthetic bacteria like *Rhodospirillum* and *Desulforibrio*, also fix nitrogen.

Azotobacter is the most thoroughly studied bacterium among the free-living nitrogen fixers. The cells are large, gram-negative, pleomorphic, ovoid rods, measuring 1.2 to 2.0 um in diameter. It has a high respiration rate, which may help in retaining low O_2 levels in the cell. As mentioned earlier, nitrogenase is not functional in the presence of high levels of oxygen. Members of the genus *Azotobacter* are able to form a type of resting cell called *cyst*. Each needed as energy source. These energy sources are probably derived from the decomposition of cellulose and starch by other micro-organisms of the soil. Carbohydrates added to the soil in the form of molasses and starch wastes stimulate *Azotobacter* and other non-symbiotic nitrogen fixing organisms.

Beijerinckia is another organism which fixes nitrogen non-symbiotically. It is found in acidic soils, particularly of tropical regions. *Beijerinckia* spp. form straight or curved rods with rounded ends. The cells are often with polar lipid bodies surrounded by membranes. The lipid bodies are composed of poly-B hydroxybutyric acid. Nitrogen fixing colonies form copious quantities of elastic slime. Fixation is enhanced by reduced O_2 tension. Cyst containing one cell may occur in Beijerinckia, however during growth on nitrogen – free medium, several individual cells will be enclosed in a common capsule, *Beijerinckia* spp. Use glucose, fructose sucrose as the sole source of carbon, but grow poorly on glutamate.

Associative Nitrogen Fixation

In associative nitrogen fixation the genus *Azospirillum* is of considerable importance. It is associated with the roots of grasses and is capable of fixing atmospheric nitrogen. Crops grown with pretreated seeds give increased yields up to about 10 to 30 percent. *Azospirillum* participates in all steps of the N_2-cycle except nitrification. It can fix atmospheric N_2 in pure culture and under microaerophilic conditions. A cluster of nif-genes has been identified in *Azospirillum*, which is considered to be homologous to those of *Klebsiella pneumonia*. *Azospirillum* spp. Have been isolated from the rhizosphere of a large number of monocotyledons and a few dicotyledonous plants. In some cases there has been root hair deformation due to the association with the bacterium. The bacterium could invade the cortical and vascular tissues of the host, and lead to enhancement of the number of lateral roots and of root hairs, which helps in increasing the mineral uptake by the plant. Though there is no evidence of host specificity in the *Azospirillum* spp. Studied, some affinity in the strains for C_4 as different from C_3 plants has been suggested. Plasmids isolated from *Azospirillum* spp. Seem to be associated with nitrogen fixation, but more studies are required to bring out genetics of nitrogen fixation in the group. In the flooded rice fields where micro aerobic conditions prevail, some bacteria associated with the rice rhizosphere have been found to help nitrogen fixation. The amount of nitrogen thus fixed has been estimated by the acetylene reduction test at about 42 to 80 kg/hectare per crop season.

The main bacterial genera encountered are *Beijerinckia, Azotobacter, Pseudomonas* and *Flavobacterium.* The tropical soils have much higher *Azotobacter* population than those found under temperate climates. In soil, their numbers vary from a few to a few hundred per gram of soil. Application of nitrogenous fertilizers drastically reduced the *Azotobacter* population in soil. Inoculation of soil or seed with *Azotobacter* is effective in increasing yields of crops in well manure soil with high organic matter content. Besides the ability to fix atmospheric nitrogen, *Azotobacter* is also known to synthesize some biologically active substances such as B-vitamins, IAA and gibberellins. The organism possesses fungistatic properties even on certain pathogenic ones, such as *Alternaria* and *Fusarium.* These attributes of *Azotobacter* explain the observed beneficial effects of the bacteria in improving seed germination and plant growth.

Several experiments conducted in temperate regions show that nitrogen fixation in *Azotobacter* inoculated soils is not more than 10 to 15 kg of N/hectare per year, depending on the availability of carbon sources. Bacterial inoculants containing Azotobacter cells under the name 'azotobacterin' are being produced and used to USSR and East European countries such as Czechoslovakia, Romania, Poland, Bulgaria and Hungary, where bacterization of seeds with azotobacterin has proved beneficial in increasing yields of crops like wheat, barley, maize, sugar beet, carrot, cabbage and potato. The increase in the yields of field crops was not more than 12 per cent over corresponding uninoculated controls. Repeated application of *Azotobacter* during different stages of growth of a crop is now being recommended with the object of increasing the number of bacteria in soil. Recent experiments on inoculation of soil with Azotobacter along with different doses of inorganic N_2 fertilizer have demonstrated the possibility of saving considerable amounts of N_2 fertilizer while still attaining desired yields of rice. Field experiments with *Azotobacter* in India have revealed that yield increases are not always reproducible. Regarding nitrogen-fixing bacteria, *Beijerinnckia* and *Derxia,* occurring in tropical soils, field experiments are needed before any conclusion could be made, to assess their utility and benefits to crop production.

During the last decade, Brazilian workers observed specific and abundant colonization of the rhizosphere of the grass *Paspalum* with *Azotobacter paspali,* a powerful nitrogen fixer. More recently *Azospirillum lipoferum* has been observed to fix atmospheric nitrogen in the cortical cells of the roots of maize. Substantial increases in yield were reported following the inoculation of sorghum and pearl millet with *Azospirillum brasilense* under several agro-climatic conditions in India. In addition to nitrogen fixation, hormonal effects have also been shown to be responsible for at least part of yield increases following inoculation with *Azospirillum.* Such responses to *Azospirillum* inoculations have been confirmed in Israel. It has also been shown that *Azospirillum* and *Azotobacter,* besides enhancing nitrogen uptake by plants, increase the number of root hairs and root hormone exudation. When *Azotobacter* is grown in association with cellulose decomposing bacteria, nitrogen fixation by *Azotobacter* is stimulated. *Azotobacter chroococcum* is known to fix more nitrogen when grown in association with a capsulated bacterium, *Enterobacter aerogenes.*

In order to maintain their large population these bacteria obtain their growth factors from the plants when they are released into the soil.

Bacteria in the soil are significant for the following reasons:

☆ Heterotrophic bacteria are able to metabolize and mineralize many organic compounds of low molecular weight.

☆ They carry out several biochemical reactions and decomposition processes under a wide range of conditions.

☆ They destroy many natural organic materials and also decompose synthetic organic compounds.

☆ They are also significant in oxidizing and reducing ions of nitrogen, sulphur, iron and manganese.

☆ They are the dominant members of microflora under anaerobic conditions.

Actinomycetes

Actinomycetes are generally considered as the intermediate group of organisms between Eubacteria and fungi. Taxonomically, the actinomycetes are bacteria, but they are considered separately in soil microbiology because of their morphological characteristics. They are like fungi in producing hyphae and conidia or sporangia. The width of the mycelium is about the same as that of bacterial cell. Certain actinomycetes, whose hyphae undergo segmentation, resemble bacterial cells. Both morphologically and physiologically, actinomycetes differ from fungi in the composition of their cell wall. They do not have chitin and cellulose which are usually found in the fungi. They are Gram-positive and grow on agar media very slowly forming powdery colonies.

Actinomycetes make up approximately 20-60 per cent of the total microbial population of the soil. The so called "earth smell" of soil is due to the production of perpenoids and extracellular enzymes by actinomycetes. They produce a multitude of extracellular enzymes that are capable of degrading plant and animal organic matter. Actinomycetes are intolerant to acidity and their number declines at pH 5.0. The optimum growth temperature ranges form 20-30°C. Some thermophilic species grow between the temperature ranges of 55-65°C. The commonest genera of actinomycetes are *Streptomyces, Nacordia, Micromonospora, Actinomyces, Actinoplanes, Microbispora* and *Streptosporangum*. Among these species of *Streptomyces* are known for antibiotic production. The importance of actinomycetes in the soils because of the following reasons:

☆ Actinomycetes play an important role in the decomposition of resistant components of organic matter such as lignocelluloses and in the formation of humus.

☆ Thermophilic actinomycetes dominate and actively participate in the transformations taking place in compost piles, manure pits and hay.

☆ Many members of Actinomycetes are pharmaceutically important for the production of antimicrobial substances.

Frankia

Species of *Frankia* are slow growing actinomycetes. There are four media in which *Frankia* can grow *in vitro*. Two of the media are for isolation of *Frankia* from root nodules and its cultivation. They are complex ones having mannitol solution. The other three ones are also complex. One is a sugar medium used for cultivation of *Frankia* spp. from root nodules. The Albumin Fatty Acid supplement is used for cultivation and maintenance. The other one is a basal medium which also contains Albumin Fatty Acid supplemented which is also used for cultivation and maintenance.

Sarma *et al.,* reported a method based on calcium alginate beads for the isolation and genetic purification of *Frankia* strains. Besides the biochemical tests prescribed in Bergey's Manual (9th edn.) the only confirmatory test is by reinoculating *Frankia* strains on seedlings of their respective hosts. Effective nodulation should occur. Chinese report says that *Frankia* can best be preserved in sterile sand for eight years and in lyophilized milk for three and half years retaining their nitrogenase activity as determined by acetylene reduction tests. David Labeda maintains 75 *Frankai* strains in ARS Actinomycetes culture collection (N.R.R.L) in Peoria. Plants produce oxygen radicals such as oxygen and hydrogen peroxide (Levene *et al.,* 1994) to protect themselves against pathogens. These radicals can kill the symbiont also. To protect itself, *Frankia* produce certain enzymes which reduce the effect of these radicals. One such enzyme is superoxide dismutase (SOD). *Frankia* shows highest level of SOD as reported by Puppo *et al.* (1989).

In *Casuarina* it was found that before the onset of uptake hydrogenase actively, nitrogen fixation was occurring. This indicates that hydrogen has to be evolved from the nitrogenase before uptake hydrogenase starts functioning. *Frankia* can recycle the hydrogen generated as a byproduct of the nitrogenase activity and has a potential to improve the energy efficiency of the symbiosis. This is important since uptake of hydrogenase is uncommon in many rhizobia used as inoculatns of legumes. However, detailed studies made in depth of the hydrogen uptake (Hup) system of *Rhizobium leguminosarum b.v. viciae* and *Bradythizobium japonicum* show multigenic (18-24) cluster that is responsible for synthesis of an active hydrogenase. This gene cluster has been isolated from these two organisms. This information along with gene transfer technology will help in biotechnological exploitation of the Hup system which will design and generate more energy efficient strains of Rhizobia. The same can be done with *Frankia* also.

Dobritsa (1998) clustered 39 selected *Frankia* strains belonging to different genomic species on the basis of their *in vitro* susceptibility to 17 antibiotics, pigment production and ability to nodulate *Alnus, Alacaganus* and *Casuarina*. Most of them fell into three cluster groups corresponding to their hosts. Encarna *et al.* (1998), has shown that *Frankia* are even more efficient than rhizobia. They used staircase electrophoresis method to look for the low molecular ribonucleic acids of *Frankia*. *Frankia* has the greatest potential for genetic engineering of the host ranges which includes other plants other than their own hosts.

Fungi

The habitats of fungi are quite diverse. Some are aquatic and some others are marine; but most fungi are found in soil or on dead plant matter. The fungi lack chlorophyll. Being typical of eukaryotes they contain nucleus, vacuoles and mitochondria. Three groups of fungi namely the molds, yeasts and mushrooms have major practical importance. The molds are filamentous, yeasts are unicellular and mushrooms are again filamentous fungi that typically form fruiting bodies. Fungal cell walls resemble plant cell walls but differ in chemical composition. Chitin is a common constituent of fungal cell walls. They are also rich with 80-90 percent of polysaccharides with proteins, lipids, polyphosphates and inorganic ions making up the matrix of the cell wall.

Hundreds of different species of fungi inhabit the soil. They are numerous near the surface of the soil because of their aerobic nature. They are dominant in acid soils. They are also present in neutral and alkaline soils and some can even tolerate pH more than 9.0. Fungal counts range from thousands to hundreds of thousands per gram of soil. Some of the more common species of fungi in soil are *Penicillum, Mucor, Rhizopus, Fusarium, Cladosporium, Aspergillus* and *Trichoderma.* The physical and chemical factors of the soil will dictate the dominance of species.

For the enumeration and isolation of fungi the soil dilution method is usually done from soil samples. Martin's Rose Bengal Agar (MRBA) with the antibiotic streptomycin is the most widely employed medium for this purpose. The buried slide technique is another method used for the direct observation of soil fungi. The fungi in the soil are significant for the following reasons.

- ☆ They are very active in decomposing the complex organic constituents of plant tissues such as cellulose, lignin and pectin.
- ☆ The accumulation of mold mycelia improves the physical nature of soil by increasing porous crumbling structure.
- ☆ The mycelium penetrates through the so9il forming a network that entangles the small particles and this improves the soil aeration and water movement in the soil.
- ☆ They are also significant in forming humus.
- ☆ A number of fungi are prominent because of their role in causing diseases of crop plants and livestock.

Yeasts

The yeasts are unicellular fungi, and most of them are classified under the Ascomycetes. Yeast cells are usually spherical, oval or cylindrical and cell division generally takes place by budding. They do not form filaments or mycelium, and the population of yeast cells remains as collection of single cells. Yeast cells are much larger than bacterial cells and can be distinguished microscopically from bacteria by their size and cell structure. *Saccharomyces* and those belonging to Deuteromycotina *e.g. Candida* are the predominant species in acidic soils.

Arbuscular Mycorrhizal (AM) Fungi

In a way, we have rediscovered that the AM fungi do not function very well in fertile soils or soils made fertile by addition of fertilizers (Mosse, 1973), where only the symbiotically inefficient AM fungal strains get selected (Anonymous, 2005; Bagyaraj, 2001). Field evidence remains unconvincing that AM fungi substantially improving P-relations and growth of crops in most production oriented agro-ecosystems (Rilling and Mummey, 2006).

This might effectively limit the scope of AMF use and restrict the same for 'low input sustainable agricultural systems' (LISA) that have come to be recognized under various descriptions, with low or no use of synthetic chemicals (fertilizers as the common denominator. Though based on limited observations, there is some skepticism for that also. The contribution of the AMF to plant growth promotion even in soils of low P-availability, for reasons not well understood yet, seems not to be always there. Increased AM Fungal colonization by virtual exclusion of soluble P-fertilizers in low P-soils may not necessarily translate to increased yield of many crops (Cardoso *et al.,* 2006). Unless the plant physiologically suffers from phosphorus deficit – the positive balance between its potential demand for and current supply of phosphorus (Koide and Li, 1989), there may not be any sense to increase the P-supply either through mycorrhizae from the soil source or from a soluble fertilizer source through the roots.

Crop responses to AM fungal inoculation are governed by soil type, host variety, AM fungal strains, temperature, moisture, cropping practices and soil management practices. In general, field experiments with AM fungal inoculation are fewer than those reported for other biofertilizers. The major constrains for field trials with AM fungi has been the inability to produce 'clean pure' inoculums on large scale. Field trials indicated that AM fungal inoculation increased yields at certain locations and the response varied with soil type, soil fertility particularly with available P status of soil and AM fungal culture. Our understanding of the biology of mycorrhizae, though severely hampered by our inability to culture them in the laboratory has greatly increased in the last two decades. That the myhcorrhizal fungi assist in the uptake of phosphorus and trace metals and positively influence water and nutrients status via hormonal influences is not in doubt. Lack of suitable inoculums production technology is the major limitation for the commercial exploitation of this system.

Cyanobacteria and other Algae

Cyanobacteria and other groups of algae are able to carry out photosynthesis. Therefore, their grown is easily noted when soil sample is inoculated into an inorganic nutrient solution in presence of light. They grow very actively in moist soil with light. Below the soil surface, water is abundant, but the absence of light restricts the growth of algae. Their number in soil usually ranges from 100 to 10,000 per gram of soil. They may form scum on the soil surface visible to the naked eye. Green algae and diatoms are numerous in soil. The genera of green algae found in soil are *Chlorella, Chlamydomonas, Chlorococcum* and *Oedogonium.* The prominent genera of diatoms, *Achnanthes, Frangilaria, Navicula, Pinnularia* and *Synedra.* The dominant members of cyanobacteria (blue-green algae) in soil are *Oscilletoria, Cylindrospermum, Anabaena, Scytonema, Tolypothrix, Chlorococcus* and *Lyngbya.*

Cynabacteria are a group of photoautotrophic prokaryotic micro-organisms, which may be single celled or colonial or filamentous forms. Filamentous cyanobactgeria show the ability to differentiate into three different cell types. Vegetative cells are the normal photosynthetic cells formed under favourable growing conditions. Climate resistant spores may be formed when environmental conditions become harsh. A third type of cell, a thick walled "heterocyst", contains the enzyme nitrogenase, responsible for nitrogen fixation. Among the biofertilizers, cyanobacteria are capable of both nitrogen fixation and oxygenic photosynthesis. The oxygen sensitivity of nitrogenase requires that these processes be separated in nitrogen fixing cyanobacteria. Heterocystous cyanobacteria achieve this by confining nitrogen fixation to heterocysts while carbon fixation occurs in vegetative cells. Thus, a special separation exists between two processes. When heterocystous organisms such as Anabaena are grown on light/dark (L/D) cycles, both CO_2 and nitrogen fixation occur in light (Mackeras, 1990).

Cyanobacteria play a spectrum of remarkable roles in agriculture, especially in sustainable integrated agro ecosystems. As biofertilizer they can contribute around 30 kg fixed nitrogen per hectare in flooded rice fields each season. The value could be increased dramatically with up-to date biotechniques (Venkataraman and Shanmugasundaram, 1992). The cyanobacterial biomass when incorporated in the soil conserves organic matter, nitrogen, phosphorus and moisture and improves the chemical and physical structure of the soil. This characteristic of several strains of cyanobacteria may be very useful in reclaiming saline and alkaline soil. This aspect has been well substantiated and documented under different agroclimatic conditions (Kaushik, 1993 and 1989, Kaushik and Ummat 1992 and Boussiba, 1999).

Singh (1961) has indicated the possibility of reclamation of salt affected soils with native cyanobacteria. Generally, in salt affected areas the paddy yield is considerably reduced compared to normal fields. The ultimate success of a bio-ameliorant cyanobacterium in a salt affected soil would be reflected by the successful establishment of the crop with a moderately high yield (Singh, 1961). Hence, marine cyanobacterial strains were tested for amelioration as well as amendment to saline soil paddy fields. In an experiment, the application of nitrogen fertilizer alone did not improve the crop yield much, as it was only 3 and 8 per cent over the control with 25N/h and 50N/h respectively. Gypsum amendment resulted in increase in the grain yield over the control, which rose to 23 per cent and 37 per cent with 25N/h and 50N/h respectively. A similar response pattern was also observed with cyanobacterial application. The three striking findings of the experiments were (a) the effect of gypsum and cyanobacteria application were comparable. (b) the combined application of gypsum and cyanobacterial was more beneficial than the individual application of either of the two with or without nitrogen fertilizer and (c) the effect of cyanobacteria was equivalent to that of 25N/h fertilizer nitrogen (Sujatha, *et al.,* 1996).

Cyanobacteria are the largest and the most widely distributed photosynthetic group on earth, forming a prominent component of microbial populations in wet land soils, especially in rice paddy fields, significantly contributing to fertility as a natural biofertilizer (Vaishampayan, *et al.*, 1998). Many tropical paddy fields receive neither chemical fertilizers nor natural manur, yet they remain productive and capable of supporting large populations with basic food. The fertility of paddy soil is maintained by the cyanobacteria, which grow spontaneously and often luxuriously in the waterlogged field. They provide nitrogen to the plants by either fixing atmospheric nitrogen or/by both secretion of nitrogenous substances and subsequent liberalization of organic substances (Santra, 1993).

The role of nitrogen fixing cyanobacteria in the maintenance of the fertility of rice fields has been well substantiated and documented all over the world. In India alone, the beneficial effect of cyanobacteria on the yield of many rice varieties has been demonstrated in a number of localities (Saga and Mandal, 1980; Santra, 1993). In many areas, 10 to 20 per cent increase in the grain yield has been observed as a result of cyanobacterial application in the absence of any added chemical nitrogen fertilizer (Santra, 1993). Rice field ecosystem provides favourable environment for the growth of cyanobacteria with respect to their requirements of light, water, temperature and nutrient availability (Kannaiyan, 1999). Kaushik (1996) has reported that a large number of field trials were conducted in different agro climatic regions of our country to assess the effect of algalisation on rice yield.

The combination of nitrogen fixing and non-nitrogen fixing algae give more effect in terms of growth and yield of paddy. The non-nitrogen fixing algae have also independently played an important role in the growth and yield of paddy. The combined effect of fixing and non-nitrogen fixing cyanobacteria have an overall beneficial effects on soil enrichment. The non-nitrogen fixing cyanobacteria also enriched the phosphorus and potassium content in the soil. This may be due to release of growth promoting substance from cyanobacteria (Selvarani, 1983). Cyanobacteria can bring about a variety of beneficial effects on the soil. The production of polysaccharides and other gummy substances helps in improving aggregation property of soil by acting as cementing agent. This helps in improved hydraulic conductivity, aeration and water holding capacity of soil. The release of various nitrogen and carbon compounds helps in maintaining or improving soil carbon to nitrogen ration (Kaushik 1983a and Kaushik 1983 b) leading to improved microbial activity. Cyanobacteria are capable of solubilizing unavailable phosphatic deposits, causes reduction in the soil pH, electrical conductivity, and exchangeable sodium percentage.

Pot trials were conducted by Jayaraman and Shanmugusundaram (1993) to assess the efficacy of *Scytonema normanii* application as microbial inoculants in the presence and absence of urea nitrogen. The cyanobacterium was given as a single dose and also in three doses, on the 5th, 35th and 65th day after transplantation of paddy seedlings. Uniniculated, controls with and without the addition of urea were maintained. Cyanobacterial treatments were better compared to control which received neither cyanobacterium nor urea nitrogen. A definite saving of 25 per cent chemical N was found to be possible when cyanobacteria were applied along with chemical N

(Jayaraman and Shanmugasundaram, 1983). Application of heterocystous and non-heterocystous cyanobacterial biofertilizers to the rice field clearly showed increase in number of grains and weight of the grains and also enhanced the NPK content of the soil (Selvarani, 1983). Immobilized as well as free living cyanobacterial application was found to be distinctly advantageous over control, as it enhanced significantly the various parameters of growth of rice plants such as shoot and root length, fresh and dry weight of the plants, chlorophyll and protein content over a period of 30 days (Sophia Rajini, 1995).

From the above investigations, it is proved that cyanobacteria serve as natural renewable biological sources as biofertilizers for rice. In biofertilizers formulations carrier is of paramount significance as it forms the vital ingredient and also helps to spread a small volume of inoculum over a large surface area. Soil is generally used as a carrier material for cyanobacterial biofertilizer but this was often not successful due to:

☆ Decreased shelf life.

☆ Loss of viability due to the hydrophilic nature of soil that removes moisture beyond the minimum requirement for retaining the viability in cells.

Members of the blue-green algae (BGA), family Nostocaceae, genera *Nostoc, Anabaena, Aulosira* and *Cylindrospermum,* and a few belonging to the families Rivulariaceae. Stigonemataceae and Scytonemataceae have been reported to possess the power to fix atmospheric nitrogen. They are of significance in rice fields, R.N. Singh reported that the algal film that developed on paddy soils in North India belonged mainly to blue-green algae capable of fixing nitrogen, and 40 per cent of nitrogen is excreted by the algae into the substratum, which may become available to rice plants. Watanabe *et al.,* showed that species of *Tolypothrix* are active nitrogen-fixers in rice fields in Japan. In desert soils, BGA are reported to fix nitrogen in the surface layers.

The blue-green algae abundantly distributed in the tropics play important roles in agriculture. Most of the nitrogen-fixing blue green algae belong to the genera *Anabaena, Anabaenopsis, Aulosira, Cylindrospermum, Nostoc, Calothrix, Scytonema, Tolypothrix, Fischerella, Stigonema* etc. The amount of nitrogen fixed by the blue-green algae has been reported to vary from 35 to 195 kg/hectare per season. Blue-green algae also add a bulk of organic matter to the soil. Further, they synthesize and secrete several vitamins and growth substances (Vitamins B12, auxin and ascorbic acid) which improve plant growth. Field trials conducted in different parts of India and South East Asia have shown significant increases in grain yields of many rice varieties by inoculation of rice fields with blue-green algae (Wani, 1990). Algal material for field application can be continuously prepared by the farmers in their own fields. For this purpose, open air soil culture method is advocated as it is simple and could be adopted by the farmer himself. A thin layer of soil is spread in galvanized iron trays, 50x50 cm, filled with 2cm standing water. The algae are inoculated into the trays and kept in open sun for 5 to 6 days. The algal growth can be dried and stored in polythene bags or used immediately for field inoculation. The production of algae on farmer's fields can be a continuous process. The culture can be mixed with lime or in a bucket of water, broadcasted or sprinkled throughout the field, one week before transplanting rice; 10 kg of inoculums per hectare may be needed.

The blue-green alga *Anabaena azollae* is symbiotically associated with the water ferm *Azolla*. *Azolla* is being used as green compost for rice cultivation of Vietnam, Thailand and China because of its association with the blue-green alga. *A. azollae* fixes nitrogen and multiplies fast and produces higher yields of green compost (200 to 300 tons/hectare per year) than the conventional green manure plants such as *Sesbania* and *Crotalaria* which are known to yield 30 to 50 tons/hectare per year. It is reported from Vietnam that a tone of *Azolla* increases the rice yield by 10-25 per cent. It is very difficult to obtain pure cultures of algae because most of them are covered by mucilaginous matrix which harbors many contaminants. However, with some special techniques, untangle cultures can be obtained. Cyanobacteria and algae in the soil are prominent because,

☆ They are oxygen producing photosynthetic organisms.

☆ They grow on freshly exposed rock and their cells accumulate organic matter, which supports the growth of the bacterial species. Therefore, they are important in starting the life cycle on rocks, in volcanic areas and in other regions that are free from life.

☆ The acids produced during the microbial metabolism dissolve the mineral constituents of the rock leading to the formation of soil.

☆ They also fix nitrogen in soil and are especially useful in the rice fields.

Protozoa

Protozoans are mainly found in the upper layer of the soil. The application of organic manures increases their number in soil. Soil moisture is another important factor affecting soil protozoa. When the supply of water is limiting for its life processes, the protozoan encysts itself and remains in the cyst form until the environment becomes more conducive to growth. Therefore, the number of protozoans in moist soils ranges from a few hundred to several hundred thousand per gram. Most protozoa feed upon bacteria and some organic materials. They eat up the bacteria belonging to the genera *Enterobacter, Agrobacterium, Bacillus, Escherichia, Micrococcus* and *Pseudomonas.* This predatory habit of protozoa regulates biological equilibrium in soil. The enumeration and isolation of protozoa from soil is regulates biological equilibrium in soil. The enumeration and isolation of protozoa from soil is usually done by the glass ring method, where well grown *Aerobacter* colonies are placed in the ring of solidified surface of water agar in petri dish. When protozoa grow, the bacterial colonies disappear indicating the growth of protozoan.

The major genera of protozoa found in soil belong to the following three groups;

☆ Ciliates – *Paramecium, Colpoda, Pleuronema* and *Vorticella*.

☆ Flagellates – *Peranema, Astasia, Olicomonas* and *Bodo*.

☆ Rhizopods – *Amoeba, Acanthamoeba, Pelomyxa, Arcella* and *Euglypha*.

Viruses

The viruses are ultramicroscopic obligatory parasites on other soil micro-organisms such as bacteria, actinomycetes, fungi and cyanobacteria. Some plant and animal viruses also come in contact with soil. As viruses are seen only under electron microscope, one can ob observe lysogenic activity of specific phages on their hosts in the form of *plaques* on agar plates. Bacteriophages (Bacteria eaters), Actinophages (viruses attacking actinomycetes), and Cyanophages (viruses attacking cyanobacteria) are found in soil samples. Around 60 species of fungal viruses have also been reported.

Nematodes

Among the microscopic forms of animals are nematodes or eelworms. They are nonsegmented, cylindrical or spindle shaped animals about 0.5 to 1.5 mm long. They are saprobic, feeding on decaying organic matter or parasitic or predacious. Other than nematodes many other macrofauna such as ants, mites, millipedes, earthworms, larvae of beetles, slugs and snails are also found in the soil.

Other organisms in nitrogen cycle are the nitrifying bacteria which oxidize ammonia to nitrite and then nitrite to nitrate. *Nitrosomonas* and *Nitrobacter* are the two bacteria, respectively and then nitrite to nitrate. *Nitrosomanas* and *Nitrobacter* are the two bacteria, respectively responsible for the two-step reaction in soil, and their importance has already been detailed M. Alexander in 1978 showed that several heterotrophs belonging to the genera *Pseudomonas, Corynebacterium, Bacillus, Nacordia, Streptomyces, Aspergillus, etc.,* produce either nitrite or nitrate from ammonia or other reduced forms of nitrogen. Various factors like aeration, moisture, temperature, pH and organic matter influence the growth and activity of nitrifies in soil. Denitrification of bound nitrogen to gaseous nitrogen is mediated by numerous species of bacteria. Of these, *Pseudomonas* and *Achromobacter* are the predominant ones in soil. Plant Growth Promoting Rhizobacteria (PGPR) are believed to improve plant growth by colonizing the root system and pre-emptying the establishment of or suppressing Deleterious Rhizosphere Micro-organisms (DRMO) on the root (Schroth and Hancock, 1981). Incoulating planting material with PGPR presumably prevents or reduces the establishment of pathogens (Sushlow, 1982). Production of siderophores is yet another mechanism through which microbes influence plant growth. Siderophores are low molecular weight high affinity Fe^{+++} chealtors that transport iron into bacterial cells and are responsible for increased plant growth by PGPR (Kloepper *et al.,* 1980). Under Fe-deficient conditions, fluorescent pseudomonads produce a yellow-green fluorescent siderophore iron complex (Hohnadel and Meyer, 1980) which creates an iron deficient environment deleterious to fungal growth.

Role of Soil Micro-organisms in Improving P Nutrition of Plants

After nitrogen (N), phosphorus (P) is the major plant growth limiting nutrient despite being abundant in soils in both inorganic and organic forms. However, many soils throughout the world are P-deficient because the free phosphorus concentration (the form available to plants) even in fertile soils is generally not higher than 10 µM even at pH 6.5 where it is most soluble (Arnou, 1953). On an average, most mineral nutrients in soil solution are present in millimolar amounts, however, phosphorus is present only in micromolar or lesser quantities (Ozanne, 1980). These low levels of P are due to high reactivity of soluble P with Calcium (Ca), iron (Fe) or aluminum (Al) that lead to P precipitation. Inorganic P in acidic soils is associated with Al and Fe compounds (Sharpley *et al.,* 1984) whereas calcium phosphates are the predominant form of inorganic phosphates in calcareous soils. Organic P may also make up a large fraction of soluble P, as much as 50 per cent in soils with high organic matter content (Barber, 1984). Phytate, a hexaphosphate salt of inositol, is the major form of P in organic matter contributing between 50 and 80 per cent of the total organic P (Alexander, 1977). Although micro-organisms are known to produce phytases, that can hydrolyze phytate, phytate tends to accumulate in virgin soils because it is rendered insoluble as a result of forming complex molecules with Fe, Al and Ca (Alexander, 1977). Phospholipids and nucleic acids form a pool of labile P in soil that is easily available to most of the organisms present there (Molla and Chowdary, 1984).

Nature of P Biofertilizers

Many plants have shown to have benefited from the association with micro-organisms under P-deficient conditions. This association could result either in better uptake of the available P, or rendering unavailable P-sources accessible to the plant. The arbuscular mycorrhizae (AM) belong to the former category and the later category includes various bacteria and fungi isolated for their ability to solubilize insoluble mineral phosphate complexes especially those of calcium phosphate complexes.

Mycorrhizae

AM fungi are known to be ubiquitous in agricultural soils and are believed to enhance P nutrition of plants by scavenging the available P due to the large surface area of their hyphae, and by their high affinity P uptake mechanisms (Hayman, 1974, 1983; Mosse, 1980; Sanders and Tinker, 1973). There are also reports of organic acid production by AM (Lapeyrie, 1988; Paul and Sundara Rao, 1971) that could solubilize the insoluble mineral phosphates. It has also been suggested that there could be further effects on the availability of Fe phosphates (Bolan *et al.,* 1987; Cress *et al.,* 1984), but so far no alternative to the original mechanism of Sanders and Tinker (1973) has been accepted.

Production of organic acids by AM would certainly affect the availability of acid-labile insoluble phosphate and the whole issue of AM mediated increase in available P needs re-examination. Ectomycorrhizal fungi have been shown to possess P solubilizing activity (Lapeyrie *et al.,* 1991). They are capable of utilizing P from inositol phosphates and possess phosphatase activity, that could further affect their ability to release P from soil organic matter (Antibus *et al.,* 1991; Koide and Schreiner, 1992). However, the use of AM as phosphate biofertilizer is hindered by the inability to culture them *in vitro,* since they are obligate symbionts. In addition, AM infection also seems to depend on the P status of the plant (Abbott *et al.,* 1984). It is known that the AM fungi are not able to colonize plant roots strongly under P sufficient conditions (Amijee *et al.,* 1989; Jasper *et al.,* 1979; Koide and Schreiner, 1992). In certain cases, the growth rates of plants were reduced by AM coloniation in the presence of available P (Buwalda and Goh, 1982; Peng *et al.,* 1993; Son and Smith, 1995).

Phosphate-Solubilizing Micro-organisms (PSMs)

The involvement of micro-organisms in solubilization of inorganic phosphates was known as early as 1903 (Kucey *et al.,* 1989). Since then, there have been extensive studies on the solubilization of mineral phosphates by micro-organisms. These have been reviewed by Goldstein (1986), Kucey *et al.* (1989), Subba Rao (1982) and Tandon (1987). Phosphate solubilizing micro-organisms (PSMs) are ubiquitous, and their numbers vary from soil to soil. In soil, P-solubilizing bacteria constitute 1–50 per cent and fungi 0.5 per cent –0.1 per cent of the total respective population. In general, P solubilizing bacteria generally out-number P solubilizing fungi by 2–150 fold (Banik and Dey, 1982; Kucey, 1983; Kucey *et al.,* 1989). The majority of the PSMs solubilize Ca–P complexes and only a few can solubilize Fe–P and Al–P (Banik and Dey, 1983; Kucey *et al.,* 1989). Hence, these PSMs could be effective in calcareous soils in which CaP complexes are present, but not in other soils such as alfisols in which phosphates are complexed with Fe and Al ions. However, these PSMs may be effective even in these soils when supplemented with rock phosphate. Most P-solubilizing bacteria were isolated from the rhizosphere of various plants and are known to be metabolically more active than those isolated from sources other than rhizosphere (Baya *et al.,* 1981; Katznelson and Bose, 1959). The P-solubilizing ability in bacteria was lost upon repeated subculturing but no such loss has been observed in the case of Psolubilizing fungi (Kucey, 1983; Sperber, 1958). In general, fungal isolates exhibit greater P-solubilizing ability than bacteria in both liquid and solid media (Banik and Dey, 1982; Gaur *et al.,* 1973; Kucey, 1983). The P-solubilizing ability of PSMs also depends on the nature of N source used in the media, with greater solubilization in the presence of ammonium salts than when nitrate is used as N source. This has been attributed to the extrusion of protons to compensate for ammonium uptake, leading to a lowering of extracellular pH (Roos and Luckner, 1984). In some cases, however, ammonium can lead to decrease in P solubilization (Reyes *et al.,* 1999). In addition, other media components were also found to affect the P solubilization ability (Cunningham and Kuiack, 1992).

Mechanisms of P Solubilization

The ability to solubilize Ca–P complexes has been attributed to the ability of the PSMs to reduce the pH of their surroundings, either by the release of organic acids or protons. The organic acids secreted can either directly dissolve the mineral phosphate as a result of anion exchange of PO_4^{2-} by acid anion or can chelate both Fe and Al ions associated with phosphate (Bajpai and Sundara Rao, 1971; Bardiya and Gaur, 1972; Katznelson and Bose, 1959; Moghimi *et al.,* 1978; Sperber, 1957). The PSMs produce organic acids such as acetate, lactate, oxalate, tartarate, succinate, citrate, gluconate, ketogluconate, glycolate, etc. (Banik and Dey, 1982; Cunningham and Kuiack, 1992; Goldstein, 1986; Gyaneshwar *et al.,* 1998). In certain cases, P solubilization has been found to be inducible by P starvation (Goldstein and Liu, 1987; Gyaneshwar *et al.,* 1999). Many micro-organisms can reduce the pH of the medium and solubilize P from rock phosphate (Gyaneshwar *et al.,* 1998). The Psolubilizing activity of Rhizobiumwas abolished by addition of NaOH indicating that P solubilizing activity of this strain was entirely due to its ability to reduce the pH of the media (Halder and Chakrabarty, 1993).

P Assimilation in Microbes

Some non P-solubilizing bacteria can also be important as P biofertilizers as they can take up the sparingly soluble P through their high affinity transporters and this P can become available to plants through mineralization as the bacteria die. P uptake and metabolism has been extensively studied in the model bacterium Escherichia coli(Torriani-Gorini *et al.,* 1987; Wanner, 1994, 1996) and similar systems have been found in other bacteria. In E. coli, P stress can affect the expression of over 400 proteins, and this global affect is mediated by a two component regulatory system PhoR and PhoB, in which PhoR is the sensor and PhoB is the cognate positive regulator. Under P-starvation conditions, PhoR phosphorylates PhoB, that in turn binds specific DNA sequences called PHO box (Willsky and Malamy, 1976; Wanner and Chang, 1987; Wanner, 1996). Many genes that are regulated by P starvation may not be involved in P acquisition or assimilation, *e.g.* gltBandgltD (which code for glutamate synthase) andglpB (coding for anaerobic glycerol-3-phosphate dehydrogenase) (Metcalf *et al.,* 1990). Recently, it has been shown that P starvation resulted in induction of glucose dehydrogenase, an enzyme involved in oxidative glucose metabolism in Enterobacter asburiae, a related member of enterobacteraciae family (Gyaneshwar *et al.,* 1999).

In *Rhizobium,* P limitation can result in higher P transport rates and the induction of alkaline phosphatase (Al-Niemi *et al.,* 1997). Rhziobium has a functional homologue of PhoB (Wanner, 1996), but a PhoR counterpart has not yet been detected in Rhizobium (Bardin and Finan, 1998). As in *E. coli,* PHO box sequences are present upstream of the genes regulated by P starvation in *Rhizobium* (McDermott, 1999).

References

Abbott L K, Robson A D and De Bore G 1984. The effect of phosphorus on the formation of hyphae in soil by the vesicular–arbuscular mycorrhizal fungus *Glomus fasciculatum. New Phytol.* 97, 437–446.

Alexander M 1977. *Introduction to soil microbiology.* Wiley New York.

Amijee F, Tinker P B and Stribley D P 1989. Effects of phosphorus on the morphology of vesicular–arbuscular Mycorrhizal root system of leek (*Allium porrum* L). *Plant Soil* 119, 334–336.

Antibus R K, Sinsabaugh R L and Linkins A E 1991. Phosphatase activities and phosphorus uptake from inositiol phosphates by ectomycorrhizal fungi. *Can. J. Bot.* 70, 794–801.

Arnou, D. I. 1953. In: *Soil and Fertilizer Phosphorus in Crop Nutrition* (IV). Ed. WH Pierre. Noramn, AG. Acad. Press NY.

Bajpai P D and Sundra Rao W V B 1971. Phosphate solubilizing bacteria. Part 2. Extracellular production of organic acids by selected bacteria solubilizing insoluble phosphate. *Soil Sci. Plant Nutr.* 17, 44–45.

Banik S and Dey B K 1982. Available phosphate content of an alluvial soil as influenced by inoculation of some isolated phosphate solubilizing bacteria. *Plant Soil* 69, 353–364.

Banik S and Dey B K 1983. Phosphate solubilizing potentiality of micro–organisms capable of utilizing aluminum phosphate as sole phosphate source. Zentrabl. Backteriol. Prasitenkd. *Infektionskr. Hyg.* II 138, 17–23.

Barber S A 1984. *Soil nutrient bioavailability.* John Wiley, New York, USA.

Bardiya M C and Gaur A C 1972. Rock phosphate dissolution by bacteria. *Indian J. Microbiol.* 12, 269–271.

Baya A M, Robert S B and Ramos C A 1981. Vitamin productionin relation to phosphate solubilization by soil bacteria. *Soil Biol. Biochem.* 13, 527–532.

Bolan N S, Robson A D and Barrow N I 1987. Effect of vesicular arbuscular mycorrhiza on avalibility of iron phosphates to plants. *Plant Soil* 99, 401–410.

Bossiba, S. 1999. Cyanobacteria for agricultural applications. In: International workshop and training course on micro algal biology and biotechnology. Mosomagyarover, Hungary: June 13–26: pp–6–10.

Buwalda J G and Goh K M 1982. Host–fungus competition for carbon as a cause of growth depressions in vesicular–arbuscular mycorrhizal rye grass. *Soil Biol. Biochem.* 14, 103–106.

Cress W A, Johnson G V and Barton L L 1984. The role of endomycorrhizal fungi in iron uptake byHilaria jamesii. *J. Plant Nutr.* 40, 547–555.

Cunningham J E and Kuiack C 1992. Production of citric and oxalic acids and solubilization of calcium phosphates byPenicillium bilaii. *Appl. Environ. Microbiol.* 58, 1451–1458.

Gaur A C, Medan M and Ostwal K P 1973. Solubilization of phosphate ores by native microflora of rock phosphates. *Indian J. Expt. Biol.* 11, 427–429.

Golstein A H 1986. Bacterial phosphate solubilization: Historical perspective and future prospects. *Am. J. Alt. Agric.* 1, 57–65.

Gupta, A., Saxen, A.K., Gopal, M. and Tilak. K.V.B.R 1998c. Effect of plant growth promoting rhizobacteria on competitive ability of introduced Bradyrhizobium Sp. (Vigna) for nodulation. *Mocrobiol Res.*, 113–117.

Gupta, A., Saxena. A. K. Gopal. M. and Tilak. K.V.B.R. 1998a. Enhanced nodulation of greengram by introduced Bradyrhizobium when inoculated with plant growth promoting rhizobacteria. *J. Sci. Indus. Res.*, 57: 720–725.

Gupta. A., Sexena. A. K., Gopal, M. and Tilak, K.V.B.T. 1998b. Bacterization of grengram with rhizosphere bacteria for enhanced plant growth. *J. Sci. Indus. Res.*, 57: 726–731.

Gyaneshwar P, Naresh Kumar G and Parekh L J 1998. Effect of buffering on the phosphate–solubilizing ability of micro-organisms. *World J. Microbiol. Biotechnol.* 14, 669–673.

Gyaneshwar P, Parekh L J, Archana G, Poole P S, Collins M D, Hutson R A and Naresh Kumar G 1999. Involvement of a phosphate starvation inducible glucose dehydrogenase in soil phosphate solubilization byEnterobacter asburiae. *FEMS Microbiol. Lett.* 171, 223–229.

Halder A K and Chakrabarty P K 1993. Solubilization of inorganic phosphates by Rhizobium. *Folia Microbiol.* 38, 325–330.

Hayman D S 1974. Plant growth responses to vesicular-arbuscular mycorrhiza VI. Effects of light and temperature. *New Phytol.* 73,71–80.

Hohandel, O. B. and J. M. 1986. Pyoverdin–facilitated iron uptake uptake among fluorescent Pseudomonades. In: *Iron Siderophoes and Plant Diseases* (Swinburne. T. R. ed.), Plenum Press, New York USA. pp. 119–129

Jasper D A, Robson A D and Abbott L K 1979. Phosphorus and the formation of vesicular–arbuscular mycorrhizae. *Soil Biol. Biochem.* 11, 501–505.

Jayaraman, S. and Shanmugasundarm, S. 1993. Algal Biofertilizer: Choice species of Scytonema. *Proceedings of the national seminar on Cyanobacterial Research Indian Scene*, Jan, 1993. Edited by G. Subramanian, Bharathidasan University, Tiruchirapalli pp. 24–31.

Kannaiyan, S. 1999. Ammonia excretion by the polyurethane foam immobilized N2–fixing cyanobacteria for Rice. In: *Cyanobacterial and algae metabolism and environmental Biotechnology*. Edited by. Tasneem Fatma, Narosa Publishing House. New Delhi. pp. 20–27.

Kannaiyan, S., Aruna, S. J. Marina Prem kumair, S., and Hall, DoO. 1997. Immobilized cyanobacteria as biofertilizer for rice crops. *J. Appl.* 7: 1–8.

Katznelson H and Bose B 1959. Metabolic activity and phosphate dissolving capability of bacterial isolates from wheat roots in the rhizosphere and non rhizosphere soil. *Can. J. Microbiol.* 5, 79–85.

Kaushik, and Ummat, J. 1992. Reclamation of salt affected soils with blue green alge (cyanobacteria): A Tecnology development. *Proc. Natl. Seminar Biofertilizer Technology Transfer*. Edited by Gangawne. L.V. Marathwada University, Autangabad (India). 157.

Kaushik, B. D. 1983a. Effect of native algal flora on nutritional and physic–chemical properties of sodic soil. *J. Biol. Res.* 3: 99.

Kaushik, B. D. 1983b. Amelioration of salt affected so. In: Proceedings of the All India applied Phycology Congress, Kanpur, India. 60–66.

Kaushik, B. D., 1989. Management of salt affected soils with blue green algae. In: *Soil micro-organisms and crop growth*. Edited by Somani. L.L and Bhandri. S.C. Divyajyoti Prakashan, Jodhpur. 169.

Kaushik. B.D. 1993. Cyanobacterial research – An IARI Pursuit. *Proceedings of the National Seminar on Cyanobacterial Research–Indian Scene*, Jan 19–21, 1993. Edited by G. Subramanian, Bharathidasan Universtity, Tiruchirapalli pp. 32–49.

Kaushik. B.D. 1996. Use of cyanobacterial biofertilizer in rice cultivation: A technology improvement. In: *Cyanobacterial Biotechnology. Proceedings of the International Symposium*. September 18–21. Edited by G. Subramanian, B.D. Kaushik, and G.S Venkataraman Oxfor and IBH publishing Co PVT LTD. New Delhi. pp. 211–222.

Kloepper. J.W. Leong, J. Teintze, P., Arayangkool, T. Sintwongse, P., Siripaibool. C., Wadisirisuk. P. and Boonkerd, N. 1988. Nitrogen fixation (15N dilution) with soybeans under Thai field conditions. II. Effect of herbicides and water application schedule. *Plant and Soil*, 87–92.

Koide T R and Schreiner P R 1992. Regulation of Vesicular arbuscular mycorrhizal symbiosis. *Ann. Rev. Plant Physiol. Plant Mol. Biol.* 43, 557–581.

Kucey R M N 1983. Phosphate solubilizing bacteria and fungi in various cultivated and virgin Albreta soils. *Can J. Soil Sci.* 63, 671–678.

Kucey R M N, Jenzen H H and Leggett M E 1989. Microbially mediated increases in plant available phosphorus. *Adv. Agron.*, 42, 199–228.

Lapeyrie F 1988. Oxalate synthesis from soil bicarbonate by fungus *Paxillus involutus*. *Plant Soil* 110, 3–8.

Lapeyrie F, Rangers J and Vairelles D 1991. Phosphate-solubilizing activity of ectomycorrhizal fungi *in vitro*. *Can J. Bot.* 69, 342–346.

Moghimi A, Tate M E and Oades J M 1978. Characterization of rhizospheric products especially 2–ketogluconic acid. *Soil Biol. Biochem.* 10, 283–287.

Molla M A Z and Chowdary A A 1984. Microbial mineralization of organic phosphate in soil. *Plant Soil* 78, 393–399.

Mosse B 1980. Vesicular-arbuscular mycorrhiza research for tropical agriculture. Research Bulletin 194, Hawaii Institute of Tropical Agriculture and Human Resources, Honolulu, HI, USA; University of Hawaii.

Ozanne, P. G. 1980. Phosphate nutrition of plants: A general treatise. In: *The Role of Phosphorus in Agriculture*. Eds. FE Khasawneh, EC Sample and EJ Kamprath. Soil Sci. Soc. Am. Madison WI.

Paul N B and Sundara Rao W V B 1971. Phosphate dissolving bacteria and VAM fungi in the rhizosphere of some cultivated legumes. *Plant Soil* 35, 127–132.

Prasad, R and J. F Power 1997. Phosphorus. In: *Soil Fertility Management for Sustainable Agriculture*. CRC Lewis Publ, New York. pp. 171–209.

Reyes I, Bernier L, Simard R R and Antoun H 1999. Effect of nitrogen source on the solubilization of different inorganic phosphates by an isolate of *Penicillium rugulosum* and two UV induced mutants. *FEMS Microbiol. Ecol.* 28, 281–290.

Roos W and Luckner M 1984. Relationships between proton extrusion and fluxes of ammonium ions and organic acid in *Penicillium cyclopium. J. Gen. Microbiol.* 130, 1007–1014.

Sah, R. N. and Millelsen, D.S. 1986. Transformation of inorganic phosphorus during the flooding and draining cycle of soil. *Soil Sci. Soc. Am. J.* 50, 62–67.

Saha, K. C. and Mendal L.N., 1980. A green house study on the effect of inoculation of N_2 fixing blue green algae. In an alluvial soil trated with P and Mo on the yield of rice and changes in the nitrogen content of soil. *Plant and Soil* 57: 23–30.

Sanders F E and Tinker P B 1973. Phosphate flow into Mycorrhizal roots. *Pest. Sci.* 4, 385–395.

Santra, S.C. 1993. Importance of blue green algae. In: Biology of rice–fields blue green algae. Edited by S.C. Santra, Daya Publishing House, Delhi – 110035.

Saxena. A. K. Rathi, S. K. and Tilak, K.V.B.R. 1997. Differential effect of various endomycorrhizal fungi on nodulation effects of greengram by *Bradyhizobium* sp. (Vigna) strain S. 24. *Biol. Fertil. Soils.* 24: 175–178.

Schroth. M. N. and Hancock, J. G. 1981. Selected topics in biological control. *Annual Review of Microbiol.* 35: 453–476.

Sevarani, V. 1983. Studies on the influence on nitrogen fixing and non–nitrogen fixing blue-green algae on the soil, growth and yield of paddy (*Oryza sativa –* IR50). *M.Sc. thesis.*

Singh, S and K. K. Kapoor 1992. Solubilization of insoluble phosphate by bacteria and their effect on growth and phosphorus uptake by mungbean. *Intl J Tropic Agric.* 10: 209–213.

Son C L and Smith S E 1995. Mycorrhizal growth responses: interaction between photon irridiance and phosphorus nutrition. *New Phytol.* 108, 305–314.

Sophia Rajini, V. 1995. Studies on immobilization of cyanobacteria. *Ph.D Thesis,* Department of Microbiology, Bharathidasan University, Tiruchirapalli.

Sperber J I 1958. The incidence of apatite-solubilizing organisms in the rhizosphere and soil. *Aust. J. Aric. Res.* 9, 778–781.

Sri. Lavanya Priya, S. 1997. Efficacy of coir waste based Anabaena azollae ML2 as a biofertilizer for rice. *M.Sc. Thesis,* Department of Microbiology, Bharathidasan University, Tiruchirapalli.

Subba Rao N S 1982. In: *Advances in Agricultural Microbiology.* Ed. NS Subba Rao. pp. 229–305. Oxford and IBH Publ. Co.

Subba Rao. N. S. 1995. *Biofertilizers in Agriculture and Forestry.* Oxford and IBD Publ Co, New Delhi.

Sujatha, V., Kaushik, B.D. and Venkataraman. G.S 1996. Agronomic potential marine Nostoc calcicola BDU 40302. In *Cyanobacterial Biotechnology. Proceedlings of the International Symposium.* September 18–21. Edited by G. Subramanian,B.D Kaushik and G. S Venkatraman. Oxford and IBH Publishing Co. Pvt Ltd. New Delhi, pp. 245–247.

Suslow. T. V. 1982. Role of root colonizing bacteria in plant growth. In: *Phytopathogenic Prokaryots.* (eds. M. S. Mount, and and G. H. Lacy, G.H. Academic Press, London. pp. 187–223.

Tandon H L S 1987. *Phosphorus Research and Production in India.* Fertilizer Development and Consultation Organization New Delhi. 160 pp.

Thomas, G.V., M. V. Shantaram and N. Sarawathi 1985. Occurrence and activity of phosphate solubilizing fungi in coconut plantation soils. *Plantation and Soil,* 87: 357–364.

Thopate, A. M., R. R. More and S. B. Jadhav 1998. Interaction effect of Azospirillum and PSB culture on yield and quality of sugar cane. In: *Biofertilizers and Bipesticides*. A. M. Deshmukh (ed)., Technosci Publ, Jaipur, 56–60

Tilak, K.V.B.R. 1991. *Bacterial Fertilizer Tech. Bull.*, ICAR, New Delhi, pp. 1–66.

Tilak, K.V.B.R. and Saxena, A. K. 1994. Competition among Rhizobium strains for nodulation of legumes. In: *Biology and Application of Nitrogen-Fixing Organisms: Problems and Prospectes*. (Eds A.B. Prasad and A. Vaishampayan), Scientific Publshihers. India, pp. 241–256.

Vaishampayan. A.R., Sinha, P. and Harder, D. P. 1998. Use of genetically improved nitrogen fixing cyanobacteria in rice paddy fields. Prospects as a source material for engineering herbicide sensitivity and resistance in plants. *Bot. Acta* 111: 176–190.

Venkataraman G.S and Shanmugasundaram S. 1992. Algal Biofertlizer Technology for rice Sankar Printing Press, 271, Goods shed Street, Madurai –1.

Wani. S.P. 1990. Inoculation with associative nitrogen fixing bacteria: Role of cereal grain production improvement. *Indian J. Microbiol.*, 30: 363–393.

2015, Recent Trends in Microbiology, Mycology and Plant Pathology   *Pages* 27–38
Editor: **Dr. H.C. Lakshman**
Published by: **DAYA PUBLISHING HOUSE, NEW DELHI**

Chapter 2

A Protocol for DNA Extraction from Rhizosphere of Field Growing Plant, Harboring Arbuscular Mycorrhizal Fungi

Ashok Shukla, Deepak Vyas*, Keerti Dehariya,
Vandana Bharti Ahirwar, Onkar Salunkhe,
Daulat Ram Gwalwanshi and Anuradha Jha

Laboratory of Microbial Technology and Plant Pathology,
Department of Botany, Dr. H.S. Gour Central University,
Sagar – 470 003, Madhya Pradesh, India
*E-mail: dvyas64@yahoo.co.in

ABSTRACT

Present study was carried out to develop a phenol-free and cost-effective protocol for extraction of metagenomic DNA directly from rhizosphere of field growing *Pongamia pinnata* L. plants, harboring arbuscular mycorrhizal fungi (AMF). The protocol consisted of three steps *i.e.* homogenization, lysis and PEG purification. To standardize this protocol, two different amounts of soil (2.5 and 5.0 g), and two extraction buffers (EB-I and EB-II) with different compositions, were tested. One set of experiment was performed by vortexing, followed by sonication, precipitation in ethanol/isopropyl alcohol and PEG purification, while in another set, sonication was replaced by grinding in liquid nitrogen. Sonication gave better results with 2.5 g soil. Further, results also suggested that EB-II [3.5 per cent cetyltrimethyl-ammonium bromide (CTAB)] was better than EB-I (3.0 per cent CTAB). Moreover, ethanol was preferred agent for precipitation of DNA.

The ratio of A_{260}/A_{280} was used for the assessment of DNA quality. Result showed that optical density ratio was 1.90 which suggested that protocol gave good quality DNA. The downstream processing of DNA was carried out by amplification of ITS region of rDNA by ITS primer pair (ITS4/ITS5). Thus, this method yielded a PCR-quality DNA that may be used to study the AMF diversity, directly from field.

Keywords: *Arbuscular mycorrhiza, DNA extraction, Metagenome, Sonication, PCR.*

Introduction

Arbuscular mycorrhizal fungi (AMF) are important component of soil ecosystem (Oehl *et al.*, 2005; Vyas *et al.*, 2011; Shukla *et al.*, 2012a). Roots of more than 90 per cent terrestrial plant species associate with AMF and establish symbiotic relationships (Graham 2008; Vyas *et al.*, 2011). AMF integrates with plants and add dimensions to the plant-soil-microbe systems (Smith and Read 2008; Mishra *et al.*, 2012; Shukla *et al.*, 2012b). AMF association not only exists in angiosperms (Wang and Qiu 2006) but has also been reported in lower plants like bryophytes and pteridophytes (Ligrone *et al.*, 2007; Vyas *et al.*, 2008). Its presence has also been reported in early land plants (Vyas *et al.*, 2007).

The traditional AMF classification was mainly based on morphological characteristics of asexual spores which had some limitations in taxonomy (Jha *et al.*, 2011; Kamalvanshi *et al.*, 2012), because morphological features allow the identification up to the family level (Redecker *et al.*, 2003; Oehl *et al.*, 2011). Identification at species or isolate level can be possible by using molecular methods (Redecker and Raab 2006). Molecular techniques makes the AMF classification more accurate and scientific, and can improve the AMF taxonomy (Kruger *et al.*, 2009) but the prime requirement for any such identification is the isolation of a good quality genomic DNA (Manian *et al.*, 2001).

Most of the methods used for DNA isolation (which harbors AMF) are mainly focused on trap culturing (Di Bonito *et al.*, 1995; Manian *et al.*, 2001). Several workers have isolated DNA from trap cultures, established by field collected soil (Lanfranco 1995; Simon *et al.*, 1992). However, these procedures have been often criticized because the trap cultures are always different from the field soil (Jansa *et al.*, 2002). AMF strains trapped in cultures might be different from those occurred in field soil (Shukla *et al.*, 2013). So, when most effective AMF strains in field has low tolerance of greenhouse conditions, may not be trapped in trap cultures and can never be observed, which often result in misinterpretation of population distribution in fields (Shukla *et al.*, 2012c). Thus, molecular studies can provide a more exact and rapid method for identification of desired strains (Sanders *et al.*, 1995). A metagenomic approach can be used to get the insight of AMF diversity.

Metagenomics is the analysis of DNA of micro-organisms recovered from an environment, without the need for culturing them (Tringe and Rubin 2005). Various workers have suggested that metagenomic DNA extracted from soil represents the DNA of all indigenous microbes (Saano *et al.*, 1995; Handelsman *et al.*, 1998; Schneegurt *et al.*, 2003). Thus, metagenomics coupled with PCR techniques and bioinformatics can be a powerful tool for studying phylogeny and taxonomy of particular gene in a community (Theron and Cloete, 2000; Torsvik and Ovreas, 2002). One of the important requirements for metagenomic is a good method for DNA isolation, capable of molecular studies (Wyss and Bonfonte, 1993). Therefore, to investigate the molecular diversity of AMF in rhizosphere of field growing *Pongamia pinnata* L., present study was carried out to develop a protocol for extraction of metagenomic DNA. To standardize this protocol, several combinations for cost-effective and phenol-free extraction of PCR quality DNA, were made. Moreover, the quality and quantity of DNA was confirmed and the applicability of isolated DNA for downstream molecular processing was also assessed by PCR amplification, cloning and sequencing of ITS region of mycorrhizal DNA.

Materials and Methods

Sampling

Root and soil samples were collected from rhizosphere of *P. pinnata*, growing at botanical garden, Dr HS Gour Central University, Sagar (Madhya Pradesh), India. The region (23° 10'–24° 27' N latitude and 78° 4'–79° 21' E longitude) lies in Agro-ecoregion 10; Central Highlands, Hot Sub-humid Ecoregion with medium and deep black soils (Sehgal *et al.,* 1990). Sampling was done as close as possible to the base of plant (five inches from stem) by removing the litter from the topsoil.

Extraction of DNA

Buffers

Various buffer solutions *i.e.* wash buffer, TE buffer and two extraction buffers (EB-I and EB-II) were used in this study. The compositions of different solutions were: a. Wash buffer: 50 mM Tris-HCl (pH 8.0), 50 mM EDTA (pH 8.0) and 50 mM NaCl; b. TE buffer: 50 mM Tris-HCl (pH 8.0) and 50 mM EDTA (pH 8.0); c. EB-I: 100 mM Tris-HCl (pH 8.0), 20 mM EDTA (pH 8.0), 3M NaCl, 3 per cent cetyltrimethyl-ammonium bromide (CTAB), 2 per cent Sarkosyl, 2 per cent Polyvinyl Pyrrolidone (PVP) and 1 per cent 2-mercaptoethonal; and d. EB-II: 100 mM Tris-HCl (pH 8.0), 20 mM EDTA (pH 8.0), 3M NaCl, 3.5 per cent CTAB, 2 per cent Sarkosyl, 2 per cent PVP and 1 per cent 2-mercaptoethonal.

Homogenization

Soil samples were weighed in two sets *i.e.* 2.5 and 5.0 g, and replicated eight times. Tubes were labeled as A_1 to A_8 for 2.5 g soil and B_1 to B_8 for 5.0 g soil. Samples were transferred to 50 ml centrifuge tubes and suspended in equal volume (w/v) of buffer *i.e.* 2.5 g soil in 2.5 ml and 5.0 g soil in 5.0 ml. Samples (A and B) were homogenized on vortex mixer for 5 and 10 min, respectively. The homogenate was centrifuged at 6000 rcf for 15 min and supernatant containing impurities was decanted retaining the pellets. These pellets were subjected to two washes with wash buffer by centrifugation at 6000 rcf which were subjected to lysis.

Cell Lysis and Proteinase K Treatment

Cell lysis was performed by two different procedures *i.e.* sonication and grinding in liquid nitrogen. The pellets obtained from tubes A_1 to A_4 and B_1 to B_4 were suspended in 2.5 ml of TE buffer and subjected to sonication for 10 and 15 min, respectively. While, pellets of remaining samples (A_5 to A_8 and B_5 to B_8) were dried in oven, ground in liquid nitrogen and suspended in 2.5 ml of TE buffer. After sonication and grinding, all the samples were centrifuged at 6000 rcf for 15 min for removing the soil particles. Supernatant was collected in fresh 50 ml tubes where 3 mg/ml proteinase K was added and incubated at 37 °C for 30 min. Further, samples were taken for treatment with extraction buffers and chloroform: isoamyl alcohol.

Treatment with Extraction Buffers and Chloroform: Isoamyl Alcohol

Two extraction buffers *i.e.* EB-I and EB-II were tested for DNA extraction. Equal volume (2.5 or 5.0 ml) of pre-warmed EB-I was added to the samples (A_1, A_2, A_5, A_6, B_1, B_2, B_5 and B_6), while equal volume of pre-warmed EB-II was added to another set of samples (A_3, A_4, A_7, A_8, B_3, B_4, B_7 and B_8). Samples were incubated at 65 °C for 1 h. Then, suspension was allowed to cool at room temperature. At this stage, approximately 5 and 10 ml suspensions were obtained from A_1 to A_8 and B_1 to B_8, respectively. Then, equal volume of chloroform: isoamyl alcohol (24:1) was added and tubes were subjected to gentle rolling (to prevent shearing of DNA) until homogenous suspension formed. These tubes were centrifuged at 9200 rcf for 20 min which gave two distinct layers. Upper layer, containing nucleic acid was transferred to new 50 ml tubes.

Precipitation of DNA and PEG Purification

DNA was precipitated either in isopropanol (IPA) or in ethanol. The samples were distributed in two groups: one with odd numbers and another with even numbers. Samples with odd number (A_1, A_3, A_5, A_7, B_1, B_3, B_5 and B_7) were precipitated with 1/10 of total volume of 3M sodium acetate (pH 5.2) and double volume of pre-chilled ethanol, while in set of even numbers (A_2, A_4, A_6, A_8, B_2, B_4, B_6 and B_8), ethanol was replaced by IPA, and incubated overnight at -20 °C. To pellet the DNA, tubes were centrifuged at 20,000 rcf for 15 min. Then, pellets were taken in 1.5 ml micro centrifuge tubes and washed with 70 per cent ethanol. To remove the excess of ethanol, samples were dried in vacuum dessicator. The dried DNA pellets were re-suspended in 50 µl sterile water and purified with double volume of 30 per cent PEG (8000) solution and 1.6 M NaCl. Then, tubes were incubated for 1 h at room temperature, followed by centrifugation at 17,000 rcf at 4 °C for 15 min. The pellets were washed with 70 per cent ethanol, vacuum dried and dissolved in 50 µl sterile water.

Quantification of DNA

The concentration of aqueous solution of DNA was derived from widely accepted assertion. For a double stranded DNA solution, an optical density of 1.0 at 260 nm (A_{260}) corresponds to DNA concentration of 50 µg/ml. An optical density at 280 nm (A_{280}) was also measured. The ratio of A_{260}/A_{280} was used for the assessment of DNA quality. The absorbance was measured using a UV visible double beam spectrophotometer.

Restriction Digestion and Agarose Gel Electrophoresis of DNA

Restriction digestion was performed by using 250 ng of DNA and 10 U of *Bam*H-I, *Eco*R-I (**Fermentas International, Canada**). The isolated genomic DNA and its restriction digestion product was electrophoresed in 0.8 per cent agarose gel containing 0.5 µg/ml ethidium bromide, using TAE buffer and photographed by using a gel documentation system (Vilber Lourmart, France).

Polymerase Chain Reaction (PCR)

The ITS primer pair (ITS4/ITS5) was used to amplify ~600 bp region specific to rDNA. The amplification was performed on thermocycler in a total volume of 50 µl containing 0.04 U of *Taq* DNA polymerase (Sigma-Aldrich, USA), 5 µl of 10 X *Taq* polymerase reaction buffer (Sigma-Aldrich, USA), 0.2mM of dNTPs (Genei, India), 5 pmol of each of the two primers and 50 ng of genomic DNA. The reaction was performed with initial denaturation at 94 °C for 5 min and 35 cycles of 94 °C for 1 min, 58 °C for 1 min, 72 °C for 2 min, a final elongation of 10 min at 72 °C and storage at 4 °C.

Electrophoresis of PCR Product

25 µl of PCR product was mixed with loading dye and electrophoresed on 1.2 per cent agarose gel at 50 v/cm for 1½ h and visualized on UV trans-illuminator (Genei, India). 100 bp DNA ladder plus (Fermentas International, Canada) was used as size marker. The remaining PCR product (25 µl) was taken further for cloning.

Cloning, Sequencing and Sequence Analysis

PCR products were cloned into pJET 1.2 vector following the manufacturer's instruction of clone JET PCR cloning kit (Fermentas International, Canada) and transformed into competent *E. coli* (DH5α) cells. Colony PCR was carried out using ITS4/ITS5 primer to check the positive clones. Sequencing was done and all the sequences were blast in NCBI public database (http://www.ncbi.nlm.nih.gov/) to find the homology with AMF sequences.

Results and Discussion

Our study suggested a protocol for the isolation of metagenomic DNA directly from rhizosphere of field growing *P. pinnata*, colonized by AMF. Initially, 5.0 g soil sample was used for this purpose but due to inappropriate breakage, good quality DNA could not be extracted. Then, protocol was tested with 2.5 g soil (Table 2.1 and Figure 2.1). Here, sonication gave better quality of DNA (confirmed by agarose gel electrophoresis) as compared to grinding in liquid nitrogen. Similar observations were found by Frostegard *et al.* (1999) and de Lipthay *et al.* (2004). We deduced that this can be due to the greater physical forces applied during grinding. Moreover, sonication reduces the use of manual labour required for grinding (Volossiouk *et al.*, 1995; Krsek and Wellington 1999). Since, in sonication, larger amount of soil (5.0 g) hampers the breakage of spores. Thus, it may be stated that larger amount of soil can be a limiting factor in sonication.

After cell lysis, the samples were subjected to proteinase K treatment for removal of protein impurities, because proteins are known to hamper the amplification process (Sorensen *et al.*, 2002). Thereafter, samples were incubated in extraction buffers where EB-II (3.5 per cent CTAB) gave better results. According to Saano *et al.* (1995), CTAB helps in breakage of cell membrane and forms complex with nucleic acid. Then, equal volume of chloroform: isoamyl alcohol (24:1) was added to the samples. To

Table 2.1: Different Combinations Used to Standardize the Protocol and their A_{260}/A_{280} Ratio

Amount of Soil Samples (g)	Tube Labels	Amount of Buffer (ml) Used for Dissolving the Samples	Duration (min) of Homogenization	Procedures Used for Cell Lysis	Amount and Type of Extraction Buffers (EB) Used for DNA Precipitation	Solvent used for DNA Precipitation	A_{260}/A_{280} Ratio
2.5	A_1	2.5	5.0	Sonication	EB-I (2.5 ml)	Ethanol	1.28
	A_2			Sonication	EB-I (2.5 ml)	IPA[b]	1.06
	A_3			Sonication	EB-II (2.5 ml)	Ethanol	1.90
	A_4			Sonication	EB-II (2.5 ml)	IPA	1.13
	A_5			Grinding in LN[a]	EB-I (2.5 ml)	Ethanol	1.25
	A_6			Grinding in LN	EB-I (2.5 ml)	IPA	1.17
	A_7			Grinding in LN	EB-II (2.5 ml)	Ethanol	1.26
	A_8			Grinding in LN	EB-II (2.5 ml)	IPA	1.27
5.0	B_1	5.0	10.0	Sonication	EB-I (5.0 ml)	Ethanol	1.30
	B_2			Sonication	EB-I (5.0 ml)	IPA	1.11
	B_3			Sonication	EB-II (5.0 ml)	Ethanol	1.41
	B_4			Sonication	EB-II (5.0 ml)	IPA	1.27
	B_5			Grinding in LN	EB-I (5.0 ml)	Ethanol	1.15
	B_6			Grinding in LN	EB-I (5.0 ml)	IPA	1.13
	B_7			Grinding in LN	EB-II (5.0 ml)	Ethanol	1.29
	B_8			Grinding in LN	EB-II (5.0 ml)	IPA	1.28

a: LN–Liquid Nitrogen; b: IPA–Isopropyl Alcohol.

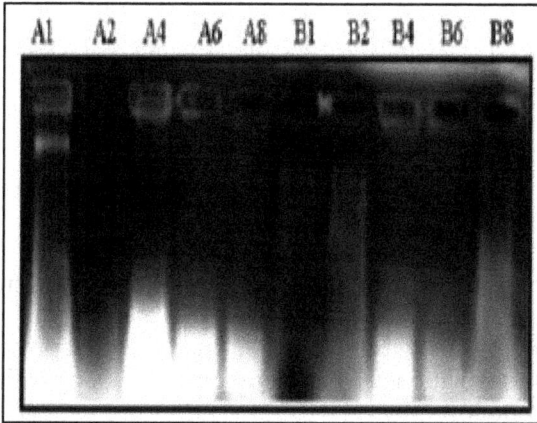

**Figure 2.1: DNA Isolated with different Combinations.
DNA was resolved in 0.8 per cent agarose gel.**

prevent the shearing of DNA and to get a homogenous suspension, samples were mixed gently. Then, tubes were centrifuged and upper aqueous layer was taken out (avoiding contamination with lower layer) which was subjected to DNA precipitation. Here, ethanol gave better results. During precipitation of DNA in ethanol with 3M sodium acetate, polysaccharides remained dissolved (Harini *et al.,* 2008) and DNA molecule was precipitated out. Since, DNA isolated from soil contains lots of polysaccharide; therefore to remove impurities, a final step (PEG purification) was also included in our protocol (Fang *et al.,* 1992). The quality of DNA was tested with three parameters *viz.* spectrophotometer, restriction digestion and PCR amplification,

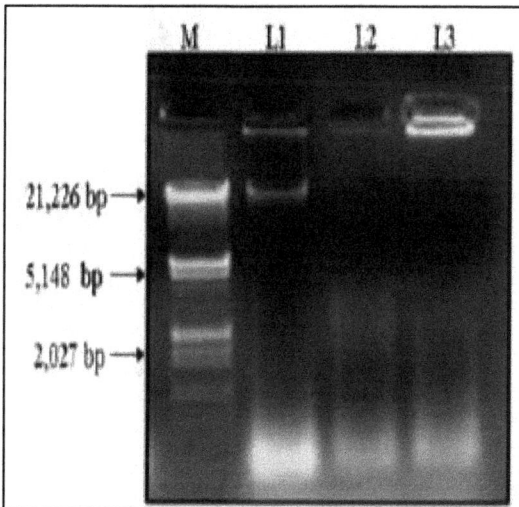

**Figure 2.2: Restriction Digestion of DNA. Restricted DNA was resolved in 0.8 per
cent agarose gel. M, *Eco*R I/*Hind*III double digested λ DNA size marker;
L1, Unrestricted DNA; L2, DNA digested with *Bam*H I; L3, DNA digested with *Eco*RI.**

and cloning and sequencing. A complete smear, followed by incubation of genomic DNA with *Bam*HI and *Eco*RI, attested the high quality of DNA (Figure 2.2). Optical density ratio of A_{260}/A_{280} was 1.9 for 2.5 g soil subjected to sonication however it was 1.27 for 2.5 g soil ground in liquid nitrogen. Further, it was evident by PCR that this protocol gave good quality DNA.

The successful PCR amplification of metagenomic DNA using ITS primers gave the amplification of ~600 bp band (Figure 2.3), which was cloned. This indicates that DNA samples were free of any PCR inhibitors. From the randomly selected colonies, colony PCR was carried out (Figure 2.4) and positive clones were sequenced. The sequences were tested for homology with AMF sequences in NCBI database (http://www.ncbi.nlm.nih.gov/), for more than 20 clones (data not presented here). Thus, this method *i.e.* direct isolation of DNA from soil (2.5 g) along with PCR can be used to identify the presence of endomycorrhizal fungi.

Conclusions

☆ In this study, PCR quality DNA was extracted directly from field grown plants. Thus, our protocol eliminates the need of trap culturing because

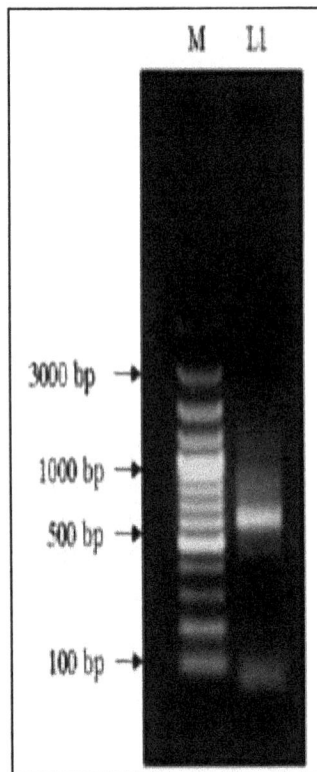

Figure 2.3: Amplified Product after PCR with ITS5/ITS4 Primers of DNA Extracted directly from *Pongamia* Rhizosphere Harboring AMF. The amplified product was resolved on 1.2 per cent gel. M, 100 bp ladder plus; L1, approximately 600bp PCR product.

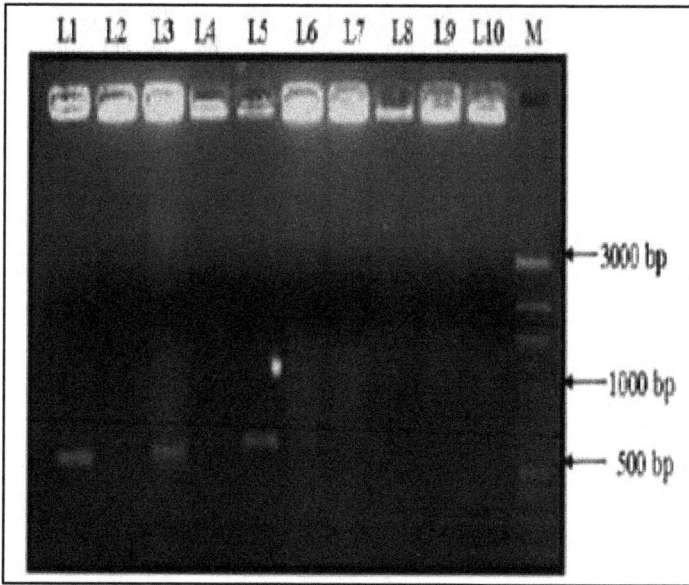

Figure 2.4: Colony PCR using ITS5/ITS4 Primers to Screen Positive Clones. M, 100 bp ladder plus; L1, L3 and L5 (positive clones) approximately 600bp.

researchers usually go for trap culturing to get good percentage of colonization.

☆ Further, metagenomic DNA can be used for PCR to detect the AMF in rhizosphere, as there is no established procedure for the isolation of DNA from *Pongamia* rhizosphere, harboring AMF.

☆ For cell breakage, use of expensive enzymes *viz.*, lysozyme and novozyme, can be avoided thereby making it a safe and cost-effective protocol.

Acknowledgments

The authors are thankful to Head, Department of Botany, Dr. H.S. Gour Central University, Sagar, India for facilitating the research programme and constant encouragement during the study. Ashok Shukla acknowledges funding through University Grants Commission's Dr. DS Kothari Post Doctoral Fellowship Scheme [sanction no. 4-2/2006 (BSR)/13-255/2008(BSR)], New Delhi, India.

References

de Lipthay JR, Enzinger C, Johnsen K, Aamand J and Sorensen SJ. 2004. Impact of DNA extraction method on bacterial community composition measured by denaturing gradient gel electrophoresis. *Soil Biol Biochem* 36: 1607–1614.

Di Bonito R, Elliotts ML and Des Jardin EA. 1995. Detection of an arbuscular mycorrhizal fungus in roots of different plant species with the PCR. *Appl Environ Microbiol* 61: 2809–2810.

Fang SG, Hammer S and Grumet R. 1992. A quick and inexpensive method for removing polysaccharides from plant genomic DNA. *Biotechniques* 13: 52–57.

Frostegard A, Courtois S, Ramisse V, Clerc S, Bernillon D, Le Gall F, Jeannin P, Nesme X and Simonet P. 1999. Quantification of bias related to the extraction of DNA directly from soil. *App Environ Microbiol* 65: 5409–5420.

Graham JH. 2008. Scaling–up evaluation of field functioning of arbuscular mycorrhizal fungi. *New Phytol* 180: 1–2.

Handelsman J, Rondon MR, Brady SF, Clardy J and Goo–dman RM. 1998. olecular biological access to the chemistry of unknown soil microbes: a new frontier for natural products. *Chem Biol* 5: 245–249.

Harini SS, Leelambika M, Kameshwari MNS and Sathyanarayana N. 2008. Optimization of DNA isolation and PCR–RAPD methods for molecular analysis of *Urginea indica* Kunth. *Int J Integ Biol* 2: 138–144.

Jansa J, Mozafar A, Anken T, Ruh R, Sanders IR and Freossard E. 2002. Diversity and structure of AMF communities as affected by tillage in a temperate soil. *Mycorrhiza* 12: 225–234.

Jha A, Kumar A, Kamalvanshi M and Shukla A. 2011. Occurrence of arbuscular mycorrhizal fungi in rhizosphere of selected agroforestry tree species of Bundelkhand region. *Ind Phytopathol* 64: 186–188.

Kamalvanshi M, Kumar A, Jha A and Dhyani SK. 2012. Occurrence of arbuscular mycorrhizal fungi in rhizosphere of *Jatropha curcas* L. in arid and semi arid regions of India. *Ind J Microbiol* 52: 492–494.

Krsek M and Wellington EMH. 1999. Comparison of different methods for the isolation and purification of total community DNA from soil. *J Microbiol Methods* 39: 1–16.

Kruger M, Stockinger H, Kruger C and Schubler A. 2009. DNA–based species level detection of *Glomeromycota*: one PCR primer set for all arbuscular mycorrhizal fungi. *New Phytol* 183: 212–223.

Lanfranco L, Wyss P, Marzachi C and Bonfante P. 1995. Generation of RAPD–PCR primers for the identification of isolates of *Glomus mosseae* and arbuscular mycorrhizal fungus. *Mol Ecol* 4: 61–68.

Ligrone R, Carafa A, Lumini E, Bianciotto V, Bonfante P and Duckett JG. 2007. *Glomeromycotean* associations in liverworts: a molecular, cellular, and taxonomic analysis. *Am J Bot* 94: 1756–1777.

Manian S, Sreenivasaprasad S and Mills PR. 2001. DNA extraction method for PCR in mycorrhizal fungi. *Lett Appl Microbiol* 33: 307–310.

Mishra M, Singh PK and Vyas D. 2012. Arbuscular mycorrhizal fungi as symbiotic bioengineers. In: *Microbes Diversity and Biotechnology* SS Sati and M Belwal (Eds). Daya Publishing House, New Delhi, India. 119–135 pp.

Oehl F, Sieverding E, Ineichen K, Ris EA, Boller T and Wiemken A. 2005. Community structure of arbuscular mycorrhizal fungi at different soil depths in extensively managed agroecosystems. *New Phytol* 165: 273–283.

Oehl F, Sieverding E, Palenuela J, Ineichen K and da Silva GA. 2011. Advances in Glomeromycota taxonomy and classification. *IMA Fungus* 2: 191–199.

Redecker D, Hijri I and Wiemken A. 2003. Molecular identification of arbuscular mycorrhizal fungi in roots: perspective and problems. *Folia Geobot* 38: 113–124.

Redecker D and Raab R. 2006. Phylogeny of the Glomeromycota (arbuscular mycorrhizal fungi): recent developments and new gene markers. *Mycologia* 98: 885–895.

Saano A, Tas E, Pippola S, Lindstrom K and Van Elsas JD. 1995. Extraction and analysis of microbial DNA from soil. In: *Nucleic Acids in the Environment: Methods and Applications* JD Van Elsas and JT Trevors (Eds). Springer–Verlag, Heidelberg, Germany. 49–67 pp.

Sanders IR, Alt M, Groppe K, Boller T and Wiemken A. 1995. Identification of ribosomal DNA polymorphisms among and within spores of the glomales: application to studies on the genetic diversity of arbuscular mycorrhizal fungal communities. *New Phytol* 130: 419–427.

Schneegurt MA, Dore SY and Kulpa CF. 2003. Direct extraction of DNA from soils for studies in microbial ecology. *Curr Issues Mol Biol* 5: 1–8.

Sehgal JL, Mandal DK, Mandal C and Vadivelu S. 1990. Agroecological Regions of India. National Bureau of Soil Survey and Land Use Planning, Tech Bull, Nagpur, India. 25 pp.

Shukla A, Kumar A, Jha A, Ajit and Rao DVKN. 2012b. Phosphorus threshold for arbuscular mycorrhizal colonization of crops and tree seedlings. *Biol Fertil Soil* 48: 109–116.

Shukla A, Kumar A, Jha A, Dhyani SK and Vyas D. 2012a. Cumulative effects of tree based intercropping on arbuscular mycorrhizal fungi. *Biol Fertil Soil* 48: 899–909.

Shukla A, Kumar A, Jha A, Salunkhe O and Vyas D. 2012c. Soil moisture levels affect growth and mycorrhization of agroforestry plants. *Biol Fertil Soil* (doi: 10.1007/s00374–012–0744–8).

Shukla A, Vyas D and Jha A. 2013. Soil depth: an overriding factor for distribution of arbuscular mycorrhizal fungi. *J Soil Sci Plant Nutr* Vol. No. In press.

Simon L, Lalonde M and Bruns T. 1992. Specific amplification of 18 S fungal ribosomal genes from vesicular–arbuscular endomycorrhizal fungi colonizing roots. *App Environ Microbiol* 58: 291–295.

Smith SE and Read DJ. 2008. *Mycorrhizal symbiosis*. Academic Press, New York.

Sorensen SJ, de Lipthay JR, Muller AK, Barkay T, Hansen LH and Rasmussen LD. 2002. Molecular methods for assessing and manipulating the diversity of microbial populations and processes. In: *Enzymes in the Environment* RG Burns and RP Dick (Eds). Marcel Dekker, New York. 363–389 pp.

Theron J and Cloete TE. 2000. Molecular techniques for de–termining microbial diversity and community structure in natural environments. *Crit Rev Microbiol* 26: 37–57.

Torsvik V and Ovreas L. 2002. Microbial diversity and function in soil: from genes to ecosystems. *Curr Opin Microbiol* 5: 240–245.

Tringe SG and Rubin EM. 2005. Metagenomics: DNA sequencing of environmental samples. *Nat Rev Genet* 6: 805–814.

Volossiouk T, Robb EJ and Nazar RN. 1995. Direct DNA Extraction for PCR–Mediated Assays of Soil Organisms. *App Environ Microbiol* 61: 3972–3976.

Vyas D, Dubey A, Soni A, Mishra MK and Singh PK. 2007. Arbuscular mycorrhizal fungi in early land plants. *Mycorr News* 19: 22–24.

Vyas D, Dubey A, Soni A, Mishra MK, Singh PK and Gupta RK. 2008. Vesicular arbuscular mycorrhizal association in bryophytes isolated from eastern and western Himalayas. *Mycor News* 19: 16–18.

Vyas D, Shukla A and Jha A. 2011. Occurrence and diversity of arbuscular mycorrhizal fungi in rhizosphere of medicinal plants. In: Proc Nat Conf on Biodiversity and Biotechnology for Sustainable Development, 21–22 March, 2011. Karnataka University, Dharwad, India, pp. 26–41.

Wang B and Qiu YL. 2006. Phylogenetic distribution and evolution of mycorrhizas in land plants. *Mycorrhiza* 16: 299–363.

Wyss P and Bonfonte P. 1993. Amplification of genomic DNA of arbuscular–mycorrhizal (AM) fungi by PCR using short arbitrary primers. *Mycol Res* 97: 1351–1357.

2015, Recent Trends in Microbiology, Mycology and Plant Pathology   *Pages* **39–45**
Editor: **Dr. H.C. Lakshman**
Published by: **DAYA PUBLISHING HOUSE, NEW DELHI**

Chapter 3

Mycorrhizae and Molecular Approaches to its Functioning*

M.N. Sreenivasa

Department of Agricultural Microbiology,
University of Agricultural Sciences, Dharwad – 580 005, Karnataka, India
E-mail: sreenivasamn@gmail.com

Certain beneficial fungi form mutualistic association with the roots of higher plants. Such symbiotic associations are often referred to as "Mycorrhizae" (In Greek, Mykes mean mushroom or fungus; rhiza mean root).

The term mycorrhizae was coined by a German botanist Albert Bernard Frank in 1885 which literally mean "Fungus – roots". Generally agricultural and horticultural crops in addition to certain tropical tree species form mycorrhizal association. Fossil and molecular data indicate that these organisms evolved about 400 million years ago.

Mycorrhizal fungi are broadly classified in to two groups:

1. Ectomycorrhizae
2. Endomycorrhizae

1. Ectomycorrhizae

This is otherwise referred to as ectotrophic mycorrhizae. They are formed by the invasion of actively growing absorbing roots of Gymnospermous and Angiospermous plants by Basidiomycetes or Ascomycetes fungi. A compact mantle of mycelium is

---

* Paper presented in DBT sponsored training programme on 15-9-2009 at GKVK, Bangalore.

formed on the surface of roots. The infested roots become morphologically different from normal roots. Ectomycorrhizal roots are short, fleshy, dichotomously branched and devoid of root hairs.

The ectomycorrhizae are characterized by the presence of an intercellular fungal network in the root cortex termed as "Hartig net". The Basidiomycetous genera, *Amanita, Boletus, Russula, Laccaria,* and *Pisolithus* etc form ectomycorrhizal association in tree species belonging to Pinaceae, Fagaceae, and Juglandaceae etc. Among Ascomycetes, Truffles (Tuberaceae) are known to form ectomycorrhizal association with trees like oak and beech. Each genus show a wide host range, inturn, a single tree species may harbour different mycorrhizal fungi.

The ectomycorrhizal fungi can be easily isolated on a laboratory medium (Melin Norkran's medium) using a carbon source and certain growth factors such as thiamine, aminoacids etc. They play a key role in afforestation programmes.

2. Endomycorrhizae

The endomycorrhizae includes smaller groups like Arbutoid, Ericoid and Orchid mycorrhizae in addition a predominant group, vesicular-arbuscular mycorrhizae (VAM). VAM fungi are formed by zygomycetous fungi belonging to the genera *Glomus, Gigaspora, Acaulospora, Sclerocystis, Entrophospora* and *Scutellospora* in the family Glomaceae of the order Glomales. All these genera produce azygospores as they are produced with out sexual reproduction. VAM fungi occur in soil universally and are very well established globally as plant growth promoting fungi.

The AM fungi penetrate the epidermis of the root and produce intercellular non septate hyphae. The hyphae produce minutely branched structures that appear intracellularly soon after the entry of the fungus to root cortex. These structures are called as "arbuscules" which are involved in bidirectional transfer of nutrients. The hyphae also produce thin walled spherical to oval bladder like structures called as "vesicle" which act as storage organ of extra nutrients and release them under nutrient deficient conditions.

Of late, VAM associations are called as Arbuscular mycorrhizal (AM) associations as vesicle formation is not seen in few genera.

Several scientists characterized the symbiosis between the two biotrophic organisms by bidirectional transfer of nutrients which gives access for the plant to low mobile elements like phosphorus. AM fungi may also influence plant health and development through other non nutritional mechanisms such as production of growth regulating substances or by increasing the resistance to root pathogens. (Barea and Azcon Aguilar 1982; Azcon – Aguilar and Barea, 1996). Many times the structure of plant communities are modified in mycorrhizal plants.

Beginning of Molecular Work on Mycorrhizae

AM symbiosis is the result of a very dynamic process in both space and time. AM fungi being obligate symbiont always multiply in presence of the host plant roots. Key developmental steps are the germination of spores and development of hyphae in the absence of the host (Hepper, 1983) but further hyphal branching and subsequent formation of appressoria will take place in the presence of roots. Many saprophytic bacteria and fungi are found in the vicinity of mycorrhizal spores (Lee and Koske, 1994) which may support the growth of hyphae and formation of mycorrhizae.

Recently, the cytoplasm of spores was found to contain bacteria like organisms (BLO's) belonging to the genus *Burkoldheria*. Interestingly a sequence cloned from this endosymbiont showed similarity to nif D which belongs to the nitrogen fixing gene cluster of *Rhizobium*.

Two main questions posed in respect to understanding of AM fungal biology are on their biodiversity and phylogeny which are dependent on their identification and differentiation.

Until recently the evaluation has relied mainly on the morphology of the asexual spores which depends on environmental factors and physiological state. Molecular methods provide an alternative for precise analysis. Few scientists have tried isozyme analysis while others have tried the use of nucleic acids. However, in case of AM fungi, the amount of material available is very limited, most methods are based on enzymatic amplification of DNA by polymerase chain reaction (PCR). Lanfranco *et al.* (1995) showed that the RAPD approach can lead to clear and reproducible identification of certain AM fungus in low amounts and complex samples of DNA. PCR normally needs information of at least short sequence stretches for the design of oligonucleotides. RAPD is not dependent on this because short randomly chosen primers are used for this technique.

A modification of the RAPD technique for finding polymorphic loci is the AFLP method. This AFLP was originally developed to detect polymorphism in bacteria and plants. Rosendahl and Taylor (1997), showed the possibility of application of this technique to the DNA of a single AM fungal spore. The extracted DNA was digested with a restriction enzyme, the resulting fragments were coupled to a linker and amplified with oligonucleotides binding to this linker. This mixture of PCR products served for further amplification obtaining reproducible polymorphic patterns characteristic of each starting spore. The results showed that there are differences between single spores of certain isolates and suggested that AM fungi reproduce clonally without showing recombination.

Redecker *et al.* (1997) used a technique combining PCR and RFLP to generate specific fragment patterns from spore extracts of AM fungi. With the universal primers ITS1 and ITS4, DNA fragments were amplified from species of *Scutellospora* and *Gigaspora* that were approximately 500 bp long. The apparent lengths of the corresponding fragments from *Glomus* sp. varied between 580 and 600 bp. The restriction enzymes Mbo1, Hinf1 and Taq 1 were used to distinguish species with in genus *Glomus*. Depending upon the restriction enzyme used, groups of species with common fragment patterns could be found. The variation of internal transcribed spacer sequences among the *Gigaspora* species under study was low. Fragment patterns of *Scutellospora* spp. showed that phylogenetic relationship with *Gigaspora* and revealed only a slightly higher degree of variation.

Redecker (2000) designed a set of PCR primers targeted at five major phylogenetic sub groups of AM fungi to facilitate specific amplification of internal transcribed spacers and 18S r RNA gene from colonized roots in the absence of spores.

Dickie and Fitzjohn (2007), used terminal restriction fragment length polymorphism (T-RFLP) to identify mycorrhizal fungi. The other techniques used in this regard are sequencing, denaturing gradient gel electrophoresis (DGGE) and clone libraries (Renker *et al.,* 2006).

Ma *et al.* (2005) standardized PCR-PGGE procedure for the detection of AM fungal 18S r RNA gene in cultivated soils.

Many scientists claim that T-RFLP is more sensitive than DGGE for fungi although obtaining sequences directly from samples is easy with DGGE. Clone library (Renker *et al.,* 2006) is most accurate for identifying species although T-RFLP is cost effective. Hence, both these techniques can be used together. T-RFLP can be applied to process large number of samples and clone libraries on selected samples to obtain identities of key species (Widmer *et al.,* 2006).

PCR on r RNA Genes

Ribosomal RNA (r RNA) genes are common to all organisms and contain very conserved sequences as well as highly variable regions. Hence, it is possible to design universal primers for the amplification of DNA from a group of organisms. It is possible to compare the fragments obtained between groups.

More than any thing since r RNA genes are present in relatively high copy numbers, only small amount of DNA are necessary for PCR. For these reasons they are one of the main targets for molecular studies with respect to biodiversity and phylogeny.

The first gene that was amplified from AM fungi was 18S r RNA gene (Simon *et al.,* 1992). Sequencing or analysis of restriction enzyme digestion of PCR products from r RNA genes has now been used to answer doubts concerning biodiversity and phylogeny. Simon *et al.* (1993) constructed phylogenetic tree for AM fungi based on molecular approach for the first time and set the origin of these organisms to the time where plants start to colonize terrestrial ecosystems confirming the results of fossil data. Thus many scientists working on the aspect of evolution found rRNA – PCR technology convenient.

Signaling between the Partners

The interaction between an AM fungus and the plant root starts even before both partners physically contact each other.

Giovannetti and Citernesi (1993), detected an increased hyphal growth and branching when the fungus was in close proximity to the root which indicate the importance of root exudates. Analysis of the effect of root exudates on the fungus showed that it was most pronounced when these exudates were derived from phosphate starved plants. That may be the reason for better colonization in phosphate starved soils than in phosphate fertilized soils.

Since, there is considerable amount of similarities between the nodule symbiosis between legume- *Rhizobium* and mycorrhizal symbiosis, many scientists started working on flavonoid compounds present in the root exudates and are known to modify bacterial nod – gene expression. Baptista and Sequeira (1994) found the effect of quercetin and naringenin on presymbiotic development of *Gigaspora.*

Later research carried out by Becard *et al.* (1995) revealed the influence of additional compounds like polyamins and salicylic acid present in root exudates which influence the development of mycosymbiont through an experiment using maize mutant completely lacking flavonoid compounds that showed same degree of colonization as their corresponding wild type plants Nagahashi *et al.* (1998) developed a method where by they injected root exudates close to developing hyphae in to the water agar. At these points, fungi showed branching after 6-12 hrs of injection. Any way the final identification of the responsive substances has still to be ascertained.

Molecular Approaches in Mass Production of AM Fungi

Since AM fungi are obligate symbionts, they can be mass produced as pot cultures using suitable host and substrate (Sreenivasa and Bagyaraj, 1989). Several other techniques are viz Hydroponic system (Hung and Sylvia, 1988) and root organ culture (Mosse and Hepper, 1975). Later several scientists obtained similar results using carrot roots genetically transformed by *Agrobacterium rhizogenes.* This system of dual culture allowed abundant production of spores (Average 9500 spores/Petridish). The root organ culture exhibited higher inoculum potential due to the numerous vesicles and extensive intraradical mycelium (Tahir, 2003).

Pure culture of AM fungi is still a big challenge for microbiologists. The pre symbiotique growth of AM is characterized by formation of running hyphae. After few weeks, growth of germinated AM propagulas stops and the cytoplasm is retracted without host partner. Hildebrant *et al.* (2002) mentioned the presence of slime forming bacteria identified as *Paenibacillus validus* on surface sterilized spores of *Glomus intraradices.* These bacteria stimulate the growth of *Glomus intraradices* up to spore formation in the absence of any plant tissue. This is attributed to a chemical component secreted by the bacteria.

It is better and necessary to evaluate the inoculum potential of AM propagules in continuous monoxenic cultures.

Conclusion

At present, the research on AM fungi is at the cross roads. The beneficial role of AM fungi in over all plant growth have been thoroughly investigated. However, the biosystematics, culturability etc are still lagging behind for want of suitable techniques which have become hurdles for further progress in AM research. The work on molecular methods for the investigation of biodiversity and for identification of AM fungi is in progress. In few laboratories, the work on cloning and analysis of protein encoding genes is going on since 5-6 years. However, it may not be just sufficient to isolate such genes but also we need to analyze their regulation and their function. The molecular work on AM fungi may help to resolve many confronted issues in ensuing years.

References

Azcon–Aguilar C and Barea J. M., 1996. Arbuscular mycorrhizas and biological control of soil–borne plant pathogens–an overview of the mechanisms involved. *Mycorrhiza*, **6**: 457–464.

Baptista M. J. and Sequeira J. O., 1994. Effect of flavonoids on spore germination and a symbiotic growth of the arbuscular mycorrhizal fungus, *Gigaspora gigantea*. *Review of Plant Pathology*, 76: 10–28.

Barea J. M. and Azcon–Aguilar C. 1982. Production of plant growth–regulating substances by the vesicular arbuscular mycorrhizal fungus, *Glomus mosseae. Applied and Environmental Microbiology* **43**: 810–813.

Becard G, Taylor L. P., Douds D. D., Pfeffer P.E. and Doner L. W., 1995. Flavonoids are not necessary plant signal compounds in arbuscular mycorrhizal symbioses. *Molecular Plant–Microbe Interaction*, **8**: 252—258.

Dickie I. A. and Fitzjohn R. G., 2007. Using terminal restriction fragment length polymorphism (T–RFLP) to identify mycorrhizal fungi: a methods review. *Mycorrhiza*. 17: 259–270.

Giovannetti M and Citernesi 1993. Time–course of appresorium formation on host plant by arbuscular mycorrhizal fungi. *Mycological Research.*, **97**: 1140–1142.

Hepper C. M. 1983. Limited independent growth of vesicular arbuscular mycorrhizal fungus *in vitro. New Phytologist.* **93**: 537–542.

Hildebrant U, Janetta K and Bothe H., 2002. Towards growth of arbuscular mycorrhizal fungi independent of a plant host. *Applied and Environmental Microbiology.* 68: 1919–1924.

Hung, L. L., and Sylvia, D. M. 1988. Production of vesicular– arbuscular mycorrhizal fungus inoculum in aeroponic culture. *Applied and Environmental Microbiology.* 54: 353–357.

Lanfranco L., Wyss P, Marzachi C and Bondante P. 1995. Generation of RAPD–PCR primers for the identification of isolates of *Glomus mosseae,* an arbuscular mycorrhizal fungus. *Molecular Ecology.* **4**: 61–60.

Lee P. J. and Koske R. E., 1994. *Gigaspora gigantea*: Parasitism of spores by fungi and actinomycetes. *Mycological Research*, **98**: 458–466.

Ma W. K., Siciliano S. D and Germida J. J., 2005. A PCR–DGGE method for detecting arbuscular mycorrhizal fungi in cultivated soils. *Soil Biology and Biochemistry.* **37**: 1589–1597.

Mosse, B and Hepper, C. M., 1975. Vesicular–arbuscular mycorrhizal infection in root organ culture. *Physiological Plant Pathology.* **5**, 215–223.

Nagahashi G., Douds, D. D. and O' Connor J, 1998. Fractioning of AM fungal branching signals from aqueous extracts of Ri T–DNA transform carrot roots. In: *Proceedings of 2nd International Conference on Mycorrhizae.* Uppsala, 1998, 125pp.

Redecker D, 2000. Specific PCR primers to identify AM fungi within colonized roots. *Mycorrhiza*, 10: 73–80.

Redecker D, Thiergelder H, Walker, C and Werner D, 1997. Restriction analysis of PCR–amplified internal transcribed spacers of ribosomal DNA as a tool for species identification in different genera of the order Glomales. *Applied and Environmental Microbiology.* 63: 1756–1761.

Renker C. WeiBhuhan K., kellner H. and Buscot F., 2006, Rationalizing molecular analysis of field collected roots for assessing diversity of AM fungi: to pool or not to pool, that is the question. *Mycorrhiza,*16: 525–531.

Rosendahl S and Taylor J., 1997. Development of multiple genetic markers for studies of genetic variation in arbuscular mycorrhizal fungi using *AFLP. Molecular Ecology.* **6:** 821–829.

Simon L., Bousquet J., Levesque R. C. and Lalonde M., 1993. Origin and diversification of endomycorrhizal fungi and coincidence with vascular land plants. *Nature,* 363: 67–69.

Simon, L., Lalonde and Mand Bruns, T. D., 1992, Specific amplification of 18S fungal ribosomal genes from VAM fungi colonizing roots. *Applied and Environmental Microbiology.* 58: 291–295.

Sreenivasa, M. N. and Bagyaraj D. J. 1989. Mass producing plant inoculum. *Appropriate. Technology* (UK): 16: 4.

Tahir, A., 2003. *In vitro* culture of AM fungi: advances and future prospects. *African Journal of Bio–technology.* 2: 692–697.

Widmer F., Hartmann M, Frey B. and Kolliker R., 2006. A novel strategy to extract specific phylogenetic sequence information from community T–RFLP. *Journal of Microbiological Methods.* 66: 512–529.

2015, Recent Trends in Microbiology, Mycology and Plant Pathology   *Pages* **47–59**
Editor: **Dr. H.C. Lakshman**
Published by: **DAYA PUBLISHING HOUSE, NEW DELHI**

Chapter 4

Recent Techniques in Arbuscular Mycorrhizal Fungal Mass Production

P. Savitha, K.S. Jagadeesh, D.K. Dushyantha
and P. Nirmalnath Jones*

*Department of Agricultural Microbiology,
University of Agricultural Sciences, Dharwad – 580 005, Karnataka, India
E-mail: abdhi2050@gmail.com

ABSTRACT

Different cultivation and inoculum production techniques of arbuscular mycorrhizal (AM) fungi have been developed in the past decades. Among many techniques, pot culture, aeroponics, hydroponics and *in vitro* culture techniques are used in large scale. In this review,an attempt is made to describe these techniques along with their advantages and disadvantages.

Introduction

Mycorrhizal fungi, especially those that are arbuscular (AM), are ubiquitous soil inhabitants forming symbioses with most naturally growing terrestrial (Jeffries, 1987) and most aquatic (Khan and Belik, 1995) plants and have been shown to enhance the growth of numerous plants of economic importance (Jeffries, 1987) including native plants used for revegetation of disturbed sites such as renourished coastal beaches (Sylvia, 1989). Arbuscular mycorrhizal fungi (AMF) are obligate symbionts and need a compatible plant host to complete their life cycle and produce spores which are the main source of propagules used for application in crop production. Major benefits of AM fungi to plant are increased uptake of phosphorus and other poorly mobile nutrients and tolerance of water stress. Their potential to enhance plant growth is well documented and recognized but not fully exploited. Nowadays, they are increasingly considered in agriculture, horticulture, and forestry programs, as well as for environmental reclamation, to increase crop yield and health and to limit the application of agrochemicals (Gianinazzi *et al.,* 2002; Johansson *et al.,* 2004). The potential advantages of inoculation of nursery plants with AM fungi (AMF) in horticulture, agriculture and forestry is not perceived by these industries as a significant issue. This is partially due to inadequate methods for large-scale inoculum production. Thus, the exploitation of AM fungi still presents a great challenge.

Materials and Methods

Production systems of AM fungi have evolved considerably during recent years, from relatively simple technologies to more complex ones, for example, *in vitro* methods (Jarstfer and Sylvia, 1994). At present, inoculum is produced for commercial purposes in the following ways:

1. Containers (pots) with different substrates (Feldmann and Idczak, 1994; Feldmann and Grotkass, 2002). Advantages: low technology input, undesirable contaminations fairly easily eliminated, reasonable costs; disadvantages: not pure, limited in its industrial development.

2. Aeroponic systems (Jarstfer and Sylvia, 1994), where pre-inoculated plant roots are continuously misted with nutrient solution sprayed within cultivation boxes. Advantages: easier control of contaminants, carrier-free inoculum, adapted for microplants; disadvantages: relatively complicated technological setup.

3. *In vitro* on roots transformed with *Agrobacterium rhizogenes* (Becard and Fortin 1988; Declerck *et al.,* 1996). Advantages: pure cultures, permits industrial development; disadvantages: high technological investment, high costs, not all AM fungi successfully culturable in this system, and suitability of inoculum produced *in vitro*, in particular its competitive ability toward other microbes in field soil, is yet to be tested.

Production of Monoxenic Cultures

The details are furnished in Figure 4.1. They are multiplied in association with a mycotrophic plant under greenhouse conditions [Figure 4.1(2)]. Isolated spores or vesicles can be monoxenically cultivated in presence of isolated roots [Figure 4.1(3,4)] or on entire plant host [Figure 4.1(5)]. Intraradical forms of AM fungi [Figure 4.1(6,7)] are used as starter inocula to have monoxenic cultures [Figure 4.1(8)] on M medium (Becard and Fortin, 1988) or on MSR medium (Diop, 1995). Regular subcultures every 4 months with total transmission of AM symbiosis [Figure 4.1(9)] allowed to obtain different fungal generations in continuous cultures. Probable loss of infectivity is avoided by introducing plant in the *in vitro* system [Figure 4.1(10)]. The use of Sunbags [Figure 4.1(11)] in *in vitro* system is another alternative to obtain large scale AM inoculum without contaminations. The axenic AM propagules are conserved at 4°C in the dark for several months or use for fundamental or inoculation practices.

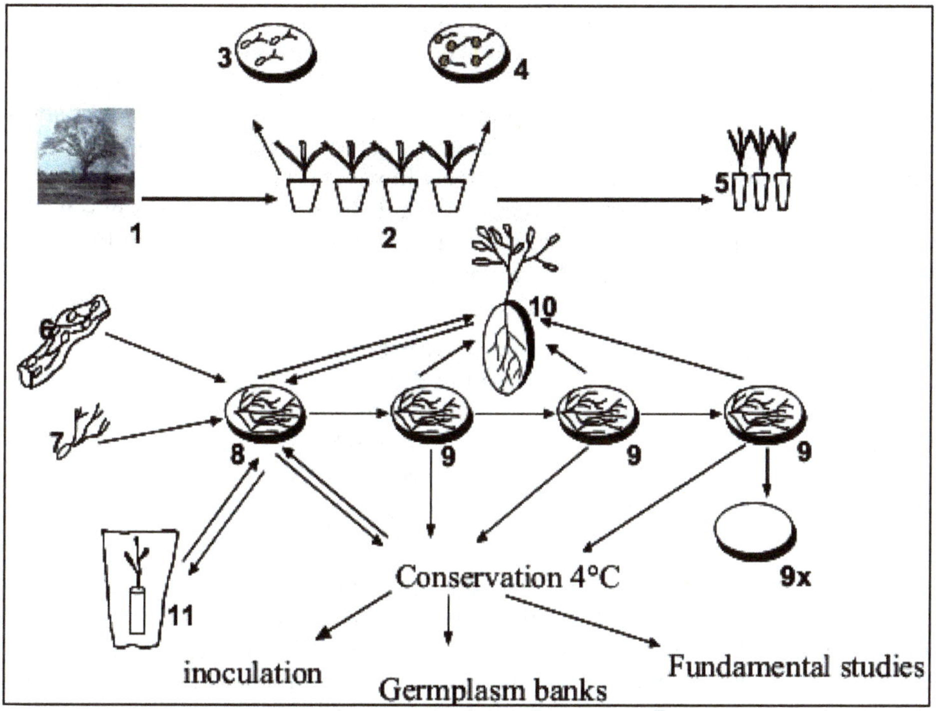

Figure 4.1: Protocol of Extraction, Purification, Cultivation and Preservation of AM Fungi.

Pot Culture

Cultivation of AM fungi in pots, bags or beds is the most widely adopted technique for AM fungal inoculum production because relatively low technical support is needed and consumables are cheap. Substrate based production systems are considered as a convenient system for large-scale production that is able to reach inoculum densities set for mass production of 80–100 propagules per cubic centimetre (Feldmann and Grotkass, 2002). Large-scale production may be achieved in single pots of various materials (earthenware or plastic) and sizes (Millner and Kitt 1992; Sylvia and Schenck 1983) or scaled up to medium-size bags and containers and to large raised or grounded beds (Douds *et al.,* 2005, 2006; Gaur and Adholeya, 2002). The production process is often conducted under controlled or semi controlled conditions in greenhouses or performed in growth chambers for the easy handling and control of parameters such as humidity and temperature.

To initiate production of AM fungi, starter inoculum may be isolated spores (Douds and Schenck, 1990a, b) or mixture of spores and mycorrhizal root pieces (Gaur and Adholeya, 2000). Spores can be obtained by wet sieving and decanting. Mixed inoculum is obtained by drying and chopping the roots into fine pieces. The soil containing AM fungal hyphae may also be used in a mixed inoculum (Gaur and Adholeya, 2000). Plantlets can be precolonised before their transplantation into beds (Douds *et al.,* 2005, 2006) or containers.

Host dependent sporulation of AM fungal species (Dodd *et al.,* 1990; Strubble and Skipper, 1988) is an important determinant for inoculum production. Large scale production of AM fungi is done by using host plants like Onion, Leek, Maize, and Bahia grass. Because these offer some of the advantages like they have very short life cycle, adequate root system development, a good colonisation level by a large range of AM fungi and tolerance to relatively low levels of phosphorus. AM fungi inoculum type partly determines the host plant/fungus association chosen. For example high intraradical colonisation levels are important for the production of mixed spore-root inoculum, while this might not always be needed for the achievement of a spore inoculum.

Inert substrates (vermiculite and perlite) can be used as substitutes for soil, sand and substrate amendments, as they used to dilute nutrient rich soil and compost (Douds *et al.,* 2005, 2006). Inert substances have also been used as carrier medium to support roots and fungal growth under conditions where plant feeding was mainly provided by a nutrient solution (Lee and George, 2005). Compost or other organic substrates such as peat can be added to nutrient deficient soils (Gaur and Adholeya, 2002; Ma *et al.,* 2007).

Substrate particle size influences drainage, humidity and aeration. These parameters have been shown to influence sporulation of AM fungi (Gaur and Adholeya, 2000). Substrate used for cultivation is pre-treated to avoid contamination by plant pathogens. Substrate used in pots, containers and bags can be treated by steam or heat sterilisation or by irradiation. The substrate in raised beds can be either fumigated or left untreated (Douds *et al.,* 2005, 2006). AM fungal colonisation is favoured under low nutrient (mainly P) conditions (Amijee *et al.,* 1993; Smith and Read, 2008).

Aeroponic and Hydroponic System

In substrate-free production systems (*i.e.,* hydroponics and aeroponics) pre-colonized plants are produced prior to their introduction into the systems (Figure 4.2). For preinoculation, plant seedlings and AM fungal propagules (both preferably surface-sterilized) are usually pre-cultured in pots containing a substrate (*e.g.,* mixture of sand and perlite) for several weeks. The container in which the roots (and AM fungus) develop is usually protected from light to prevent the development of algae (Jarstfer and Sylvia, 1995). Aeroponics and hydroponics mainly differ in the mode of aeration and application of the nutrient solution. Aeroponics is a form of hydroponics in which the roots (and AM fungus) are bathed in a nutrient solution mist (Zobel *et al.,* 1976). Spraying of micro droplets increases the aeration of the culture medium, and in addition, the liquid film surrounding the roots allows gas exchange. This mist can be applied by various techniques that differ mainly in the size of the fine droplets produced.

In vitro Culture

In vitro culture of AM fungi was achieved for the first time in the early 1960s (Mosse, 1962). It is a valuable tool for the study of AM fungi. Different production systems have been derived from the basic root organ culture (ROC) in Petri plates. For

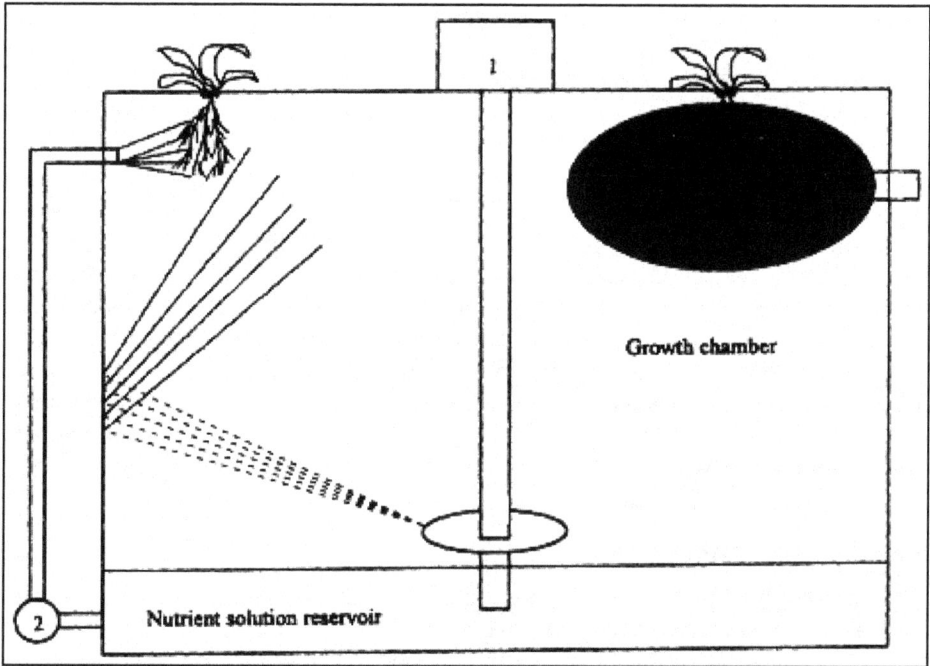

Figure 4.2: Three Ways of Producing Atomised Nutrient Solution for the Growth of Mycorrhizal Plants in a Flat-bed Aeroponic Chamber (1) Impeller System for Distributing Nutrient Spray (2) System with Nozzles for Producing a Pressurised Spray (after Jarstfer and Sylvia, 1994).

example, Tiwari and Adholeya (2003) cultured root organs and AM fungi in small containers, by which large-scale production was obtained (Adholeya *et al.,* 2005). Large-scale cultivation of AM fungi has also been performed in an airlift bioreactor (Jolicoeur *et al.,*1999), in a mist bioreactor with perlite as a substrate (Jolicoeur 1998), and in a bioreactor containing solid (*i.e.,* gelled medium) support elements (Fortin *et al.,* 1996). In the patented container-based hydroponic culture system of Wang (2003), the root organs and AM fungus were periodically exposed to a liquid culture medium. Gadkar *et al.* (2006) further developed a container, in which a Petri plate containing a ROC was used to initiate fungal proliferation in a separate compartment filled with sterile expanded clay balls.

Root Organ Culture

The use of excised roots as host partner in AM symbiosis was first proposed by Mosse and Hepper (1975). Isolated root can be propagated continuously in different solid and liquid media with high reproducibility. Initiation of isolated roots requires pregermination of seeds previously surface sterilized with classical disinfectants (sodium hypochlorite, hydrogen peroxide), then thoroughly washed in sterile distilled water. Germination of seeds occurs after 48 h at 28°C in the dark on water agar or moistened filter papers. The tips (2 cm) of emerged can be transferred to a rich medium such as modified White medium (Bécard and Fortin, 1988) or Strullu and Romand medium. The pH of the medium is adjusted to 5.5 before autoclaving. Fast-growing roots are cloned by repeated subcultures. Transformation of roots by *Agrobacterium rhizogenes* has provided a new way to obtain mass production of roots in a very short time. A two day old loopful of a bacterial suspension is used to inoculate sections of root organs. Genetically modified carrot (*Daucus carrota L.*) roots by *A. rhizogenes* show profuse roots two to four weeks later. Then, the tips are aseptically cultivated on rich medium (Bécard and Fortin, 1988). Several subcultures (3 to 4) are necessary in this medium enriched with antibiotics such as carbenicillin or ampicillin, to obtain free living roots without bacteria. A clonal culture derived from a single root is then established. *A. rhizogenes* inserts in transformed root copies of tDNA (transfer DNA) which occurs in a large plasmid of *A. rhizogenes* (Chilton *et al.,* 1982). Therefore, transformed roots have a quick, vigorous and homogenous growth in relative poor substrates without supplementation of hormonal substances. The negative geotropism of transformed roots facilitates contacts with hyphae of AM fungi (Bécard and Fortin, 1988; Mugnier and Mosse, 1987). Tepfer (1984) indicated that they can survive for a long time without subculture.

Results and Discussion

Different host plants are used for mass production of AM fungi in pot culture and Table 4.1 represents arbuscular mycorrhizal colonization (per cent) in four host plants. Pot culture is the most widely used method for producing AMF inoculum but it is time consuming, bulky, and often not pathogen free. Besides, these methods are often space consuming and need pest control. Harvesting is usually performed by wet sieving and decanting, which can be followed by centrifugation. Different substrates used in pot cultures have effect on colonisation and spore population of AMF at different stages of plant growth (Table 4.2). When the substrate is not used as a carrier, the final inoculum can be difficult to prepare due to the attachment of clay particles and organic debris (Millner and Kitt 1992). Technical adaptations such as the addition of glass beads, river sand, or vermiculite seem to have limited this problem and facilitate harvesting of relatively clean AM fungal spores and roots that can be chopped into pieces. The presence of a substrate, however, provides an inoculum which is not directly suitable for mechanical application, as is the case for substrate-free production methods (Mohammad *et al.,* 2004).

Table 4.1: Arbuscular Mycorrhizal Colonization (per cent) in Four Hosts Plant Used for Mass Production of AM Fungi (Chaurasia and Khare, 2005)

Host Plant	Period (days)			
	15	30	45	60
H. vulgare	54	60	72	92
T. aestivum	40	49	59	71
P. vulgaris	37	44	50	62
P. mungo	38	41	52	66

Table 4.2: Effect of Various Substrates on Colonization and Spore Population of *G. mosseae* in Onion at different Stages of Plant Growth (Sunita *et al.,* 2011)

Days	30					60					90					120				
Substrates	%age	H	A	V	SC	%age	H	A	V	SC	%age	H	A	V	SC	%age	H	A	V	SC
Soil-FYM (3:1)	*75.8±4.33	+	+	+	14.0±3.26	87.98±5.78	+	–	–	48.33±2.49	83.5±2.48	+	+	+	76.3±2.86	100±0	+	–	+	103.0±1-63
Soil-VC (3:1)	55.0±10.66	+	+	+	46.7±1.24	61.78±1.47	+	+	+	54.3±2.49	72.5±0.79	+	–	+	64.33±2.86	91.5±1.57	+	–	+	84.7±2.05
Soil-OM (3:1)	83.4±4.71	+	+	+	45.0±4.98	71.78±1.10	+	+	+	51.7±2.86	82.7±1.97	+	+	+	60.4±1.24	94.6±4.12	+	–	+	103.0±1.63
Soil-sand (3:1)	24.2±4.23	+	–	–	9.7±2.05	25.78±3.18	+	–	–	21.4±1.24	51.0±0.81	+	+	+	26.3±8.49	63.6±7.43	+	+	+	54.3±2.62

%age: Percentage mycorrhizal root colonization (H: Hyphael, A: Arbuscular; V: Vesicular); SC : Mycorrhizal spore count.

* Mean of three replicates; ±: Standard deviation; FYM: Farmyard manure; VC: Vermicompost; OM: Organic manure.

To eliminate contamination from soil organisms and for better control of physical and chemical properties, AM fungi have been cultured in soil-less media; such techniques provide greater uniformity in the composition of the product and facilitate better aeration than soil. Some species of AM fungi from four genera have been cultured in soil less media and are listed in Table 4.3.

Table 4.3: Soil Less Media Used in the Culture of Arbuscular Mycorrhizal Fungi (Jarstfer and Sylvia, 1992)

Materials	AM Fungi
Bark-shredded douglas fir	*Glomus fasciculatum*
Calcined montmorillonite clay	*G. macrocarpum, G. monosporum, G. versiforme, Scutellospora calospora*
Expanded clay aggregates	*Acaulospora laevis, G. constrictum, G. fasciculatum, G. intraradix, G. mosseae*
Peat	
Hypnum	*G. fasciculatum*
Sphagnum	*A. spinosa, G. fasciculatum+ perlite and G. mosseae,*
Vermiculite + pumice (2:1)	*G. intraradix, G. fascicultum*
Perlite + soilrite (1:1)	*G. fasciculatum*
Vermiculite	*A. spinosa, G. fasciculatum, G. mosseae*

To overcome these problems, various soil-free methods, namely aeroponic system (Hung *et al.,* 1991; Jarsfter and Sylvia 1995), hydroponic (Elmes and Mosse 1984; Mosse and Thompson 1984), and use of excised roots (*in vitro*) as host partner in AM symbiosis (Mosse and Hepper,1975) have been used successfully to produce AMF-colonized root inoculum.

The main advantage of the aeroponic cultivation system is the production of inoculum, free from attached substrate particles. Sheared-root inoculum (roots chopped up in a food processor and washed over sieves) with high propagules density can directly be used for application or can be processed for storage (Sylvia and Jarstfer 1992; Jarstfer and Sylvia, 1995). Spores can be easily separated from the roots in absence of debris on the root material (Millner and Kitt 1992). Samples of clean roots can also be harvested and analyzed without roughly interrupting the cultivation of the AM fungus. Moreover, the risk of cross-contamination by other AM fungi is low in such systems. In addition, nutrient supply and pH can be monitored and/or manipulated in substrate-free cultivation systems to optimize cultivation settings for a particular host/AM fungus association. As a disadvantage, liquid nutrient solutions are prone to the multiplication and dissemination of microbial contaminants as well as the development of algae (Elmes and Mosse 1984). Covering channels, addition of clean P sources, and the utilization of soil-free substrate in the preinoculation phase will solve part of the problems. Dugassa *et al.* (1995) for example pre-cultured plants in a sand substrate before transferring them into the common nutrient solution, while Voets *et al.* (2009) successfully transplanted autotrophic *in vitro* produced mycorrhizal plants.

The most obvious advantage shared by all *in vitro* cultivation systems is the absence of undesirable micro-organisms, which makes them more suitable for large-scale production of high-quality inoculum. Factors that influence optimal production (*e.g.*, nutrient availability, presence of contaminants) can be more easily detected and controlled in (liquid) *in vitro* cultures. As a disadvantage, the diversity (in terms of genera) of AM fungi that have been grown *in vitro* is lower than under pot cultivation systems. Another disadvantage of *in vitro* production is the costs associated with the production systems, requiring skilled technicians and laboratory equipments such as sterile work flows, controlled incubators for ROC, and growth chambers for plant systems.

In India also, the technology standardised by The Energy and Resources Institute (TERI), New Delhi, has been transferred to many private entrepreneurs for commercial production of VAM biofertilizer. For example, M/s. Cosme Biotech Pvt. Ltd., Goa, India, has started commercial production and diversification of this biofertilizer since 2005, under the brand name Shubhodaya and the mass production technology involves genetically stable, pure culture of *Glomus* sp. of mycorrhizae. The field trials conducted by the company in collaboration with UAS Dharwad, Karnataka, have indicated that this bioformulation could substitute for 50 per cent NPK in many crops such as sugarcane, cotton, chilli, maize and banana without sacrificing plant growth and crop yields.

Conclusion and Future Prospectus

As production is a prerequisite to fundamental research as well as for application purposes, numerous methods have been developed for decades for large scale production of AM fungi. It is expected that in the future, new cultivation technology will emerge, taking into consideration of several aspects of quality control problem of commercial inoculum. As an example, the production of AM fungi on plants under *in vitro* conditions has been recently proposed (Voets *et al.*, 2005) and extended to hydroponic systems (Declerck *et al.*, 2009). Other *in vitro* methods might come up, which could involve spore production on callus or sporulation in sterile alginate beads or fully closed hydroponic plant cultivation suitable for the production of AM fungi (upscaling the system of Dupré de Boulois *et al.*, 2006). However, other relatively clean methods (*e.g.*, in aeroponics) also have a strong developmental potential and could be further developed in the future. New and advanced methods for AM fungal large-scale inoculum production methods are expected to emerge in the close future.

References

Adholeya A, Tiwari P and Singh R. 2005. Large-scale production of arbuscular mycorrhizal fungi on root organs and inoculation strategies. In: Declerck S, Strullu DG, Fortin JA (eds) *In vitro* culture of mycorrhizas. Springer, Heidelberg, pp. 315–338

Amijee F, Stribley DP and Tinker PB. 1993. The development of endomycorrhizal root systems. VIII effects of soil phosphorus and fungal colonization on the concentration of soluble carbohydrates in roots. *New Phytol* 123: 297–306.

Bécard G and Fortin JA. 1988. Early events of vesicular–arbuscular mycorrhiza formation on Ri T-DNA transformed roots. *New Phytol* 108: 211–218.

Chilton MD, Tepfer DA, Petit A, David C, Casse-Delbart F and Tempé J. 1982. *Agrobacterium rhizogenes* inserts T-DNA into the genomes of the host plant root cells. *Nature* 295: 432–434.

Chaurasia B and Khare PK. 2005. *Hordeum vulgare:* a suitable host for mass production of arbuscular mycorrhizal fungi from natural soil. *Appl Eco Environ Res* 4: 45–53.

Declerck S, Strullu DG and Plenchette C. 1996. *In vitro* mass production of the arbuscular mycorrhizal fungus, *Glomus versiforme*, associated with Ri T-DNA transformed carrot roots. *Mycol Res* 100: 1237–1242.

Dodd JC, Arias I, Koomen I and Hayman DS. 1990. The management of populations of vesicular–arbuscular mycorrhizal fungi in acid infertile soils of savanna ecosystem. I. The effect of pre-cropping and inoculation with VAM–fungi on plant growth and nutrition in the field. *Plant Soil* 122: 229–240.

Douds DD, and Jr Schenck NC. 1990b. Relationship of colonization and sporulation by VA mycorrhizal fungi to plant nutrient and carbohydrate contents. *New Phytol* 116: 621–627.

Douds DD and Jr, Schenck NC, 1990a. Increased sporulation of vesicular–arbuscular mycorrhizal fungi by manipulation of nutrient regimens. *Appl Environ Microbiol* 56: 413–418.

Douds DD, Jr Nagahashi G, Pfeffer PE, Kayser WM and Reider C. 2005. On-farm production and utilization of arbuscular mycorrhizal fungus inoculum. *Can J Plant Sci* 85: 15–21.

Douds DD, Jr Nagahashi G, Pfeffer PE, Reider C and Kayser WM. 2006. On-farm production of AM fungus inoculum in mixtures of compost and vermiculite. *Biores Tech* 97: 809–818.

Dugassa DG, Grunewaldt-Stöcker G and Schönbeck F. 1995. Growth of Glomus intraradices and its effect on linseed (*Linum usitatissimum* L.) in hydroponic culture. *Mycorrhiza* 5: 279–282.

Dupré de Boulois H, Voets L, Delvaux B, Jakobsen I and Declerck S. 2006. Transport of radiocaesium by arbuscular mycorrhizal fungi to *Medicago truncatula* under *in vitro* conditions. *Environ Microbiol* 8: 1926–1934.

Elmes RP and Mosse B. 1984. Vesicular-arbuscular endomycorrhizal inoculum production II Experiments with maize (*Zea mays*) and other hosts in nutrient flow culture. *Can J Bot* 62: 1531–1536.

Feldman F and Idczak E. 1994. Inoculum production of VA mycorrhizal fungi. *In* Techniques for mycorrhizal research. *Edited by* J.R. Norris, D.J. Read, and A.K. Varma. Academic Press, San Diego. pp. 799–817.

Feldmann F and Grotkass C. 2002. Directed inoculum production— shall we be able to design AMF populations to achieve predictable symbiotic effectiveness? In: Gianinazzi S, Schüepp H, Barea JM, Haselwandter K (eds) Mycorrhizal technology in agriculture: from genes to bioproducts. Birkhauser, Basel, pp 261–279.

Fortin JA, St-Arnaud M, Hamel C, Chaverie C and Jolicoeur M. 1996. Aseptic *in vitro* endomycorrhizal spore mass production. US Pat. No. 5554530.

Gadkar V, Driver JD and Rillig MC. 2006. A novel *in vitro* cultivation system to produce and isolate soluble factors released from hyphae of arbuscular mycorrhizal fungi. *Biotechnol Lett* 28: 1071–1076.

Gaur A and Adholeya A. 2002. Arbuscular-mycorrhizal inoculation of five tropical fodder crops and inoculum production in marginal soil amended with organic matter. *Biol Fertil Soils* 35: 214–218.

Gaur A and Adholeya A. 2000. Effects of the particle size of soil-less substrates upon AM fungus inoculum production. *Mycorrhiza* 10: 43–48.

Gianinazzi S, Schüepp H, Barea JM. and Haselwandter K. 2002. Mycorrhizal technology in agriculture: from genes to bioproducts. Birkhauser, Basel.

Hung LL, O'Keefe DM and Sylvia DM. 1991. Use of a hydrogel as a sticking agent and carrier of vesicular-arbuscular mycorrhizal fungi. *Mycol Res* 95: 427–429.

Jarstfer AG and Sylvia DM. 1994. Aeroponic culture of VAM fungi. *In* Mycorrhiza: structure, function, molecular biology and biotechnology. *Edited by* A.K. Varma and B. Hock. Springer-Verlag, Berlin. pp. 427–441.

Jarstfer AG and Sylvia DM. 1995. Aeroponic culture of VAM fungi. In: Varma A, Hock B (eds) Mycorrhiza. Springer, Heidelberg, pp. 427–441.

Jarstter AG and Sylvia DM. 1992. Inoculum production and inoculation strategies for vesicular arbuscular mycorrhizal fungi; in Soil Microbial Technologies: Applications in Agriculture, Forestry and Environmental Management pp. 349–377 ed. B. Meeting. Marcel (New York: Decker Inc.).

Jarstter AG and Sylvia DM. 1994. Aeroponic culture of VAM fugi; in Mycorrhiza: Structure, function, molecular biology and bioechnology, pp.427–41 eds. A K Varma and Hock (Berlin: Springer-Verlag).

Jeffries P. 1987. Use of mycorrhizae in agriculture. CRC *Critical Reviews in Biotechnology* 5(4): 319–357.

Johansson JF, Paul LR and Finlay RD. 2004. Microbial interactions in the mycorrhizosphere and their significance for sustainable agriculture. FEMS *Microbiol Ecol* 48: 1–13.

Jolicoeur M. 1998. Optimisation d'un procédé de production de champignons endomycorhiziens en bioréacteur. Dissertation, École Polytechnique de Montréal

Jolicoeur M, Williams RD, Chavarie C, Fortin JA and Archambault J. 1999. Production of Glomus intraradices propagules, an arbuscular mycorrhizal fungus, in an airlift bioreactor. *Biotechnol Bioeng* 63: 224–232.

Khan AG and Belik M. 1995. Occurrence and ecological significance of mycorrhizal symbiosis in aquatic plants. In: Varma A, Hock B (eds.) Mycorrhiza – structure, function, molecular biology and biotechnology. Springer-Verlag Heidelberg, pp 627–665.

Lee YJ and George E. 2005. Development of a nutrient film technique culture system for arbuscular mycorrhizal plants. *Hort Science* 40: 378–380.

Ma N, Yokoyama K and Marumoto T. 2007. Effect of peat on mycorrhizal colonization and effectiveness of the arbuscular mycorrhizal fungus *Gigaspora margarita. Soil Sci Plant Nutr* 53: 744–752.

Millner PD and Kitt DG. 1992. The Beltsville method for soilless production of vesicular–arbuscular mycorrhizal fungi. *Mycorrhiza* 2: 9–15.

Mohammad A, Mirta B and Khan AG. 2004. Effects of sheared-root inoculum of *Glomus intraradices* on wheat grown at different phosphorus levels in the field. *Agric Ecosyst Environ* 103: 245–249.

Mosse B. 1962. The establishment of vesicular–arbuscular mycorrhiza under aseptic conditions. *J Gen Microbiol* 27: 509–520.

Mosse B and Hepper CM. 1975. Vesicular-arbuscular infections in root–organ cultures. *Physiol Plant Pathol* 5: 215–233.

Mosse B and Thompson JP. 1984. Vesicular arbuscular endomycorrhizal inoculum production. I. v Exploratory experiments with beans (*Phaseolus vulgaris*) in nutrient flow culture. *Can J Bot* 62: 1523–1530.

Mugnier J and Mosse B. 1987. Vesicular–arbuscular mycorrhizal infection in transformed root-inducing T-DNA roots grown axenically. *Phytopathology* 77: 1045–1050.

Smith SE and Read DJ. 2008. Mycorrhizal symbiosis, 3rd edn. Academic, London.

Struble JE and Skipper HD. 1988. Vesicular–arbuscular mycorrhizal fungal spore production as influenced by plant species. *Plant Soil* 109: 277–280.

Sunita, K., Aditya, K. and Ashok, A. 2011. Influence of hosts and substrates on mass multiplication of *Glomus mosseae. African J. Agril. Res.*, 6: 2971–2977.

Sylvia DM. 1989. Nursery inoculation of sea oats with vesicular–arbuscular mycorrhizal fungi and outplanting performance on Florida beaches. *J Coastal Res* 5: 747–754.

Sylvia DM and Jarstfer A.G. 1992. Sheared roots as a VA-mycorrhizal inoculum and methods for enhancing growth. US Pat. No 5096481.

Sylvia DM and Schenck NC. 1983. Application of superphosphate to mycorrhizal plants stimulates sporulation of phosphorus tolerant vesicular–arbuscular mycorrhizal fungi. *New Phytol* 95: 655–661.

Tepfer D. 1984. Transformation of several species of higher plants by *Agrobacterium rhizogenes:* sexual transmission of the transformed genotype and phenotype. *Cell* 37: 959–967.

Tiwari P and Adholeya A. 2003. Host dependent differential spread of *Glomus intraradices* on various Ri T-DNA transformed root *in vitro. Mycol Prog* 2: 171–177.

Voets L. Dupré de Boulois H, Renard L, Strullu DG and Declerck S. 2005. Development of an autotrophic culture system for the *in vitro* mycorrhization of potato plantlets. *FEMS Microbiol Lett* 248: 111–118.

Voets L, de la Providencia IE, Fernandez K, IJdo M, Cranenbrouck S and Declerck S. 2009. Extraradical mycelium network of arbuscular mycorrhizal fungi allows fast colonization of seedlings under *in vitro* conditions. *Mycorrhiza* 19: 347–356.

Wang WK. 2003. Method of facilitating mass production and sporulation of arbuscular mycorrhizal fungi aseptic. US Pat. No. 6759232.

Zobel RW, Del Tredici P and Torrey JG. 1976. Method for growing plants aeroponically. *Plant Physiol* 57: 344–346.

2015, Recent Trends in Microbiology, Mycology and Plant Pathology   *Pages* **61–81**
Editor: **Dr. H.C. Lakshman**
Published by: **DAYA PUBLISHING HOUSE, NEW DELHI**

Chapter 5

Bioremediation: Basic Concepts, Approaches and Applications

M. Nagalakshmi Devamma*[1], Kavya Deepthi M.[1]
and Shaik Thahir Basha[2]

[1]Department of Botany,
[2]Microbiology Laboratory, Department of Virology,
Sri Venkateswara Univeristy, Tirupati – 517 502, A.P., India
*E-mail: devi.bot@gmail.com

Introduction

Approximately 6×10^6 chemical compounds have been synthesized, with 1,000 new chemicals being synthesized every year. Almost 60,000 to 95,000 chemicals are in commercial use. According to Third World Network reports, more than 45 crore kilograms of toxins are released globally into air and water. Usually the contaminated sites are treated by traditional methods like physical, chemical and thermal processes. By these methods, the cost of removal of 1 cubic meter soil from 1acre contaminated site is estimated to be INR 3.6–15 crores. Bioremediation technology is cost effective, eco-friendly and alternative to conventional treatments, which rely on incinerations, volatilization or immobilization of the pollutants. The conventional treatment technologies simply transfer the pollutants, creating a new waste such as incineration residues and they do not eliminate the problem. Bioremediation is one such option which offers the possibility to destroy or render harmless various contaminants using natural biological activity. As such, it uses relatively low-cost, low-technology techniques, which generally have a high public acceptance and can often be carried out on site. Compared to other methods, bioremediation is a more promising and less expensive way for cleaning up contaminated soil and water.

Defining Bioremediation

Bioremediation, the most effective innovative technology to come along that uses biological systems for treatment of contaminants. This novel and recent technology is a multidisciplinary approach.

Bioremediation can be defined in the following ways:

1. Bioremediation is the use of micro-organisms for the degradation of hazardous chemicals in soil, sediments, water or other contaminated materials.
2. Bioremediation is the process whereby organic wastes are biologically degraded under controlled conditions to an innocuous state, or to levels below concentration limits established by regulatory authorities.
3. Bioremediation is the application of biological treatment to the cleanup of contaminants in ground water.
4. Bioremediation is the use of micro-organism metabolism to remove pollutants.
5. The productive use of biodegradative processes to remove or detoxify pollutants that have found their way into the environment and threaten public health usually as contaminants of soil, water or sediments is bioremediation.

How does Bioremediation Occur?

Biological processes, which occur in the natural environment, can modify organic contaminant molecules at the spill location or during their transport in the subsurface. Such biological transformations, which involve enzymes as catalysts, frequently bring about extensive modification in the structure and toxicological properties of the contaminants. In case of organic compounds, biodegradation frequently leads to the conversion of much of the carbon, nitrogen, phosphorus, sulfur, and other elements in the original compound to inorganic end products. Such a conversion of an organic substrate to inorganic end products is known as mineralization. Thus, in the mineralization of organic C, N, P, S, or other elements, CO_2 or inorganic forms of N, P, S, or other elements are released by the organisms and enter the surrounding environment.

The available information suggests that the major agents causing the biological transformations in soil, sediment, surface water, and groundwater are the indigenous micro-organisms that inhabit these environments. Natural communities of micro-organisms present in the subsurface have an amazing physiological versatility. In bioremediation processes, micro-organisms use the contaminants as nutrient or energy sources. Bioremediation activity through microbes is stimulated by supplementing nutrients (nitrogen and phosphorus), electron acceptors (oxygen), and substrates (methane, phenol, and toluene), or by introducing micro-organisms with desired catalytic capabilities. Plant and soil microbes develop a rhizospheric zone which is highly complex symbiotic and synergistic relationships which is also used as a tool for accelerating the rate of degradation or to remove contaminants.

Micro-organisms can carry out biodegradation in many different types of habitats and environments, both under aerobic and anaerobic conditions. Communities of bacteria and fungi can degrade a multitude of synthetic compounds and probably every natural product. Most bioremediation systems are run under aerobic conditions, but running a system under anaerobic conditions may permit microbial organisms to degrade otherwise recalcitrant molecules.

Bioremediation process depends on several parameters like

1. The nature of pollutants,
2. The soil structure, pH, Moisture contents and hydrogeology,
3. The nutritional state, microbial diversity of the site and
4. Temperature and oxidation-reduction (redox- Potential).

Bioremediation process can be categorized into two types *i.e.*, *in situ* and *ex situ* bioremediations. If the process occurs in the same place affected by pollution then it is called *in-situ* bioremediation and when contaminated material (soil and water) is relocated to a different place to accelerate biocatalysis it is referred to as *ex-situ* bioremediation. These types are elaborated in detail in the following paragraphs.

Where is this Bioremediation Process Applied?

Bioremediation has been successfully applied for cleanup of soil, surface water, groundwater, sediments and ecosystem restoration. It has been unequivocally demonstrated that a number of xenobiotics including nitro-glycerine (explosive) can be cleaned up through bioremediation. Bioremediation is generally considered to include natural attenuation (little or no human action), bio-stimulation or bio-augmentation, the deliberate addition of natural or engineered micro-organisms to accelerate the desired catalytic capabilities to contribute significantly to the fate of hazardous waste and can be used to remove these unwanted compounds from the biosphere.

Types of Bioremediation

In situ Bioremediation

Bioremediation is the application of biological treatment to the cleanup of hazardous chemicals present in the subsurface. Hazardous compounds persist in the subsurface because environmental conditions are not appropriate for the microbial activity that results in biochemical degradation. The optimization of environmental conditions is achieved by understanding the biological principles under which these compounds are degraded, and the effect of environmental conditions on both the responsible micro-organisms and their metabolic reactions. The "biodegradation triangle" for understanding the microbial degradation of any natural or synthetic organic compound consists of knowledge of the microbial community, environmental conditions, and structure and physicochemical characteristics of the organic compound to be degraded.

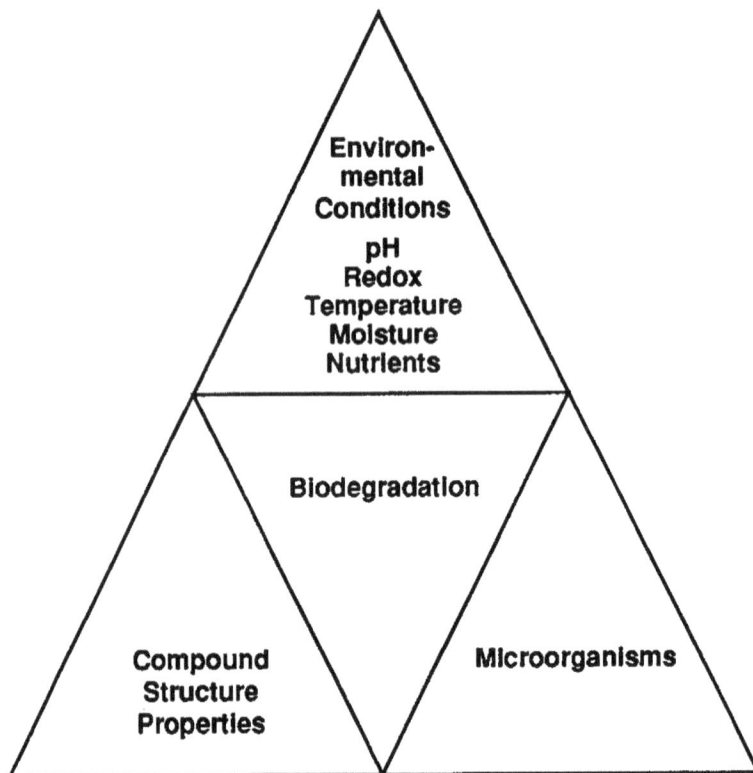

Figure 5.1: Biodegradation Triangle.

Microbial Metabolism on Site

During the process of bioremediation, micro-organisms use the organic contaminants for their growth. In addition, compounds providing the major nutrients such as nitrogen, phosphorus, and minor nutrients such as sulfur and trace elements are also required for their growth. In most cases, an organic compound that represents a carbon and energy source is transformed by the metabolic pathways that are characteristic of heterotrophic micro-organisms. However, an organic compound need not necessarily be a substrate for growth in order for it to be metabolized by micro-organisms.

Biological Transformations that occur during Microbial Metabolism

1. Growth linked transformation – In this process biodegradation provides carbon and energy to support growth.
2. Cometabolic transformations – In this process biodegradation is not linked to multiplication, but carbon is obtained for respiration to maintain cell viability. This maintenance metabolism may take place only when the organic carbon concentrations are very low.

It has been observed that the number of microbial cells or the biomass of a species acting on the compound of interest increases as degradation proceeds.

Acclimatization Period

Prior to the degradation of many organic compounds, a period is observed where no degradation of the chemical occurs. This time interval is known as the acclimatization period or, sometimes, as adaptation or lag period. The length of the acclimatization period varies and may be less than 1 h or many months. The duration of acclimatization depends upon the chemical structure, subsurface biogeochemical environmental conditions, and concentration of the compound. Once acclimatization period is over, the microbial community will retain its higher level of activity for some time.

Modes of Microbial Transformation

The design of bioremediation processes requires determination of the degradation reactions to which the target compounds are subjected to; this involves selecting the metabolism mode that will occur in the process.

The metabolism modes are broadly classified as aerobic and anaerobic. Aerobic transformations occur in the presence of molecular oxygen, with molecular oxygen serving as the electron acceptor. This form of metabolism is known as aerobic respiration. Anaerobic reactions occur only in the absence of molecular oxygen and the reactions are subdivided into anaerobic respiration, fermentation, and methane fermentation. This classification is demonstrated pictorially in Figure 5.2.

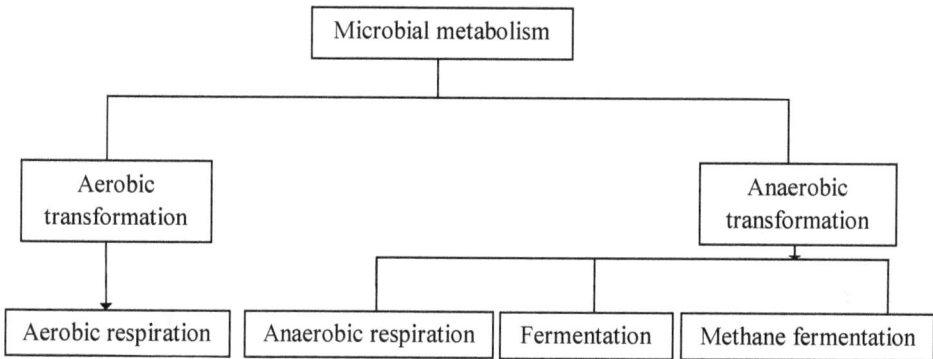

Figure 5.2: Flowchart Showing Microbial Mode Classification.

The metabolism modes that utilize nitrate, sulfate, thiosulfate, and CO_2 as electron acceptors can be used to biodegrade various organic contaminants. Recently, chlorinated organic compounds are also used as electron acceptors during anaerobic respiration.

Hydrocarbons Degradation

Hydrocarbons are compounds containing carbon and hydrogen. The most frequent and earliest application of bioremediation has been to remediate hydrocarbons present in the subsurface as a result of petroleum spills. Soil contains significant microbial populations that can use hydrocarbons as sole sources of carbon and energy and these populations constitute around 20 per cent of all soil microbes.

Bioremediation of aliphatic hydrocarbons should be performed as an aerobic process. Aerobic biodegradation of aliphatic hydrocarbons involves the incorporation of molecular oxygen into the hydrocarbon structure. This is performed by oxygenase enzymes. Unsaturated straight-chain hydrocarbons are generally less readily degraded than saturated ones. Micro-organisms capable of degrading cyclic aliphatic hydrocarbons are not as predominant in soils as those for the degradation of aliphatic alkane and alkene hydrocarbons. Hydroxylation is vital to initiate the degradation of cycloalkanes.

Aromatic hydrocarbons like benzene, toluene, ethyl benzene, and the three isomers of xylene are frequently referred to as BTEX (Benzene, toluene, ethylbenzene and xylenes) compounds and are one of the most heavily regulated groups of compounds. Aromatic hydrocarbons can be transformed under various anaerobic conditions such as denitrifying, manganese reducing, iron reducing, sulfate reducing, and methanogenic conditions. At any given location, the benzene biodegradation sequence will depend on the availability of electron acceptors and the redox potential of the environment. Under methanogenic fermentative conditions, several aromatic hydrocarbon compounds, including benzene, have been shown to transform into CO_2 and methane. Higher rates of degradation are reported under denitrifying conditions than under methanogenic conditions.

Polynuclear Aromatic Hydrocarbons (PAHs)

Polynuclear aromatic hydrocarbons (PAHs) are compounds that have multiple rings in their molecular structure. They include naphthalene and anthracene and the more complex compounds such as pyrene and benzo(a)pyrene. Biodegradation of polynuclear aromatic hydrocarbons depends on the complexity of the chemical structure and the extent of enzymatic adaptation. In general, PAHs which contain two or three rings such as naphthalene, anthracene, and phenanthrene are degraded at reasonable rates when O_2 is present. Compounds with four rings such as chrysene, pyrene, and pentacyclic compounds, in contrast, are highly persistent and are considered recalcitrant. Aerobic biodegradation of the two- and three-ring PAHs is accomplished by a number of soil bacteria. As the number of fused rings and the complexity of the substituted groups increase, the relative degree of degradation decreases. Co-metabolism may be the only metabolism mode for degradation of the heavier PAHs. Many fungal species are known to degrade PAHs under aerobic conditions. The fungus *Phanerochaete chrysosporium*, also known as white rot fungus, degrades many PAHs including benzo(a)pyrene, pyrene, fluorene, and phenanthrene. Nitrogen-limiting conditions and lower pH (around 4.5) are favorable for this degradation. However, the transformations by the fungus are slow.

Chlorinated Organics Degradation

Transformations of Chlorinated Aliphatic Hydrocarbons (CAHs) in the subsurface environment can occur both chemically (abiotic) and biologically (biotic).

The major abiotic transformations include hydrolysis, substitution, dehydrohalogenation, coupling, and reduction reactions. Abiotic transformations generally result in only a partial transformation of a compound and may lead to the formation of an intermediate that is either more readily or less readily biodegraded by micro-organisms. Biotic transformation products are different under aerobic than anaerobic conditions. Microbial degradation of chlorinated aliphatic compounds can use one of several metabolism modes. These include oxidation of the compound for an energy source, co-metabolism under aerobic conditions, and reductive dehalogenation under anaerobic conditions. The higher chlorinated aliphatic compounds, where all available valences on carbon are substituted, such as tetrachloroethylene, have not been transformed under aerobic systems. Dichloromethane can be used as a primary substrate under both aerobic and anaerobic conditions and completely mineralizes. Dichloroethane and vinyl chloride can be used as a primary energy source under aerobic conditions. The microbial transformation of most of the CAHs depends upon co-metabolism. However, with co-metabolism partial transformation of CAHs occurs.

Chlorinated aromatic hydrocarbons include a wide range of compounds present in the subsurface as contaminants and thus require remediation. Few examples of chlorinated aromatic hydrocarbons include chlorophenols, chlorobenzenes, chloronitro benzenes, chloroaniline, polychlorinated biphenyls (PCBs), and many pesticides. Methanogenic metabolism has successfully dechlorinated many aromatic organic compounds such as 3-chlorobenzoate, 2,4-dichlorophenol, and 4-chlorophenol. Anaerobic dehalogenation and the final mineralization may require multiple species of micro-organisms and reduction pathways. For example, 2,4-dichlorophenol was mineralized to CH_4 and CO_2 by as many as six species of micro-organisms. Polychlorinated biphenyls (PCBs) are chlorinated aromatic compounds that are designated by numbers that represent the number of carbon atoms and the percentage of chlorine by weight. PCBs are also known under their trade name Aroclor™. For example, Aroclor 1252 contains 12 carbon atoms and has 52 per cent chlorine by weight. PCBs are very insoluble in water and are mostly found only in soils and sediments.

Environmental Factors

Microbial populations capable of degrading contaminants in the subsurface are subjected to a variety of physical, chemical, and biological factors that influence their growth, their metabolic activity, and their very existence. The properties and characteristics of the environments in which the micro-organisms function have a profound impact on the microbial population, the rate of microbial transformations, the pathways of products of biodegradation, and the persistence of contaminants.

Microbial Factors

The microbial population of the soil is made up of five major groups: bacteria, actinomycetes, fungi, algae, and protozoa. Bacteria are the most abundant group, usually more numerous than the other four combined. Although transformations similar to those of the bacteria are carried out by the other groups, the bacteria stand out because of their capacity for rapid growth and degradation of a variety of contaminants. Measurement of the indigenous microbial activity is one method for evaluating potential toxic or inhibitory conditions at a site. Low bacteria counts can indicate a potential toxicity problem or a stressed microbial population. Groundwater bacterial counts range from 10^2 to 10^5 colony forming units (CFU) per milliliter of sample. Typical soil microbial counts range from 10^3 to 10^7 CFUs per gram of soil. Higher counts indicate a healthy microbial population. Counts below 10^3 organisms per gram of soil at contaminated sites may indicate a stressed microbial population.

Nutrients

Carbon makes up a large fraction of the total protoplasmic material of a microbial cell. Carbohydrates, proteins, amino acids, vitamins, nucleic acids, purines, pyrimidines, and other substances constitute the cell material. In addition to carbon, cell material is mainly composed of the elements hydrogen, oxygen, and nitrogen. These four chemical elements constitute about 95 per cent by weight of living cells. Two other elements phosphorus and calcium contribute to 70 per cent of the remainder. Carbon is usually supplied by organic substrates organic contaminants in the case of bioremediation—for the heterotrophic micro-organisms. Autotrophic micro-organisms obtain their carbon supply from inorganic sources such as carbonates and bicarbonates. Hydrogen and oxygen are supplied by water. Usually, the nutrients in short supply are nitrogen, phosphorus, or both. Nearly always, the supply of potassium, sulfur, magnesium, calcium, iron, and micronutrient elements is greater than the demand. These micronutrients are present in most soil and aquifer systems.

Physical–Chemical Factors

The activities of micro-organisms are markedly affected by their physical–chemical environment. Environmental parameters such as temperature, pH, moisture content, and redox potential will determine the efficiency and extent of biodegradation.

In situ Bioremediation Systems

The most significant challenge in bioremediation is introducing into the subsurface environment the reagents needed by micro-organisms and mixing them with the contaminants to be degraded. Much of the methodology usually associated with bioremediation can be attributed to the pioneering research and development carried out by Richard L. Raymond and Sun Tech in the 1970s. By the mid-1980s, the potential of bioremediation was widely accepted in the remediation industry. In the last few years, there has been an explosion of activity in bioremediation which now incorporates a wide range of processes in the environment.

At this point the two concepts to be focused mainly include biostimulation and bioaugmentation. Biostimulation consists of adding nutrients, such as nitrogen and phosphorus, as well as oxygen and other electron acceptors, to the microbial environment to stimulate the activity of micro-organisms. Bioaugmentation involves adding exogenous microbes to the subsurface where organisms able to degrade a specific contaminant are deficient. Microbes may be "seeded" from populations already present at a site and grown in an above-ground reactor, or specially cultivated strains having known capabilities for degrading a specific contaminant. Most bioremediation systems employ some form of biostimulation. However, there is a significant resistance in the industry to use bioaugmentation. This resistance stems from the ubiquity principle, which states that all micro-organisms are ever-present in the subsurface environment. Another argument against bioaugmentation is that indigenous organisms already present at the contaminated site would have developed the enzyme systems to degrade the target contaminants. Furthermore, the limitation of distributing the exogenous microbial cultures in the subsurface and the question of long-term survivability of these lab-grown cultures under field conditions also discourage bioaugmentation.

In situ Bioremediation System Evaluation Process

Prior to designing a bioremediation system, the feasibility of biodegradation should be carefully evaluated. This evaluation should include the ease or difficulty of degrading the target contaminants, the ability to achieve total mineralization, and the environmental conditions necessary to implement the process.

There are various factors that should be incorporated into this evaluation process.

1. Biodegradability of contaminants
2. Mineralization potential of the compounds
3. Specific microbial, substrate conditions
4. Availability of nutrients
5. Site's hydrogeologic characteristics
6. Extent and distribution of contaminants

This evaluation along with site hydro geologic parameters contributes for developing the engineering design of the "subsurface bioreactor."

Biogeochemical Parameters

Measurements of various biogeochemical parameters such as dissolved oxygen (DO), redox potential, CO_2, and other parameters such as NH_4^+, NO_3^-, SO_4^{2-}, S^{2-}, and Fe^{2+} will give an indication of the existing (natural or intrinsic) microbial metabolic activity at the site.

Denitrification-Based *In situ* Bioremediation

One promising alternative to the saturation limitations or high costs of the major alternative forms of oxygen involves the use of nitrate as the oxygen acceptor. In this process, the biodegradative activities of denitrifying organisms are enhanced, resulting in biodegradation of the target organic contaminants along with the transformation of NO_3^- to N_2. Nitrate feedstocks can thus be substituted for oxygen feedstocks in the groundwater manipulation system described in the previous section.

During biosparging, the consumption of O_2 is relatively fast and the rate of O_2 transfer from the injected air to the aqueous phase is slow, due to low solubility of O_2 in water. Expansion of the aerobic zone is limited by the rate of O_2 supply to the aqueous phase. Anaerobic conditions are expected to persist within aerobically treated aquifers, especially in relatively impermeable zones and zones further away from the injection wells. The overall degradation efficiency can be increased by using nitrate, which is much more water-soluble than O_2 (9200 mg/l as $NaNO_3$ vs. 8 to 10 mg/l as O_2). The reducing equivalents that can be introduced into an aquifer using saturated sodium nitrate solution is approximately 50 times higher than with a saturated oxygen solution. However, due to regulatory and microbial toxicity considerations, the nitrate feedstock solution concentration should be significantly lower than saturated concentration.

In situ Bioremediation Treatments

The most commonly used *In situ* treatments include bioventing, enhanced in-situ bioremediation of hydrocarbons in groundwater, natural attenuation of soils, natural attenuation of ground water and natural attenuation of chlorinated aliphatic hydrocarbons (CAHs). These treatments are schematically represented in Figure 5.3.

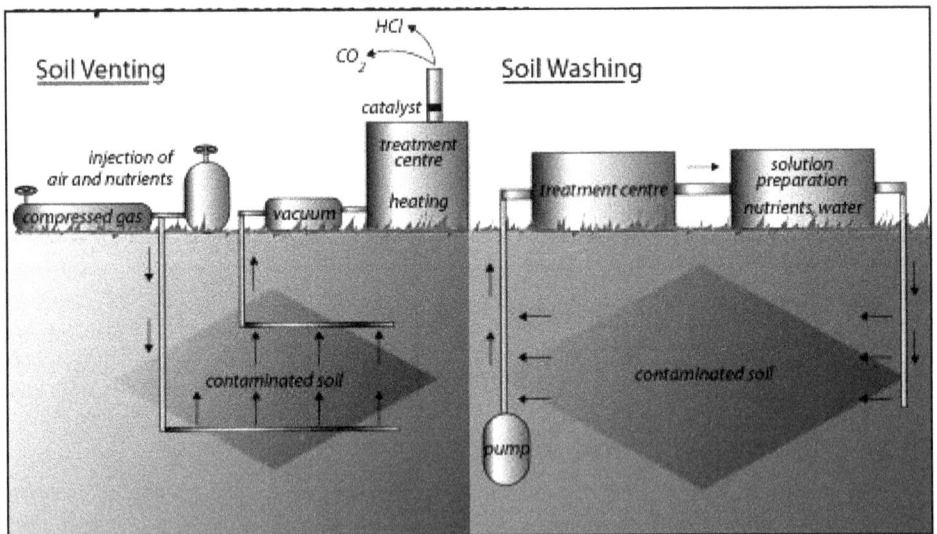

Figure 5.3: *In situ* Bioremediation Treatment Process.

Advantages of *In situ* Bioremediation

1. Capability to degrade chlorinated aliphatic hydrocarbons to relatively less toxic products.

2. Generation of relatively small amounts of remediation wastes, compared to *ex situ* technologies.

3. Reduced potential for cross-media transfer of contaminants commonly associated with *ex situ* treatment.

4. Reduced risk of human exposure to contaminated media, compared to *ex situ* technologies.

5. Relatively lower cost of treatment compared to excavation and disposal *ex situ* treatment or conventional pump and treat systems.

6. Potential to remediate a site faster than with conventional technologies.

Disadvantages of *In situ* Bioremediation

1. A perceived lack of knowledge about biodegradation mechanisms.
2. Specific contaminants at a site may not be amenable to *in situ* bioremediation.
3. Enhanced technologies when needed may be costly or their implementation may be technologically challenging.
4. The toxicity of transformation products may exceed that of parent compounds.
5. Could take longer to remediate site than a conventional technology.

Ex situ Bioremediation

Ex situ bioremediation involves excavation of the contaminated soil and treating in a treatment plant located on the site or away from the site. This approach can be faster, easier to control, and used to treat a wider range of contaminants and soil types than in-situ approach. *Ex situ* bioremediation can be implemented as slurry-phase bioremediation, or solid-phase bioremediation.

Slurry-Phase Bioremediation

In slurry-phase bioremediation, the contaminated soil is mixed with water to create slurry. The slurry is aerated, and the contaminants are aerobically biodegraded. The treatment can take place on-site, or the soils can be removed and transported to a remote location for treatment. The process generally takes place in a tank or vessel (a "bioreactor"), but can also take place in a lagoon (Figure 5.4A). Figure 5.4B presents a schematic of the process. Contaminated soil is excavated and then screened to remove large particles and debris. A specific volume of soil is mixed with water, nutrients, and micro-organisms. The resulting slurry pH may be adjusted, if necessary. The slurry is treated in the bioreactor until the desired level of treatment is achieved. Aeration is provided by compressors and air spargers. Mixing is accomplished either by aeration alone or by aeration combined with mechanical mixers. During treatment, the oxygen and nutrient content, pH, and temperature of the slurry are adjusted and maintained at levels suitable for aerobic microbial growth. Natural soil microbial populations may be used if suitable strains and numbers are present in the soil. More typically, micro-organisms are added to ensure timely and effective treatment. The micro-organisms can be seeded initially on start-up or supplemented continuously throughout the treatment period for each batch of soil treated. When the desired level of treatment has been achieved, the unit is emptied. The treated soil is then dewatered and backfilled in excavations. The wastewater is treated and disposed or recycled, and a second volume of soil is treated.

Solid-Phase Bioremediation

In solid-phase bioremediation, soil is treated above ground treatment areas equipped with collection systems to prevent any contaminant from escaping the treatment. Moisture, heat, nutrients, or oxygen are controlled to enhance bioremediation for the application of this treatment. Solid-phase systems are relatively

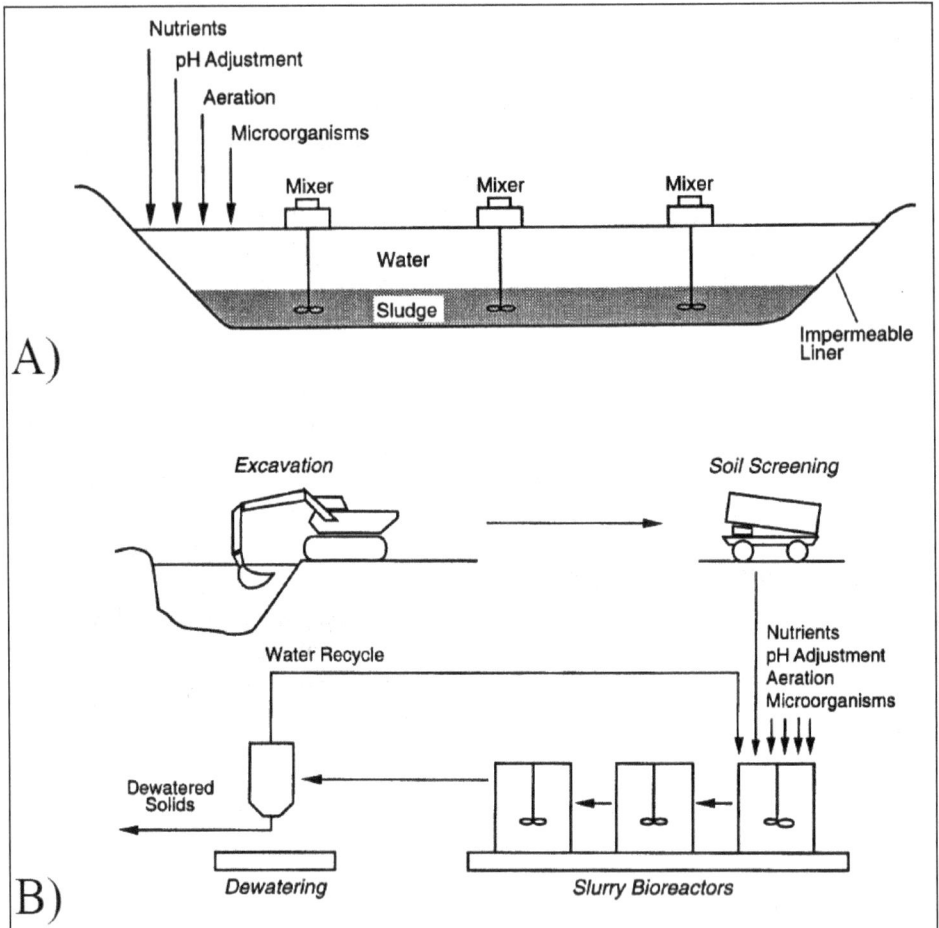

Figure 5.4: *Ex situ* **Slurry-Phase Bioremediation in (A) Lagoons, and (B) Above-ground reactors.**

simple to operate and maintain, require large amount of space, and cleanups require more time to complete than slurry-phase processes. There are three different ways of implementing solid-phase bioremediation: contained solid phase bioremediation, composting, and land farming.

Contained Solid Phase Bioremediation

In contained solid phase bioremediation, the excavated soils are not slurried with water; the contaminated soils are simply blended to achieve a homogeneous texture. Occasionally, textural or bulk amendments, nutrients, moisture, pH adjustment, and microbes are added. The soil is then placed in an enclosed building, vault, tank, or vessel (Figure 5.5). The temperature and moisture conditions are controlled to maintain good growing conditions for the microbial population. In addition, since the soil mass is enclosed, rainfall and runoff are eliminated, and VOC

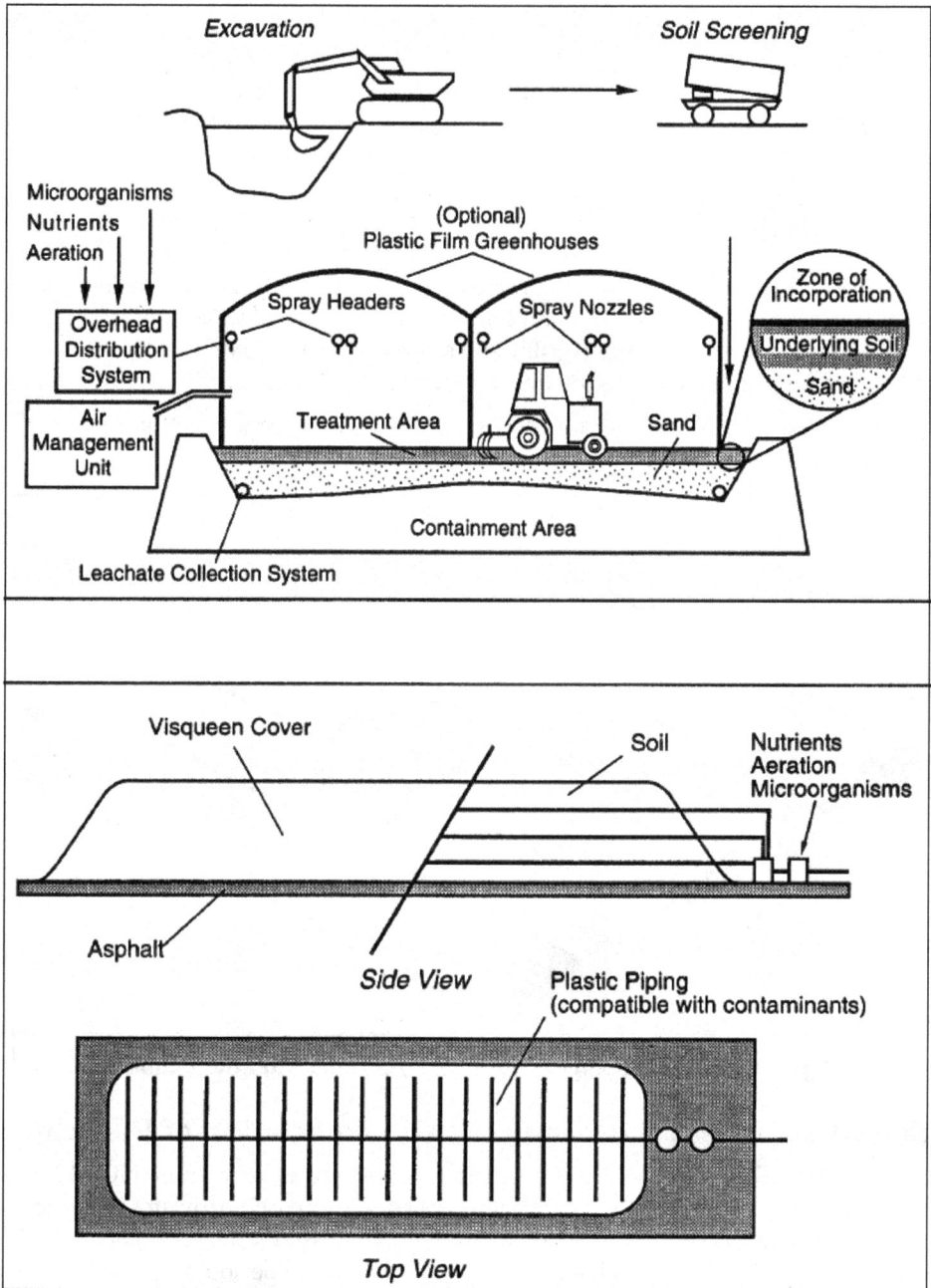

Figure 5.5: Contained *Ex situ* Solid Phase Bioremediation.

emissions can be controlled. Mechanisms for managing/controlling flammable or explosive atmospheres and special equipment for blending and aeration of the soil may be required.

Ex situ bioremediation treatments include landfarming, composting, biopiles, bioreactors-fixed film, bioreactors–suspended growth and usage of white rot fungi.

In the case of *Ex situ* bioremediation, a bioreactor with adequate volume to hold the contaminated soil is needed or adequate space is needed to spread the contaminated soil volume of bioreactor is roughly estimated using:

$$dC/dt = r_0/k_{sd} X_s + 1 \tag{1}$$

$$dM/dt = V r_0 \tag{2}$$

Where V=volume of the reactor, r_0 the rate of contaminant biodegradation, C=the mass concentration of soluble contaminant, M=the mass of contaminant to be put in the reactor, t=the time in days, K_{sd}=the soil distribution coefficient, and X_s=the mass concentration of solids (contaminated soils).

Excavating equipment is also necessary for excavating the contaminated soil from the site for the *Ex situ* treatment.

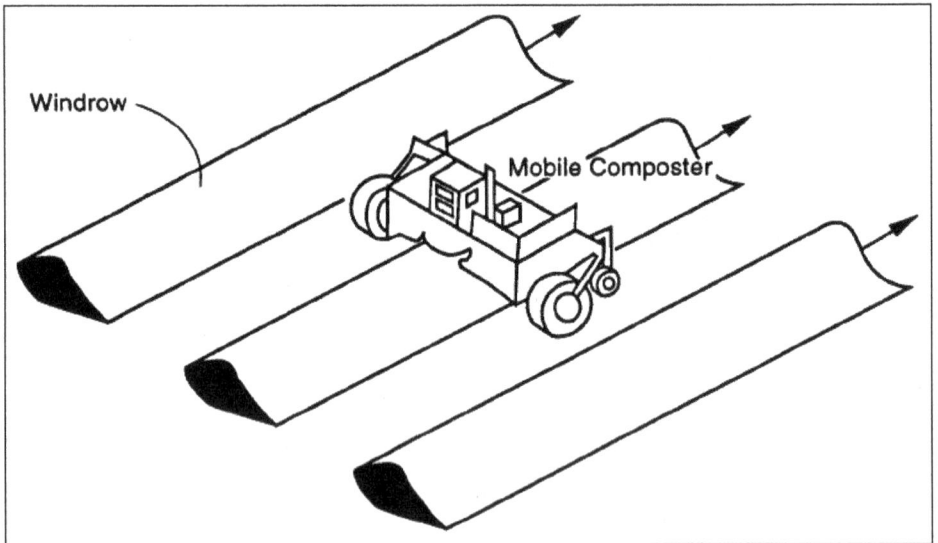

Figure 5.6: *Ex situ* Composting: Open and Static Windrow Systems.

Scale Up Process Bioreactors in Bioremediation of Pesticides

To assess the bioremediation potential of *Pseudomonas aeruginosa* (NCIM 2074) by improving its adaptability to increasing concentration of chlorpyrifos using scale up process. The biodegradation of chlorpyrifos, as assessed by GC-MS, showed that chlorpyrifos at 10, 25, 50 mg/l degraded completely over a period of 1, 5 and 7 days, respectively. The intermediate 3, 5, 6 trichloro-2-pyridion, 2, 4-bis (1, 1 dimethyiethyl) phenol and 1, 2 zenedicarboxylic acid persisted during bioremediation, but in the long run these convert to CO_2, biomass and nutrients. *Pseudomonas aeruginosa* (NCIM 2074) has been of potential use in bioremediation of chlorpyrifos at concentrations up to 50 mg/l, but the organism is inhibited by higher concentrations.

Bioremediation of Benzene using Partitioning Bioreactor

A bioreactor has been designed and developed for partitioning of aqueous and organic phases with a provision for aeration and stirring, a cooling system and a sampling port. The potential of a cow dung microbial consortium has been assessed for bioremediation of phenol in a single-phase bioreactor and a two-phase partitioning bioreactor. The *Pseudomonas putida* IFO 14671 has been isolated, cultured and identified from the cow dung microbial consortium as a high-potential phenol degrader. This study presents an advance in bioremediation techniques for the biodegradation of organic compound such as phenol using a bioreactor.

Current Practices in Bioremediation

Mycoremediation

Mycoremediation refers to the use of fungal mycelia in bioremediation. One of the primary roles of fungi in the ecosystem is decomposition, which is performed by the mycelium. The mycelium secretes extracellular enzymes and acids that break down lignin and cellulose, the two main building blocks of plant fiber. These are organic compounds composed of long chains of carbon and hydrogen, structurally similar to many organic pollutants.

Phytoremediation

Phytoremediation uses plants to clean up contaminated soil and ground water, taking advantage of plants natural abilities to take up, accumulate, and/or degrade constituents of their soil and water environments. Phytoremediation involves the use of certain plants to cleanup soil and water contaminated with inorganics and/or organics. It is strongly believed that there are three dimensions for the effectiveness of vital bioremediation process, *i.e.*, chemical landscape, abiotic landscape and catabolic landscape of which only the catabolic landscape is "genuinely" biological. The chemical landscape has a dynamic interplay with the biological interventions on the abiotic background of the site at stake. This includes humidity, conductivity, temperature, matrix conditions, redox status, etc. The term Phytoremediation was coined in 1991. This is an emerging technology that uses plants to remove contaminants from the soil and water.

Types of Phytoremediation

5 types of phytoremediation techniques are present. They include:

1. **Phytoextraction** or **phytoaccumulation** refers to the ability of plants to take up contaminants into the roots and translocate them to above ground shoots or leaves.

2. **Phytotransformation** or **phytodegradation** refers to uptake of contaminants with the subsequent breakdown, mineralization or metabolization by the plant itself through various internal enzymatic reactions and metabolic processes.

3. **Phytovolatilization** is the volatilization of contaminants from the plant either from the leaf stomata or from plant stems. Mercury or selenium once

taken up by the plant roots, can be converted into non-toxic forms forms and volatilized into the atmosphere from the roots, shoots or leaves.

4. **Phytostabilization**–refers to the holding of contaminated soils and sediments in place by vegetation and to immobilizing toxic contaminants in soils. Applicable to metal contaminants at waste sites.

5. **Phytodegradation** or **rhizodegradation** is enhanced breakdown of contaminant by increasing the bioactivity using the plant rhizosphere environment to stimulate the microbial populations.

6. **Rhizofiltration** – water remediation technique, involves uptake of contaminants by plant roots. This technique is used to reduce contamination in natural wetlands and estuary areas.

Bioaugmentation

Bioaugmentation is the process in which the rate of biological degradation can be increased through the addition of micro-organisms that have been shown to degrade the contaminants of concern at high rates or are particularly well suited to remain active under prevailing site conditions. Bioaugmentation is useful if the contaminants are particularly recalcitrant to degradation or if site conditions are extreme.

To be effective, the introduced organisms must become distributed throughout the contaminated matrix and compete with the indigenous micro-organisms for available nutrients. It they are not distributed throughout the matrix the positive effect will be localized on the other hand if the introduced organisms compete poorly, they will not persist and the treatment effect will be short lived.

Bioventing

It is the most common *In situ* treatment and involves supplying air and nutrients through wells to contaminated soil to stimulate the indigenous bacteria. It works for simple hydrocarbons and can be used where the contamination is deep under the surface. Bioventing is the only *In situ* bioremediation technology applicable to vadose zone contamination.

Vadose Zone/Unsaturated Zone

Vadose zone is the subsurface region that contains water at a pressure less than that of atmospheric pressure. This zone is limited from above by the land surface and from below by the zone of saturation. The unsaturated zone often contains the contaminant source that contributes contamination to saturation zone.

Biosparging

Biosparging is the injection of air under pressure below the water table to increase ground water oxygen concentrations and enhance the rate of biological degradation of contaminants by naturally occurring bacteria.

Natural Intrinsic Bioremediation

The basic concept behind "natural intrinsic bioremediation" is to allow the natural indigenous micro-organisms to biodegrade the contaminant present in the groundwater. While natural attenuation processes include biodegradation, abiotic oxidation, hydrolysis, dispersion, dilution, sorption, and volatilization, intrinsic bioremediation is the primary mechanism for the attenuation of biodegradable contaminants. Intrinsic bioremediation, abiotic oxidation, and hydrolysis are the only attenuation mechanisms that destroy the contaminants to innocuous end products. The use of intrinsic bioremediation as part of the site remediation strategy can significantly reduce cleanup costs. Acceptance of natural bioremediation as a remediation alternative will be greatly enhanced if a *zero line* can be established. Existence of a zero line can be inferred by evaluating groundwater quality data over a period of time. Three to four rounds of sampling data collected over a period of time may indicate the existence of the zero line.

Four distinct zones of biogeochemical dynamics are present in a petroleum contaminated site. They include

1. The heart of the plume
2. Anaerobic zone
3. Aerobic zone and
4. Remediated zone

Various biodegradation pathways will take place in these four zones. Almost all dissolved petroleum hydrocarbons are biodegradable under aerobic conditions, where micro-organisms utilize O_2 as the electron acceptor and the contaminants as the substrate for their growth and energy. When oxygen supply is depleted and nitrate is present, facultative anaerobic micro-organisms will utilize NO_3 as the electron acceptor. Once the available oxygen and nitrate are depleted, micro-organisms may use oxidized ferric ion (Fe(III)) as an electron acceptor.

Bio-buffering

The concept of bio-buffering is based on the premise that the degradative capacity of the aquifer is a lot more than the available (Dissolved oxygen) DO in the system. Bio-buffering can also be defined as the stability of the assimilative capacity of the natural system in response to the introduction of the contaminant mass flux into the aquifer. Among all the electron acceptors, O_2 and CO_2 are the most readily available, due to natural recharge processes and aquifer geochemistry. Sulfate, iron, and manganese also occur naturally, but are dependent on site minerology. Sulfate may be also introduced by manmade activities. The predominant sources of nitrate are anthropogenic activities such as agricultural fertilization.

Estimation of the assimilative capacity of benzene in an intrinsic bioremediation system involves 5 steps which are given below.

1. Aerobic Oxidation
2. Denitrification
3. Iron Reduction
4. Sulfate Reduction
5. Methanogenesis

Biopiling

Biopile treatment is a full-scale technology in which excavated soils are mixed with soil amendments, placed on a treatment area, and bioremediated using forced aeration. The contaminants are reduced to carbon dioxide and water. The basic biopile system includes a treatment bed, an aeration system, an irrigation/nutrient system and a leach ate collection system. Moisture, heat, nutrients, oxygen, and pH are controlled to enhance biodegradation. The irrigation/nutrient system is buried under the soil to pass air and nutrients either by vacuum or positive pressure. Soil piles can be up to 20 feet high and may be covered with plastic to control runoff, evaporation and volatilization, and to promote solar heating. If volatile organic compounds (VOCs) in the soil volatilize into the air stream, the air leaving the soil may be treated to remove or destroy the VOCs before they are discharged into the atmosphere. Treatment time is typically 3 to 6 months.

Applications of Bioremediation

Bioremediation is a rapidly establishing technology for contaminated soil and groundwater treatment. It is the best technology for treatment, particularly in sites where it is difficult to access the contamination such as in deeper aquifers. Although nearly all organic pollutants can be biodegraded in the laboratory and suitable for bioremediation at some sites, a few can only be efficiently treated by *in situ* or *ex situ* bioremediation. The key factors are ease of transport for amendments to the site of action, ease of biodegradation, low toxicity, and high bioavailability. Following these conditions, organic contaminants like chlorinated aliphatics, explosives, BTEX, and petroleum hydrocarbons can be best remediated.

One of the common applications is remediation of petrochemical compounds like gasoline, fuel oil and bitumen, using bioventing technology. Chakrabarty *et al.* (1980) developed and patented a "superbug" that degraded petroleum (camphor, octane, xylene, and naphthaline) by plasmid transfers. Expressed *Pseudomonas* or *Flavobacterium* organophosphorus hydrolase (opd) gene fused to a lipoprotein gene at the *E. coli* cell surface degrades organophosphate pesticides. *Deinococcus radiodurans* which is naturally resistant to high levels of radiation is an ideal bacterium to express bioremediating proteins in toxic, radioactive environments.

Several case studies have been reported where bioremediation has been used to remediate site contamination. A brief description of one such case study is explained in section Application of Oilzapper at oil spill sites in Gujarat and Assam, India

The Energy and Resources Institute (TERI), New Delhi, India in their Microbial Biotechnology Laboratory developed oilzapper–a crude oil and oily sludge degrading bacterial consortium after seven years of research work. This bacterial consortium was developed by mixing five bacterial strains, which could degrade aliphatic, aromatic, asphaltene, and NSO (nitrogen, sulphur, and oxygen compounds) fractions of crude oil and oily sludge. Crude oil and oily sludge degrading efficiency of the developed bacterial consortium was tested under laboratory conditions and field conditions. A feasibility study on the bioremediation of soil contaminated with crude oil/oily sludge was carried out at the Mathura oil refinery, India. The feasibility study was carried out with six different treatments in a 25 square meter land area contaminated with crude oil/oily sludge prior to full scale bioremediation.

The site was tilled thoroughly to mix the oily sludge uniformly with the soil and oilzapper applied onto it. The land was tilled again and watered to maintain proper aeration and moisture levels. The land was tilled at regular intervals to facilitate faster degradation. The problem of heterogenous distribution of the oily sludge was solved by extensive tilling prior to the application of the oilzapper. Another microbial-based product Oilivorous-S was jointly developed by TERI and IOC's R&D centre in Faridabad for treatment of oily sludge. Figure 5.7 represents schematically while oilzapper is being applied to the oil spilled site in Gujarat, whereas Figure 5.8 represents the oil spilled site after application of oilzapper in Assam.

Figure 5.7: Application of Oilzapper at Oil Spill Site in Gujarat, India.

Figure 5.8: After Application of Oilzapper at Oil Spill Site in Assam, India.

The end-users of Oilzapper and Oilivorous-S technologies are Indian Oil Corporation Ltd, India, Bharat Petroleum Corporation Ltd., India, Hindustan Petroleum Corporation Ltd, India, Oil and Natural Gas Corporation Ltd., India, Oil India Ltd, India, Indian Petrochemicals Corporation Ltd, India, Reliance Industries Ltd, India, Abu Dhabi National Oil Company, Abu Dhabi, Kuwait Oil Company, Kuwait.

Conclusion

In this chapter, a brief overview of basic concepts, practices and applications of bioremediation technology has been presented with special focus on developments occurring in this field in India. Bioremediation is a cost effective technology where biological agents are used to decontaminate soil, water and air. The general approaches used for bioremediation purpose are either *in situ* or *ex situ*. In *in situ* bioremediation the treatments used are bioventing, natural attenuation of soils or groundwater whereas in *ex situ* bioremediation the treatments include biopiles, land farming and bioreactors. Besides, there are several techniques of bioremediation based on the biological agents used that include phytoremediation, mycoremediation etc. The energy and Resources institute (TERI), New Delhi contributed a lot to this field especially by introducing oilzapper – a bacterial consortium into the contaminated sites. Even then, it is the necessity of the hour that more bioremediating products are being introduced into the market such that pollution levels can be controlled to a major extent.

References

B Pandey, MH Fulekar. 2012, Bioremediation technology: A new horizon for environmental clean–up. *Biology and Medicine*, 4 (1): 51–59pp.

B.R. Glick, J.J. Pasternak and C.L. Patten. 2010, *Molecular Biology: Principles and Applications of Recombinant DNA*, fourth edition, American Society for Microbiology press. 551–598pp.

Gopan Mukkulath and Santosh G. Thampi. 2012, Biodegradation of Coir Geotextiles Attached Media in Aerobic Biological Wastewater Treatment. *Bioremediation and Biodegradation*, 3: 11.

Karl J. Rockne and Krishna R. Reddy. 2003, Bioremediation of contaminated sites, International e-Conference on Modern Trends in Foundation Engineering: Geotechnical Challenges and Solutions, Indian Institute of Technology, Madras, India. 22pp.

Larry J. Forney, Mr. Tom Crossman, P.C., Geraghty and Miller, and Mr. John Shauver, Michigan. 1998, Fundamental principles of bioremediation, Environmental Response Division Design, O and M Unit, 24pp.

Mishra S, Sarma P M, and Lal B. 2004, Crude oil degradation efficiency recombinant lux tagged *Acinetobacter baumannii* strain and its survival in crude oil contaminated soil microcosm. *FEMS Microbiology Letters.* 235: 323–331.

Mishra S, Jyot J, Kuhad R C, Lal B. 2001, Evaluation of Inoculum addition to stimulate *In situ* bioremediation of oily sludge contaminated soil. *Applied and Environmental Microbiology* 67: 1675–1681.

Mishra S, Jyot J, Kuhad R C, Lal B. 2001, *In situ* bioremediation potential of an oily sludge degrading bacterial consortium. *Current Microbiology* 43.

Mishra S, Lal B, Jyot J, Rajan S, Khanna S. 1999 Field study: *In situ* bioremediation of oily sludge contaminated land using oilzapper In: *Proceedings of Hazardous and Industrial Wastes*, pp. 177–186, edited by D Bishop. Pennsylvania, USA: Technomic Publishing Co. Inc.

Molly Leung. 2004, Bioremediation: Techniques for Cleaning up a mess. *BioTeach Journal* Vol. 2 18 –22.

Prasad, M.N.V. 2011, A State–of–the–Art report on Bioremediation, its Applications to Contaminated Sites in India. Ministry of Environment and Forests, India. 85pp.

Sharma, Shilpi 2012, Bioremediation: Features, Strategies and applications. *Asian Journal of Pharmacy and Life Science*. Vol. 2 (2), 202–213.

Suthersan, S.S. 1999, " *In Situ Bioremediation" Remediation Engineering: Design Concepts*. Ed. Suthan S. Suthersan Boca Raton: CRC Press LLC, 36 pp.

Timothy P. Ruggaber, M. Asce and Jeffrey W. Talley. 2006, Enhancing Bioremediation with Enzymatic Processes: A Review. M. Asce Practice Periodical of Hazaroous, Mxic, and Radioactive Waste Management.

2015, Recent Trends in Microbiology, Mycology and Plant Pathology   *Pages* **83–93**
Editor: **Dr. H.C. Lakshman**
Published by: **DAYA PUBLISHING HOUSE, NEW DELHI**

Chapter 6

Improved Seedling Emergence and Biomass Production of Crop Plants by *Trichoderma* spp.

J.N. Rajkonda[1] and U.N. Bhale[2]

[1]*Department of Botany, Yeshwantrao Chavan Mahavidyalaya,*
Tuljapur Dist. Osmanabad – 413 601, Mh., India
E-mail: jnrajkonda@gmail.com
[2]*Research laboratory, Department of Botany,*
Arts, Science and Commerce College, Naldurg,
Dist. Osmanabad – 413 602, Mh., India
E-mail: unbhale2007@rediffmail.com, drunbhale2012@gmail.com

ABSTRACT

Trichoderma species are used extensively as a biocontrol agent for the management of many soil and seed borne diseases. Five *Trichoderma* species *viz. Trichoderma viride, T. harzianum, T. koningii, T. pseudokoningii* and *T. virens* were treated with seeds of various crop plants. Seed germination, biomass production and vigour index of treated seeds and seedling were evaluated and found to be significant. Biomass production and vigour index of wheat (*Triticum aestivum* L.) seedlings were enhanced by *T. virens* while the rate of germination also increased by *T. harzianum* and *T. pseudokoningii. Trichoderma viride* showed better results on the paddy (*Oryza sativa* L.) seedlings with respect to germination, biomass production and vigour index. Germination rate of pigeon pea (*Cajanus cajan* L.) seeds were enhanced by *T. harzianum* and *T. virens. Trichoderma harzianum* enhanced the biomass production and seedling vigour index of pigeon pea (*Cajanus cajan* L.). Germination, biomass production and vigour index of Gram (*Cicer aerietinum*

L.) due to *T. koningii*. Maximum germination and seedling vigour index of groundnut (*Arachis hypogea* L.) was recorded better by *T. pseudokoningii* and *T. virens*. Its biomass production was accelerated by *T. koningii*. *Trichoderma virens* was found to be superior in case of soybean (*Glycine max* L.) with respect to germination rate, biomass production and seedling vigour index. *Trichoderma harzianum* showed promoting effect on the germination rate and biomass production of spinach while maximum seedling vigour index was due to the effect of *T. koningii*. Germination rate and biomass production of Dill (*Anethum graveolens* L.) was better under the treatment of *T. koningii* and *T. virens*, while seedling vigour index was enhanced due to *T. viride*. *Trichoderma* species were responsible for healthy and disease free crops and products.

Keywords: *Trichoderma species, Crop plants, Biomass production.*

Introduction

Trichoderma species are biological control organisms against a wide range of soil borne pathogens and have plant growth promotion capacity. While using *Trichoderma* species as a biocontrol agent for the management of plant diseases, it should be necessary to determine its effects on the crop plants with respect to seed germination, biomass production and vigour index. The interaction between *Trichoderma* species and plants are intriguing. The *Trichoderma* species can be beneficial to the plants causing growth stimulation (Kleifeld and Chet,1992; Ousley, *et al.*, 1994) or through biocontrol of plant diseases (Askew and Laing, 1994; Devan and Sivasithamparam 1988; Lo *et al.*, 1996; Yang *et al.*, 1995).

The effect of *Trichoderma* species on the plant growth and their development was studied by many workers and researches regarding this aspect were revealed that *Trichoderma* species increases the growth of the plants. Windham *et al.* (1986) reported that addition of *Trichoderma* species to autoclaved soil increased the rate of emergence of tomato and tobacco seedlings over that of control. *Trichoderma harzianum* when applied to sterilized soil induced an emergence of seedlings, plant height, leaf area and dry weight (Kleifeld and Chet, 1992). Seed germination was maximum when *T. polysporum* was treated; fresh weight of eggplant was more with *T. viride* (Watanabe, 1993). Effect of biocontrol agent when primed was effective with respect to increased germination, biomass production of crop plants and healthy crop production. The effect of *T. harzianum* and *T. viride* on cumin seeds by biopriming method (Sharma *et al.*, 2009) indicated that germination was increased; shoot: root ratio and seedling height was also increased. It has been shown that *Trichoderma* species stimulated the growth of tomato plants (Datnoff *et al.*, 1995; Ozbay *et al.*, 2004).

The effects of *Trichoderma* isolates on plant growth and development have important economical implications such as shortening the plant growth period and time in nursery, as well as improving plant vigour to overcome biotic and/or abiotic stresses, resulting in increase plant productivity and yields. Therefore, the use of *Trichoderma* isolates in plant growth improvement is crucially important in sustainable agriculture system, because chemical fertilizer is not economical in the long run due to their cost and environmental pollution. Accordingly, reduction or elimination of synthetic fertilizer applications in agriculture is highly desirable. There are many abiotic and biotic factors such as plant species, the strain of *Trichoderma* used, the form of applied inoculum and its concentration and the soil environment as well as the rhizosphere microflora that may have an influence on *Trichoderma* activity. Therefore, it is very important to collect information about these factors. Recently, several attempts have been undertaken to survey *Trichoderma* spp. promotion of seedling establishment, enhancement of plant growth and elicit plant defense reaction in some crops such as cotton (Shanmugaiah *et al.,* 2009), vegetables (Celar and Valic, 2005; Rabeerdran *et al.,* 2000; Lynch *et al.,* 1991), bean (Hoyos-Carvajal *et al.,* 2009) and corn (Windham *et al.,* 1989).

The physical interaction between *Trichoderma* and the plant appears that this interaction evolves in to a symbiotic rather than a parasitic relationship between the fungus and the plant. (Yedidia *et al.,* 1999). It Elicitors from *Trichoderma* activate the expression of genes involved in the plant defence response system and promote the growth of the plant, root system and nutrient availability. This effect inturn augments the zone for colonization and the nutrients available for the biocontrol fungus, subsequently increasing the overall antagonism to plant pathogens (Yedidia *et al.,* 2003; Honson and Howell, 2004). Many species of *Trichoderma* are not only able to control the pathogens that cause plant disease, but are also able to promote plant growth and development.

Therefore in the present investigation an effort was taken to determination of effect of *Trichoderma* species on the germination, biomass production and vigour index of crop plants.

Materials and Methods

Isolation of *Trichoderma* Species

Five *Trichoderma* species *viz.,* *T. viride, T. harzianum, T. koningii, T. pseudokoningii* and *T. virens* were isolated from the rhizosphere soil of Marathwada region of Maharashtra state. Isolated *Trichoderma* species were maintained on PDA slants for the study.

Treatment with Seeds of Crop Plants

The effect of culture filtrate of *Trichoderma* species on germination, biomass production and vigour index of crop plants were determined by pot culture method. Healthy seeds of crop plants such as wheat (*Triticum aestivum* L.), rice (*Oryza sativa* L.), pigeon pea (*Cajanus cajan* L.), gram (*Cicer aerietinum* L.), groundnut (*Arachis hypogea* L.), soybean (*Glycine max* L.), spinach (*Spinacia oleraea* L.) and shepu (*Anethum graveolens* L.) were selected. Twenty healthy seeds were taken and sterilized by treatment with Mercuric chloride ($HgCl_2$) (0.1 per cent) solution. The seeds were washed with sterile distilled water and treated with culture filtrate of isolated *Trichoderma* species. The treated seeds were sown in pot containing sterilized 2 kg black soil. Seeds without treatment were sown in separate pot and considered as a control. Water was added periodically at regular time intervals for 30 days. The pots were allowed to maintain at natural conditions.

After 30 days, number of germinated seeds were counted and recorded in terms of percentage root length, shoot length, number of leaves fresh, weights, dry weight were measured. The vigour index [Vigour index = [Mean root length + Mean Shoot length] X Germination (per cent)] of crop seedlings was calculated as per Abdul Baki and Anderson (1973).

Results and Disucssion

Effect of isolated *Trichoderma* species on the growth of crop plants were tested by seed treatment with culture filtrate and growing such seeds in pot. The treatment of *Trichoderma* species with the seeds significantly increased the germination rate, biomass production and vigour index of the seedlings.

Table 6.1: Effect of *Trichoderma* Species on Germination, Biomass Production and Vigour Index of Cereal Crop Plants

Crop Plant	Treatment	Germ. (%)	No. of Leaves	Root Length (cm)	Shoot Length (cm)	Fresh Weight (gm)	Dry Weight (gm)	Vigour Index
Wheat	Control	75	3.9	7.6	6.5	0.62	0.090	980
	T. viride	50	4.0	8.3	7	0.78	0.100	765
	T. harzianum	80	4.1	8.0	9.1	0.8	0.100	1368
	T. koningii	75	4.0	7.8	6.7	0.7	0.080	1087
	T. pseudokoningii	80	4.1	8.4	7	0.72	0.060	1232
	T. virens	70	4.6	8.9	6.8	1.24	0.180	1089
Rice	Control	65	1.1	2.4	1.7	0.03	0.010	164
	T. viride	90	1.3	5.5	2.4	0.06	0.018	711
	T. harzianum	80	1.4	5.5	2.7	0.05	0.017	656
	T. koningii	50	1.2	3.2	2.1	0.04	0.013	265
	T. pseudokoningii	90	1.5	4.1	3	0.05	0.017	639
	T. virens	65	1.3	3.8	2.5	0.05	0.024	346
Mean±		72.5	2.7	6.1	5.19	0.42	0.059	775

Results from the Table 6.1 indicated that germination of wheat was maximum under the treatment of *T. harzianum* (80 per cent) and *T. pseudokoningii* (80 per cent). Number of leaves recorded more in case of all treated seeds and was maximum when *T. virens* (4.6) used. Root length of root recorded maximum in the treatment of *T. virens* (8.9 cm) and shoot length was found to be maximum by *T. harzianum* (9.1 cm). Biomass production of wheat seedlings with respect to fresh weight and dry weight was recorded maximum in the treatment of *T. virens*. However, the vigour index of wheat seedlings was recorded maximum when *T. harzianum* (1368) was treated. Germination of rice seeds was increased when they treated with *T. viride* (90 per cent) and *T. pseudokoningii* (90 per cent) followed by *T. harzianum* (80 per cent). The effect of *Trichoderma* species on seedlings with respect to number of leaves were little significant. Root length of paddy seedlings were enhanced by *T. viride* (5.5 cm) and *T. harzianum*

(5.5 cm) while shoot length was maximum due to the effect of *T. pseudokoningii* (3.0 cm) and *T. harzianum* (2.7cm). Biomass production was maximum in the treatment of *T. viride* which also increased vigour index of paddy seedlings (711).

Seeds of pulse crops were treated with *Trichoderma* species and the results were determined. The effects were found to be superior with respect to germination, biomass production and vigour index of seedlings of pulse crop plants. Table 6.2 showed that pigeon pea seeds when treated with *T. harzianum* (100 per cent) and *T. virens* (100 per cent) recorded highest percent germination followed by *T. viride* (90 per cent). Numbers of leaves were recorded maximum in the treatment of *T. harzianum* (11). Maximum root length of seedlings was observed in the treatment of *T. harzianum* (11.2 cm) and shoot length was better in *T pseudokoningii* (25.6cm). *Trichoderma harzianum* also showed maximum biomass production of pigeon pea. Effect of *T. harzianum* on the seedling vigour index was significant (3180) followed by *T. pseudokoningii* (2720) and *T. virens* (2520). Germination of gram seeds was maximum under the treatment of all *Trichoderma* species (100 per cent). Numbers of leaves were maximum by *T. koningii* (14). *Trichoderma koningii* was the species that found to be effective with respect to root length (14.70 cm), shoot length (12.50 cm), fresh weight (25.86 gm), dry weight (7.06gm) and vigour index (2720).

Table 6.2: Effect of *Trichoderma* Species on Germination, Biomass Production and Vigour Index of Pulse Crop Plants

Crop Plant	Treatment	Germ. (%)	No. of Leaves	Root Length (cm)	Shoot Length (cm)	Fresh Weight (gm)	Dry Weight (gm)	Vigour Index
Pigeon pea	Control	80	8	8.20	12.4	11.32	1.15	1648
	T. viride	90	8	9.40	16.5	11.80	1.18	2331
	T. harzianum	100	11	11.20	20.6	14.04	1.40	3180
	T. koningii	75	8	8.30	20.6	11.48	1.15	2167
	T. pseudokoningii	80	7	8.40	25.6	12.01	1.19	2720
	T. virens	100	5	8.20	17.30	11.42	1.16	2520
Gram	Control	90	8	13.15	10.3	17.70	4.58	2110
	T. viride	100	10	14.20	12.3	21.98	5.85	2660
	T. harzianum	100	11	14.47	11.5	25.55	6.96	2597
	T. koningii	100	14	14.70	12.5	25.86	7.06	2720
	T. pseudokoningii	100	11	14.10	10.0	18.93	6.12	2410
	T. virens	100	12	14.27	11.0	21.50	5.44	2527
Mean±		93	9	11.77	15.02	16.96	3.60	2465

Among the oil seed crops, seeds of groundnut and soybean were treated with the *Trichoderma* species to determine their effects on germination, biomass production and vigour index. The results were obtained more superior and recorded. Table 6.3 showed that *T. pseudokoningii* (95 per cent) was the species effective for germination of groundnut seeds followed by *T. virens* (85 per cent). *Trichoderma koningii* caused the increased number of leaves (8.33) followed by *T. pseudokoningii* (8.22). Maximum induction of root length was occurred by treatment of *T. harzianum* (11.10 cm) and *T. virens* (11.00 cm) while shoot length by *T. koningii* (7.00 cm). Maximum biomass production was observed under the treatment of *T. koningii*. Vigour index of seedlings were maximum when *T. pseudokoningii* (1520) and *T. virens* (1437) were treated. Germination of seeds of soybean were maximum under the treatment of *T. virens* (90 per cent), *T. viride* (85 per cent), *T. harzianum* (85 per cent) and *T. pseudokoningii* (85 per cent). Maximum numbers of leaves were observed when treated with *T. viride* (6.12) and *T. harzianum* (6.00). Root length and shoot length of seedlings were increased but root length was maximum due to *T. virens* (14.09 cm) while shoot length by *T. pseudokoningii* (16.65 cm). Biomass production of the seedlings was highest in the treatment of *T. virens*. However, seedling vigour index was highest when treated with of *T. virens* (2610) followed by *T. pseudokoningii* (2468) and *T. viride* (2055).

Table 6.3: Effect of *Trichoderma* Species on Germination, Biomass Production and Vigour Index of Oilseed Crop Plants

Crop Plant	Treatment	Germ. (%)	No. of Leaves	Root Length (cm)	Shoot Length (cm)	Fresh Weight (gm)	Dry Weight (gm)	Vigour Index
Groundnut	Control	45	6.67	6.52	4.81	8.37	0.32	510
	T. viride	65	6.77	7.57	5.7	8.62	0.35	862
	T. harzianum	50	7.40	11.10	6.16	8.71	0.37	863
	T. koningii	55	8.33	10.81	7	14.73	0.70	980
	T. pseudokoningii	95	8.22	9.74	6.26	9.30	0.38	1520
	T. virens	85	7.53	11.00	5.91	9.02	0.34	1437
Soybean	Control	55	5.81	09.46	11.67	6.08	0.24	1162
	T. viride	85	6.12	10.84	13.34	7.61	0.37	2055
	T. harzianum	85	6.00	10.37	13.07	6.16	0.28	1992
	T. koningii	75	5.91	10.03	13.65	7.71	0.31	1776
	T. pseudokoningii	85	5.83	12.39	16.65	7.87	0.33	2468
	T. virens	90	5.81	14.09	14.91	7.96	0.38	2610
Mean±		73	6.70	10.32	9.93	8.51	36.00	1520

Seeds of spinach treated with the culture filtrate of *Trichoderma* species and the results were recorded in the Table 6.4. The germination of the seeds were increased and maximum due to *T. harzianum* (90 per cent) followed by *T. virens* (80 per cent), *T. koningii* (85 per cent) and *T. pseudokoningii* (80 per cent). Number of leaves recorded maximum by the treatment of *T. viride* and *T. harzianum*. The root length was more under the treatment of *T. koningii* (7.01 cm) and *T. virens* (6.27 cm). The shoot length was increased due to *T. harzianum* (0.99 cm), *T. koningii* (0.98 cm) and *T. pseudokoningii* (0.94 cm). Biomass production with respect to fresh weight and dry weight was found to be maximum due to effect of *T. harzianum* and *T. viride* followed by *T. koningii* and *T. pseudokoningii*. *Trichoderma koningii* (639) and *T. virens* (609) were effective in increased seedling vigour index. The germination of shepu was observed maximum due to the effect of *T. koningii* (80 per cent) and *T. virens* (80 per cent). Numbers of leaves were increased slightly and maximum under the treatment of *T. koningii* (4.62) and *T. virens* (4.61). Maximum root length (11.01 cm) and shoot length (2.11 cm) was recorded by the effect of *T. viride*. Fresh weight of seedlings was observed in the treatment of *T. koningii* (0.055 gm) and *T. harzianum* (0.052 gm). Dry weight was found to be more under the treatment of *T. koningii* ((0.011 gm) and *T. harzianum* (0.011 gm). Significant increase in the seedling vigour index was recorded due to the treatment of *T. viride* (984).

Table 6.4: Effect of *Trichoderma* Species on Germination, Biomass Production and Vigour Index of Vegetable Plants

Crop Plant	Treatment	Germ. (%)	No. of Leaves	Root Length (cm)	Shoot Length (cm)	Fresh Weight (gm)	Dry Weight (gm)	Vigour Index
Spinach	Control	60	5.33	4.33	0.90	0.260	0.028	314
	T. viride	70	12.30	5.30	0.90	3.290	0.370	434
	T. harzianum	90	10.55	5.34	0.99	3.330	0.391	570
	T. koningii	80	8.80	7.01	0.98	1.630	0.178	639
	T. pseudokoningii	80	7.62	5.11	0.94	1.130	0.115	484
	T. virens	85	7.00	6.27	0.90	0.780	0.098	609
Shepu	Control	70	4.07	3.46	1.30	0.029	0.007	333
	T. viride	75	4.10	11.01	2.11	0.034	0.008	984
	T. harzianum	75	4.40	6.07	1.97	0.052	0.011	603
	T. koningii	80	4.62	4.31	1.90	0.055	0.011	497
	T. pseudokoningii	75	4.26	4.38	1.73	0.042	0.009	458
	T. virens	80	4.61	4.39	1.92	0.043	0.009	505
Mean±		77	6.47	5.58	1.38	0.890	0.102	536

Trichoderma species are the useful biocontrol agents against number of plant pathogens. When the bioagents are applied, their effects on the growth and development of plants should be determined. Therefore in the present investigation, effect of *Trichoderma* species on the various crop plants had been carried out. The obtained results showed that *Trichoderma* species exhibited promoting effects on the crop plants with respect to germination, seedling length, number of leaves, biomass production and vigour index. There were no negative effects of the *Trichoderma* species on the growth of crop plants.

During the development process, initial growth of seedling is necessary for their establishment in the surroundings. Many seedlings were failed to proper growth and due to lack of proper growth caused low productivity. The increased seedling length, number of leaves, biomass production and seedling vigour index also indicated that the survivals of crop plants were increased. This is because the *Trichoderma* species probably caused increased metabolism in the crop plants. The healthy seedlings were also the sign of increased immunity of the crop plant seedlings. The seeds of pulse crops, oil crops and vegetables germinated actively under the treatment of *Trichoderma* species.

Precolonization provides the biocontrol agent with a competitive advantage over attacking pathogens and often provides superior seed protection (Harman and Taylor,1988; Harman *et al.,* 1989; Harman, 1991). *Trichoderma* species can compete with other micro-organisms for key exudates from seeds that stimulate the germination of propagules of plant pathogenic fungi in the soil (Howell, 2003) Application of *T. harzianum* at the time of sowing resulted significant increased groundnut pod yield and reduced the incidence of stem and pod rot (Jadeja and Ralcholiya, 2009). Seed treatment and soil treatment with *Trichoderma* species resulted healthy and disease free crop plants. This indicated that these plants have increased disease resistance which was induced by *Trichoderma* species. However, the induction of disease resistance in plants by *Trichoderma* species has been poorly studied because earlier works focused on factors like mycoparasitism and antibiosis. The first clear demonstration of induced resistance by *Trichoderma* was published in 1997 by Bigirimana who showed that treating soil with *T. harzianum* (T-39) made leaves of bean plants resistant to diseases caused by *B. cinerea* and *Colletotrichum lindemuthlanum*. Similar studies have been carried out with wide range of plants, including both monocotyledons and dicotyledons. In cotton, *T. virens* induced disease resistance against *R. solani* by inducing fungitoxic turpenoid phytoalexins (Howell *et al.,* 2000).

Earlier research works confirmed that induced germination of seeds by *Trichoderma* species. Wright (1954) found that when *Trichoderma* added to soil, germination of tomato and eggplant occurred two to three days earlier and showed better plant growth over control. The increased growth response induced by *Trichoderma* species has been reported for many crop plants such as beans, cucumber, pepper, maize and wheat (Lo and Lin, 2002). Seed pelleting of sunflower and mung bean with *T. harzianum, T. polysporum, T. pseudokoningii* and *T. virens* enhanced the plant growth as compared to control (Yaqub and Shahzad, 2008). There are several reports where *Trichoderma* species effectively increased the plant growth, weight and root mass (Cole and Zvenyika, 1986; Paulitz *et al.,* 1985; Kumar *et al.,* 2007) that supports the results of present research work.

Recently chilli seeds coated with spore suspension of strains of *T. harzianum, T. pseudokoningii* and *T. virens* resulted in to increased germination, reduced the delay of germination and seedling parameters in both laboratory and field conditions (Islam *et al.,* 2011). Seed dressing and soil application with *T. virens, T. harzianum* and *T. viride* were found to be superior in controlling wet root rot of mungbean caused by *Rhizoctonia solani* and improvement of the yield (Dubey *et al.,* 2011). Increased growth of shoot and root caused by *Trichoderma* implied that there was beneficial effect of inoculation on plant growth and development since root collar and stem diameters were a measure of survivability of seedlings (Okoth *et al.,* 2011).

Earlier works suggested that the *Trichoderma* species are naturally occurring soil fungi that colonize roots and stimulate plant growth. Such fungi have been applied to a wide range of plant species for the purpose of growth enhancement, with a positive effects on plant weight, crop yield and disease control. Their agricultural use could be expanded if the mechanisms of growth enhancement were known.

Conclusion

It is concluded that the *Trichoderma* species likely act as plant growth promoting micro-organisms because the given experiments showed enhanced root growth along with shoot growth and biomass production of crop plants. The data obtained in the result suggested an important role of *Trichoderma* species in plant growth regulation. The results showed great promise for the use of *Trichoderma* species as inoculants for plant improvement under controlled and field conditions. The introduction of *Trichoderma* species in the soil with or without pathogen did not affect the population of existing beneficial micro-organisms. Use of *Trichoderma* species as seed treatment for better crop productivity could be suggested.

Acknowledgement

Authors are thankful to Principal, Dr. S. D. Peshwe, A. S. C. College, Naldurg and Principal, Dr. J. S. Mohite, Y. C. M. Tuljapur for providing necessary laboratory facilities. And also thankfully acknowledged to UGC (WRO) Pune and UGC, New Delhi for providing financial assistance.

References

Abdul Baki and Anderson. 1973. Vigour determination in soybean by multiple criteria. *Crop Science,* 13: 630–633.

Askew DJ and Liang MD. 1994. The *in vitro* screening of 119 *Trichoderma* isolates for antagonism to *Rhizoctonia solani* and an evaluation of different environmental sites of *Trichoderma* as sources of aggressive strains. *Plant and Soil,* 159: 277–281.

Bigirimana J. 1997. Induction of systemic resistance on bean (*Phaseolus vulgaris*) by *Trichoderma harzianum. Med. Fac. Landbouww Uni. Gent.,* 62: 1001–1007.

Celar F and Valic N. 2005. Effects of *Trichoderma spp* and *Glicladium roseum* culture filtrates on seed germination of vegetables and maize. *J. Plant Dis. Prot.,* 112(4): 343–350.

Cole JS and Zvenyika P. 1986. Integrating *T. harzianum* and triadimenol for the control of tobacco sore shin in Zimbabwe. *Bull. Inf. Coresta,* 68.

Datnoff LE, Sand N and Pernezny K. 1995. Biological control of *Fusarium* crown and root rot of tomato in Florida using *Trichoderma harzianum* and *Glomus intraradices. Biol. Cont.,* 5: 427–431.

Devan MM and Sivasithamparam K. 1988. Identity and frequency of occurrence of *Trichoderma* spp in roots of wheat and rye–grass in Western Australia and their effects on root rot caused by *Gaeumannomyces graminis* var. *tritici. Plant and Soil,* 109: 93–101.

Dubey S C, Ranganaicker B and Singh B. 2011. Integration of soil application and seed treatment formulations of *Trichoderma* species for management of wet root rot of mungbean caused by *Rhizoctonia solani. Pest Management Science, 67(9):* *1163–1168.*

Harman GE. 1991. Seed treatments for biological control of plant disease. *Crop Prot.* 10: 166–171.

Harman GE and Taylor AG. 1988. Improved seedling performance by integration of biological control agents at favorable pH levels with solid matrix priming. *Phytopathology,* 78: 520–525.

Harman GE, Taylor AG and Stasz TE. 1989. Combining effective strains of *Trichoderma harzianum* and solid matrix priming to improve biological seed treatments. *Pl. Dis.,* 73: 631–637.

Honson LE and Howell CR. 2004. Elicitors of plant defense responses from biocontrol strains of *Trichoderma virens. Phytopathology,* 94(2): 171–176.

Howell CR. 2003. Mechanisms employed by *Trichoderma* species in the biological control of plant disease; the history and evolution of current concepts. *Plant Dis.,* 87: 4–10.

Howell CR, Hanson LE, Stipanovic RD and Pukhaber LS. 2000. Induction of turpenoid synthesis in cotton roots and control of *Rhizoctonia solani* by seed treatment with *Trichoderma virens. Phytopathology,* 1990: 248–252.

Hoyos–Carvajal L, Ordua S and Bissett J. 2009. Growth stimulation in bean (*Phaseolus vulgaris* L.) by *Trichoderma. Biol. Control,* 51: 409–416.

Islam MS, Rahman MA, Bulbul SH and Alam M.F. 2011. Effect of *Trichoderma* on seed germination and seedling parameters in chilli. *Int. J. Expt. Agric.,* 2(1): 21–26.

Jadeja KB and Rakholiya KB. 2009. Efficacy of soil amendments with *T. harzianum* on stem and pod rot incident and yield of groundnut. (Abs.) *J. Mycol. Pl. Pathol.,* 39(3): 569.

Kleifeld O and Chet I. 1992. *Trichoderma harzianum* – interaction with plants and effect on growth response. *Plant and Soil,* 144: 267–272.

Kumar S, Arya MC and Singh R. 2007. Efficacy of *P. fluorescence* and *T. harzianum* as bioenhancers in tomato at high altitude in Central Himalayas. *Indian J. Crop Sciences,* 2(1): 79–82.

Lo CT and Lin CY. 2002. Screening strains of *Trichoderma* spp for plant growth enhancement in Taiwan. *Plant Pathology Bull.,* 11: 215–220.

Lo CT, Nelson EB and Harman GE. 1996. Biological control of turfgrass diseases with a rhizosphere competent strains of *Trichoderma harzianum. Plant Dis.,* 80: 736–741.

Lynch JM, Wilson KL, Ousley MA and Wipps JM. 1991. Response of lettuce to *Trichoderma* treatment. *Lett. Appl. Microbiol.,* 12: 59–61.

Okoth SA, Otaodh JA. and Ochanda JO. 2011. Improved seedling emergence and growth of maize and beans by *Trichoderma harzianum. Tropical and Subtropical Agroecosystems,* 13: 65–71.

Ousley MA, Lynch JM, and Whipps JM. 1994. Potential of *Trichoderma* spp. as consistent plant growth stimulators. *Biol. Fertil. Soils,* 17: 85–90.

Ozbay N, Newman SE and Brown M. 2004. The effect of *Trichoderma harzianum* strains in the growth of tomato seedlings. Proc. XXVI IHC Manage, *Acta Hort.* 635: 131–135.

Paulitz T, Windham M and Baker R. 1985.Enhanced plant growth induced by *Trichoderma* amendments. *Phytopathology,* 75: 1302.

Rabeerdran N, Moot DJ, Jones EE and Stewart A. 2000. Inconsistent growth promotion of cabbage and lettuce from *Trichoderma* isolates. *New Zealand Plant Prot.,* 53: 143–146.

Shanmugaiah V, Balasubramanian N, Gomathinayagam S, Monoharan PT and Rajendran A. 2009. Effect of single application of *Trichoderma viride* and *Pseudomonas fluorences* on growth promotion in cottonplants. *Afr. J. Agric. Res.,* 4(11): 1220–1225.

Sharma YK, Anwer MM, Lodha SK, Sriram S and Ramanujam B. 2009. Effect of biopriming with biological control agents on wilt and growth of cumin. *J. Mycol. Pl. Pathol.* 39(3): 567.

Watanabe N. 1993. Promoting effect of *Trichoderma* spp on seed germination and plant growth in vegetables. *Memories of the Institute of Science and Technology, Meiji Uni.,* 32: 9–17.

Windham GL, Windham MT and Williams WP. 1989. Effects of *Trichiderma* spp. on maize growth and *Meloidogyne arenaria* reproduction. *Plant Dis.,*73: 493–495.

Windham, MT, Elad Y and Baker R. 1986. A mechanism for increased plant growth induced by *Trichoderma* spp. *Phytopathology,* 76: 518–521.

Wright JM. 1954. The production of antibiotics in soil 1 production of gliotoxin by *Trichoderma viride. Ann. Appl. Boil.,* 41: 280–289.

Yang D, Bernier L and Dessureault M. 1995. *Phaeotheca dimorphospora* increases *Trichoderma harzianum* density in soil and suppresses red pine damping off caused by *Cylindrocladium scoparium. Can. J. Bot.,* 73: 693–700.

Yaqub F and Shahzad S. 2008. Effect of seed pelleting with *Trichoderma* spp and *Gliocladium virens* on growth and colonizationof roots of sunflower and mung bean by *Sclerotium rolfsii. Pak. J. Bot.,* 40(2): 947–953.

Yedidia I, Benhamou N and Chet I. 1999. Induction of defense responses in cucumber (*Cucumis sativus* L.) by the biocontrol agent *Trichoderma harzianum. Appl. Environ. Microbiol.,* 65: 1061–1070.

Yedidia I, Shoresh M, Kareem Z, Lapulink Y and Chet I. 2003. Concomitant induction of systemic resistance to *Pseudomonas syringae* pv. *lachrymans* in cucumber by *Trichoderma asperellum* (T–203) and accumulation of phytoalexins. *Applied Environmental Microbiology,* 69: 7343–7353.

2015, Recent Trends in Microbiology, Mycology and Plant Pathology   *Pages* 95–104
Editor: **Dr. H.C. Lakshman**
Published by: **DAYA PUBLISHING HOUSE, NEW DELHI**

Chapter 7

Interactions between the Arbuscular Mycorrhizae, *Glomus mosseae* and *Trichoderma* and their Significance for Enhancing Plant Growth and Suppressing Wilt Disease of *Cajanus cajan* (L).

Keerti Dehariya*, Imtiyaz Ahmed Sheikh,
Ashok shukla and Deepak Vyas

Lab of Microbial Technology and Plant Pathology, Department of Botany,
Dr. Hari Singh Gour University, Sagar – 470 003, M.P., India
*E-mail: dehariya.k@gmail.com

Introduction

In the rhizosphere microbial community carried out several function which results in plant growth promotion and disease reduction caused by soil microbes (Martínez-Medina *et al.,* 2010). Arbuscular mycorrhizae and *Trichoderma* are very cosmopolitan population of all types of soil. AMF are an important component of many agricultural and natural ecosystems, forming symbiotic associations with the majority of plants and showing multifunctional roles in such systems (Sieverding., 1991). Arbuscular mycorrhizae has been well documented to mitigate the effects of root pathogens (Hooker *et al.,* 1994) and improve host nutrients and stress tolerance (Auge´, 2001; Jeffries and Barea, 2001; Tisdall, 1994). It has also been reported that AMF alter the microbial population composition that exists in the zone of soil influenced by these communities (Linderman and Paulitz, 1990). On the other hand,

AMF population density and diversity could also be affected by the antagonistic activities of other soil inhabiting micro-organisms used as biocontrol or plant growth promoting agents (Linderman, 1992; Mar Vazquezet al., 2000). *Trichoderma* sp. have been reported to producing beneficial effects to several crop plants by promoting their growth and protecting them from disease (Koike *et al.,* 2001; Muslim *et al.,* 2003). Interactions between *Trichoderma* and AMF might stimulate or inhibit the beneficial effects of each species. Although interactions between AMF and rhizobacteria have been extensively studied (Marschner and Baumann, 2003; Medina *et al.,* 2003), interactions of AMF with saprophytic beneficial fungal species have received little attention (Fracchia *et al.,* 2000; Green *et al.,* 1999). Wilt disease caused by *Fusarium udum* significantly affects agricultural crop production and yield. This disease generally controlled by fungicides, but such increase in use of pesticides adversely affects the natural balance of soil ecosystems (Vosatka and Albrechtova, 2009), pollute soil and eliminate beneficial microflora and has favored the development of resistant to pathogens (Zeng, 2006). As such in the present context, disease management with biocontrol agents offers a great promise. Biocontrol agents are vital component of sustainable agriculture. A biological control agent colonize the rhizosphere, the site requiring protection and leaves no toxic residues as opposed to chemicals (Dubey *et al.,* 2007). Co-inoculations of AMF and *Trichoderma* might provide a higher level of protection and wider range of effectiveness by activating several different mechanisms. In this study, we initially observed the interaction between AMF *Glomus mosseae* and *Trichoderma harzianum/T. viride* with respect to their colonization of *Cajamun cajan* and their effects on plant growth. Secondly, each fungal species and their combined effects were tested for the possibility of controlling occurrence of wilt disease caused by *F. udum.*

Materials and Methods

Site Description and Host Plant

Present study was carried out in the net house at the Department of Botany, Dr HS Gour Central University, Sagar (23° 10'- 24° 27' N latitude and 78° 4'-79° 21' E longitude). Sagar situated at North of Tropic of Cancer on an average altitude of 580 m above mean sea level. The region lies in agro-ecoregion 10; central highlands, hot sub-humid ecoregion with medium and deep black soils (I5C3). *Cajanus cajan* (L.) Millsp was used as the host plant.

Fungal Inocula Preparation

The AMF *G. mosseae* (Nicol. and Gerd.) Gerd. and Trappe, *T. harzianum* and *T. viride* were isolated from selected sites and selected on the basis of their antagonistic potential against pathogen. Pathogen *i.e. Fusarium udum* was isolated from infected plant part collected from aricultural field of *Cajanus cajan* and its pathogenicity was tested on healthy plant on the basis of occurrence of wilt disease. The purified culture of *F. udum* grown on PDA was multiplied on sand-maize flour medium. Fifteen g maize flour and 85 g autoclaved sand was thoroughly mixed in 250 ml flasks (100 g/ flask) and autoclaved at 15 Lbs for 30 minutes. Then, each flask was inoculated with pure culture of *F. udum* and incubated at 25+2 °C for two weeks.

Ten g wheat bran moistened with 25 ml mineral salt solution (pH: 5 to 5.2) was taken in 500 ml flask and sterilized at 121°C and 15 psi for 15 min by autoclaving. These flasks were then inoculated with the mycelial disc (4 mm diameter) of *Trichoderma* and incubated at 28±2°C for 15 days in an incubator (Sinclair and Dhingra, 1985). The cfu of *Trichoderma* was determined by serial dilution plate method before being used.

Mycorrhizal consortium (*Mc*) comprised of equal parts in weight of *Glomus fasciculatum* Thaxter, *Glomus intraradices* Schenck and Smith, and *Glomus mosseae* (Nicolson and Gerd.) Gerd. and Trappe was used. Consortium consisting of sand, chopped root bits, spores and extrametrical mycelia was used.

Experimental Design and Biological Treatments

Two experiments were conducted to test the interaction of both *Trichoderma* species with *G. mosseae*. One experiment (Exp. 1) investigated interaction effects on fungal colonization and plant growth promotion and the other (Exp. 2) evaluated wilt disease suppression. Both experiments had the same treatments as follows. (1) *Trichoderma harzianum* only (Th) (2) *T. viride* only (Tv) (3) *G. mosseae* only (Gm); (4) *T. harzianum* and *G. mosseae* (Th + Gm) (5) *T. viride* and *G. mosseae* (Tv + Gm); and (6) uninoculated control (Cont). Each treatment had five replications. Plants were laid out in a completely randomized block design.

Exp. 1: Effect on Fungal Colonization and Plant Growth Promotion

Mycorrhizal treatments were made by hand-mixing the *G. mosseae* inoculum with the potting medium (2 per cent, w/w). Non mycorrhizal treatments received the same amount of autoclaved *G. mosseae* inoculum. Potting medium–inoculum mixtures were then added to the sterilized plastic pots (100 ml). At the same time, the filtrate of AMF inoculum (without AMF propagules) was poured into the non-mycorrhizal pots (10 ml/pot) in order to introduce other microbial communities accompanying the *G. mosseae* inoculum (Green *et al.*, 1999). Surface sterilized (2 per cent NaOCl for 3 min) seeds of *Cajanus cajan* were pre-germinated and one seed was planted per pot. At planting, *Trichoderma* inoculua were added to the pots (10g/pot) according to the treatment combinations. The colonization of *Trichoderma* was measured 4 and 6 weeks after planting. To estimate colonization, each rhizosphere soil sample (1 g) was put in a 250 ml Erlenmeyer flask containing 99 ml of sterile water. After shaking for 30 min at 150 rpm, serial dilutions were made. The growth media used was *Trichoderma* selective (Elad *et al.*, 1981) medium. After 3–5 days of incubation, colonies were identified, counted and population was quantified as colony-forming units (CFU) per gram of dry soil.

AM colonization index and spore density (50 g^{-1} sand), root and soil samples were determined in Myc inoculated pots. Fine roots were cleared with 10 per cent KOH and stained with acid fuchsin (0.01 per cent in lacto-glycerol) following the method of Phillips and Hayman (1970). Colonization in cleared root parts was determined with a microscope (Nikon Eclipse 400) at x 100 using the grid line intersect method (Giovannetti and Mosse 1980). Briefly, 50 g sand was collected from each pot and AM spores were extracted using wet sieving and decanting method (Gerdemann and Nicolson, 1963). Sand samples were taken in a substantial amount of water (1 L) and decanted through a series of sieves (mesh size: 250, 150 and 53 µ). Sievings were individually collected in separate jars. For observation under a stereomicroscope (Nikon SMZ 800), the sievings were transferred into a gridded petri dish (11 cm). Then, we isolated the AM spores (Gerdemann and Nicolson 1963) and calculated the spore population (number of AM spores in 50 g sand). To estimate plant growth promotion, shoot and root systems of uprooted plants were separated, dried for 24 h at 80 °C and dry weights were recorded.

Exp. 2: Disease Severity Index

The disease severity was determined using 0-7 point scale; odd numbers 0= no wilting, 1= approximately 10 per cent leaves showing yellowing, 3= approximately 25 per cent leaves showing yellowing, 5 = 50 per cent leaves showing yellowing as well as shoot stunted or wilted, 7= severe wilting, resulting in death of plant. Even numbers *i.e.* 2, 4, 6 were given to plants whose symptoms were between two odd numbers.

Statistical Analysis

Data on Mycorrhizal colonization were arcsine transformed to normalize their distribution and subjected to analysis of variance in which effect of various treatments were tested. For each factor analyzed, the means of the different treatments were compared and ranked using Fischer F-test (P<0.05).

Results

Among different inoculants, significantly (*P*<0.05) higher shoot and root dry weight were recorded in pots where *Th* was co-inoculated with *Gm* (*Th+Gm*), followed by *Gm* alone inoculated pots. The values in terms of shoot and root dry weight in different *Trichoderma* inoculated pots (*Th* and *Tv* alone) and *Tv+Gm* were comparable with each other. The observations on spore density/50 g soil and root colonization index were recorded from AMF-inoculated pots four and six weeks after inoculation; hence, differences were found significant between different time intervals. Among AMF-inoculated pots (*Gm*, *Th+Gm* and *Tv+Gm*), spore density was maximum in *Tv+Gm* (82.60+ 6.27/50 g soil) after four weeks inoculation. Six weeks after inoculation values of spore density were comparable with each other in all the treatments. Two way interactions between various treatments and time of inoculation were found significant for spore density. The population density *i.e.* CFU of *Trichoderma* was recorded from *Trichoderma* inoculated pots after four and six weeks of inoculation; hence, differences were found significant among various treatments and time intervals in terms of population density of *T. harzianum*. Inoculation of *Gm* significantly reduced population density of *Th,* so differences were found significant among single and co-inoculation. Presence of *Gm* significantly reduced the population density of *Th* (29.70+6.70/g of soil) whereas highest population density was recorded in *Tv+Gm* inoculated pots(50.90+8.96/g of soil).

A significant interaction was found between the factors AMF and *T. harzianum* regarding disease severity. Inoculation of *Trichoderma* spp. simultaneously or prior to pathogen infection significantly suppressed *disease* but *Gm* was efficient in disease reduction when it was inoculated prior to pathogen. Plants co-inoculated with *T. harzianum* and *Gm* showed significantly lower disease incidence than all treatments either prior or simultaneous inoculation with pathogen. However, the DS under prior inoculation was lower than those observed under simultaneous inoculation. Significantly higher DS was recorded in pots where *F. udum* was inoculated alone (6.0) and it was minimum in *Th+Gm* (2.1). However, there was no difference between the disease severity of single and dual inoculated seedlings when treated simultaneously with the pathogen.

Table 7.1: Shoot Dry Weight (SDW) and Root Dry Weight (RDW) 6 Weeks after Planting (WAP)

Treatments	SDW	RDW
Control	13.18±1.02	8.20±0.91
Th	18.01±2.39	10.94±1.50
TV	16.53±21.09	10.88±1.87
Gm	21.06±1.45	16.38±1.39
Th+Gm	24.17±1.45	18.68±1.90
Tv+Gm	17.02±2.25	12.48±1.39
Mean	18.42±4.07	12.92±3.87
LSD	2.41	2.00
T	2.06	2.06

Values are means of five replicates per treatment at P = 0.05.

Table 7.2: Number of Colony Forming Units (CFU) of *Trichoderma* in the Rhizosphere of *Cajanus cajan* Co-inoculated or Not with *G. mosseae* (*Gm*) 4 and 6 Weeks after Planting (WAP)

Treatments	Colony Forming Units of Trichoderma		
	4 Weeks	6 Weeks	Mean
Control	nd	nd	nd
Th	32.0+5.00	43.80+3.42	37.90+7.41
Th+Gm	26.40+4.16	33.00+7.51	29.70+6.70
Tv	39.20+ 4.75	56.40+6.73	47.80+10.60
Tv+Gm	43.20 + 3.27	58.60+4.67	50.90+8.96
Mean	23.46+20.41	31.96+28.41	
LSD=3.77	T=2.67	F= NS	TxF= NS

Values are means of five replicates per treatment at P = 0.05.

Discussion

The advantages of using biocontrol include environmental friendly, cost effective and extent of protection. *Trichoderma* spp. and AMF are found in almost any soil,

Table 7.3: Effect of Inoculation of *Trichoderma* Species (*T.harzianum* (*Th*) or *T. viride* (*Tv*)} and/or *G. mosseae* (*Gm*), Prior or Simultaneously with Pathogen *F. udum* on Wilt Disease in *Cajanus cajan*

Treatments	Time of Inoculation			Mean
	Prior Inoculation		Simultaneous Inoculation	
	14 Days	7 Days		
Control	5.6+ 1.14	6.40+ 0.54	6.0+0.70	6.0+0.84
Th	0.8+0.45	2.40+1.14	4.80+1.45	2.67+1.99
TV	2.2+1.09	3.6+1.81	4.6+1.67	3.47+1.56
Gm	2.4+0.55	5.6+0.89	5.2+0.83	4.40+1.63
Th+Gm	0.4+ 0.54	2.0+1.56	4.0+1.20	2.13+1.89
Tv+Gm	2.4+1.14	3.20+1.30	5.0+1.59	3.53+1.69
Mean	2.3+1.89	3.87+1.94	4.93+1.33	
LSD(0.99)	T=1.40	F=NS	T×F= NS	

Values are means five replicates per treatment at P = 0.05.

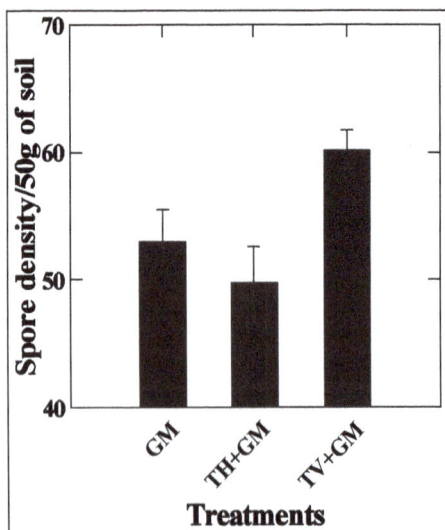

Figure 7.1: Spore Density of *G. mosseae* (*Gm*) Four Weeks after Planting in the Rhizosphere of *Cajanus cajan*.

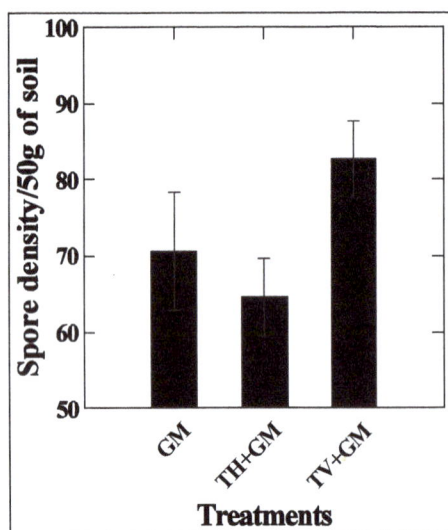

Figure 7.2: Spore Density of *G. mosseae* (*Gm*) Six Weeks after Planting in the Rhizosphere of *Cajanus cajan*.

have been investigated as potential biocontrol agents because of their ability to disease supression caused by plant pathogenic fungi, particularly many common soil borne pathogens. Present study demonstrate the interaction between two species of *Trichoderma i.e. T. harzianum/T. viride* and *G. mosseae* in terms of root colonization, *Trichoderma* population, growth promotion and disease reduction. Co-inoculation of *Th+Gm* showed significantly higher growth among all the treatments. Inoculation of

Gm was alone significant in terms of growth promotion but it showed additive effects when it was inoculated with *Th*. Calvet *et al.* (1993) has also reported a synergistic effect on the growth of marigold (Tagetes erecta) and percentage of AMF internal colonization as a result of combine-inoculation of *T. aureoviride* with *G. mosseae*. This suggest that possible mechanism for plant growth stimulation by *Trichoderma* and AM could be the solubilisation of plant nutrients and colonization (Altomare *et al.*, 1999) of *Trichoderma* to interior of roots, which increase shoot length, dry weight and P uptake (Yedidia *et al.*, 2001). Both isolates of *Trichoderma i.e. T. harzianum* and *T. viridae* showed similar results in terms of root dry weight and shoot dry weight. Growth promotion activities of *Trichoderma* have also been reported in radish, pepper and tomato. It might be due to a direct consequence of colonization, enhanced positive interaction between fungi and plant, increased nutrient uptake by plant or by reducing pathogen activity (Shoresh *et al.*, 2010). Our further results showed that spore density and root colonization were significantly increased after 14 days of inoculation as compared to 7 days and simultaneous inoculation. Among all AMF inoculated pots *Tv+Gm* showed maximum spore density whereas after 14 days spore density was comparable with each other in all the treatments. Spore density was increased with days after inoculation significantly, results obtained on population density of *Trichoderma* followed similar trends with reference to time of inoculation as in case of spore density and mycorrhization. Inoculation of *Gm* with different species of *Trichoderma* showed different response in terms of growth promotion and disease reduction. *Gm* significantly reduced population of *Th* but it enhance *Tv* population density. It might be due to the changes in the mineral composition of soil and physiology of plant tissues and in root exudation induces by AM fungi which affect microbial populations both qualitatively and quantitatively. Results obtained from experiment II demonstrate the effect of different treatments on disease severity in prior or simultaneously with pathogen. Inoculation of *Th* showed significant reduction either prior or simultaneous inoculated treatments with pathogen. *Trichoderma* function as antagonists of many phytopathogenic fungi, thus protecting plants from diseases (Vinale *et al.*, 2008) by means of ISR or localized resistance has been reported (Harman *et al.*, 2004) and evolution of positive interactions with plants (Druzhinina *et al.*, 2011) and reduces biotic and abiotic damage (Shoresh *et al.*, 2010). *Gm* was not found effective in terms of disease reduction when it was inoculated simultaneously with pathogen but it showed less disease severity when it was inoculated prior to the pathogen. The establishment of AMF symbiosis in plants is known to change physiological and biochemical properties of the host plant and these changes may alter the composition of root exudates which play a key role in the modification of the microbial population qualitatively and quantitatively in the mycorrhizosphere (Linderman, 1992). Presence of *F. udum* produced significantly higher disease severity, which is the direct consequence of pathogenic effecs of pathogen.

Acknowledgments

The authors are thankful to the Head of Department of Botany, Dr. Hari Singh Gour University, Sagar, India, for providing facilities.

References

Ainhoa Martínez–Medina, Antonio Roldán, Alfonso Albacete, Jose A. Pascual (2010). The interaction with arbuscular mycorrhizal fungi or Trichoderma harzianum alters the shoot hormonal profile in melon plants. *Phytochemistry* 72, 223–229

Altomare, C., Norvell, W. A., Bjorkman, T. and Harman, G. B. 1999. Solubilization of phosphates and micronutrients by the plant-growth-promoting and biocontrol fungus *Trichoderma harzianum* Rifai 1295–22. *Applied and Environmental Microbiology*, 65: 2926–2933.

Auge´, R.M., 2001. Water relations, drought and vesicular-arbuscular mycorrhizal symbiosis. *Mycorrhiza* 11, 3–42.

Calvet, C., Barea, J.M., Pera, J., 1992. *In vitro* interactions between the vesicular arbuscular mycorrhizal fungus *Glomus mosseae* and some saprophytic fungi isolated from organic substrates. *Soil Biol. Biochem*. 24, 775–780.

Calvet, C., Pera, J., Barea, J.M., 1993. Growth response of Marigold (*Tagetes erecta* L.) to inoculation with *Glomus mosseae, Trichoderma aureoviride* and *Pythium ultimum* in a peat–perlite mixture. *Plant Soil* 148, 1–6.

Druzhinina, I. S, Seidl, V, Estrella H., Horwitz B.A., Kenerley C. M, Monte E., Mukherjee P. K, Zeilinger S., Grigoriev I. V. and Kubicek, C. P., 2011. *Trichoderma*: the genomics of opportunistic success. *Nat Rev Microbiol* 9: 749–759

Dubey, S.C., M. Suresh and B. Singh. 2007. Evaluation of *Trichoderma* species against *Fusarium oxysporum* f. sp. *ciceris* for integrated management of chickpea wilt. Biological Control, 40(1): 118–127.

Elad, Y., Chet, I., Henis, Y., 1981. A selective medium for improving quantitative isolation of *Trichoderma* spp. from soil. *Phytoparasitica* 9, 59–67.

Fracchia, S., Garcia–Romera, I., Godeas, A., Ocampo, J.A., 2000. Effect of the saprophytic fungus *Fusarium oxysporum* on arbuscular mycorrhizal colonization and growth of plants in greenhouse and field trials. *Plant Soil* 223, 175–184.

Gerdemann JW, Nicolson TH., 1963. Spores of mycorrhizal *Endogone* species extracted from soil by wet sieving and decanting. *Trans Br Mycol Soc* 46: 235–244

Giovannetti M, Mosse B., 1980. An evaluation of techniques for measuring vesicular arbuscular mycorrhizal infection in roots. *New Phytol* 84: 489–500

Green, H., Larsen, J., Olsson, P.A., Jensen, D.F., Jakobsen, I., 1999. Suppression of the biocontrol agent *Trichoderma harzianum* by mycelium of the arbuscular mycorrhizal fungus *Glomus intraradices* in root–free soil. *Appl. Environ. Microbiol*. 65, 1428–1434.

Harman, G.E., Howell, C.R., Viterbo, A., Chet, I., 2004. *Trichoderma* spp.–opportunistic avirulent plant symbionts. *Nat. Rev*. 2, 43–56.

Hooker, J.E., Jaizme–Vega, M., Atkinson, D., 1994. Biocontrol of plant pathogens using arbuscular mycorrhizal fungi. In: Gianinazzi, S., Schuepp, H. (Eds.), *Impact of Arbuscular Mycorrhizas on Sustainable Agriculture and Natural Ecosystems*. Birkhauser, Basel, pp. 191–200.

Jeffries, P., Barea, J.M., 2001. Arbuscular mycorrhiza: a key component of sustainable plant–soil ecosystems. In: Hock, B. (Ed.), *The Mycota: Fungal Associations*, vol. IX. Springer, Berlin/Heidelberg/New York, pp. 95–113.

Koike, N., Hyakumachi, M., Kageyama, K., Tsuyumu, S., Doke, N., 2001. Induction of systemic resistance in cucumber against several diseases by plant growth promoting fungi: lignification and superoxide generation. *Eur. J. Plant Pathol.* 107, 523–533.

Linderman, R.G., 1992. Vesicular–arbuscular mycorrhizae and soil microbial interactions. In: Bethlenfalvay, G.J., Linderman, R.G. (Eds.), *Mycorrhizae in Sustainable Agriculture*. ASA Special Publication No. 54, Madison, WI, pp. 45–70.

Linderman, R.G., Paulitz, T.C., 1990. Mycorrhizal–rhizobacterial interactions. In: Hornby, D., Cook, R.J., Henis, Y., Ko, W.H., Rovira, A.D., Schippers, B., Scott, P.R. (Eds.), *Biological Control of Soil–borne Plant Pathogens*. CAB International, Wallingford, UK, pp. 261–283.

Mar Vazquez, M., Cesar, S., Azcon, R., Barea, J.M., 2000. Interactions between arbuscular mycorrhizal fungi and other microbial inoculants (*Azospirillum, Pseudomonas, Trichoderma*) and their effects on microbial population and enzyme activities in the rhizosphere of maize plants. *Appl. Soil Ecol.* 15, 261–272.

Marschner, P., Baumann, K., 2003. Changes in bacterial community structure induced by mycorrhizal colonization in split-root maize. *Plant Soil* 251, 279– 289.

Medina, A., Probanza, A., Manero, F.J.G., Azco´n, R., 2003. Interactions of arbuscular mycorrhizal fungi and *Bacillus* strains and their effects on plant growth, microbial rhizosphere activity (thymidine and leucine incorporation) and fungal biomass (ergosterol and chitin). *Appl. Soil Ecol.* 22, 15–28.

Muslim, A., Horinouchi, H., Hyakumachi, M., 2003. Control of *fusarium* crown and root rot of tomato with hypovirulent binucleate *Rhizoctonia* in soil and rock wool systems. *Plant Dis.* 87, 739–747.

Phillips JM, Hayman DS., 1970 Improved procedure for clearing roots and staining parasitic and vesicular arbuscular mycorrhizal fungi for rapid assessment of infection. *Trans Br Mycol Soc* 55: 158–161

Shoresh, M., Harman, G. E. and Mastouri, F., 2010. Induced systemic resistance and plant responses to fungal biocontrol agents. *Annu Rev. Phytopathol* 48, 21–43

Sieverding E 1991. Vesicular arbuscular mycorrhiza management in tropical agrosystems. GTZ, Rossdorf.

Sinclair J.B., Dhingra O.D. 1985. *Basic Plant Pathology Method*. CRC Press, Inc. Corporate Blud, M.W. Boca Rotam, Florida: 295–315.

Tisdall, J.M., 1994. Possible role of soil micro-organisms in aggregation in soils. *Plant Soil* 159, 115–121.

Vinale, F., Sivasithamparam, K., Ghisalberti, E.L., Marra, R., Barbetti, M.J., Li, H., Woo,S.L., Lorito, M., 2008. A novel role for Trichoderma secondary metabolites in the interactions with plants. *Physiol. Mol. Plant Pathol.* 72, 80–86.

Vosátka M., Albrechtová J., 2001. Benefits of Arbuscular Mycorrhizal Fungi to Sustainable Crop Production. In: Khan M., Zaidi A., Musarrat J. (Eds.). *Microbial Strategy for Crop Improvement*. Springer. ISBN 978–3–642–01987–4. Pp. 205–225.

Yedidia, I., A. K. Srivastva, Y. Kapulnik and I. Chet., 2001. Effect of *Trichoderma harzianum* on microelement concentrations and increased growth of cucumber plants. *Plant Soil*, 235: 235–42.

Yedidia, I., Behamou, N., Kapulnik, Y., Chet, I., 2000. Induction and accumulation of PR proteins activity during early stages of root colonization by the mycoparasite *T. harzianum* strain T–203. *Plant Physiol. Biochem.* 38, 863–873.

Zeng RS., 2006. sease resistance in plants through mycorrhizal fungi induced allelochemicals. In: Inderjit, Mukerji KG (eds) *Allelochemicals: biological control of plant pathogens and diseases*. Springer, Netherlands, pp 181–192.

2015, **Recent Trends in Microbiology, Mycology and Plant Pathology**   *Pages* **105–113**
Editor: **Dr. H.C. Lakshman**
Published by: **DAYA PUBLISHING HOUSE, NEW DELHI**

Chapter 8

Role of Cyanobacteria in Phosphate Nutrition to Plants

Ratna V. Airsang and H.C. Lakshman

Microbiology Laboratory,
Post Graduate Department of Studies in Botany, Karnataka University,
Pavate Nagar, Dharwad – 580 003, Karnataka, India

Introduction

During the last 30 years, a fundamental shift has taken place in the strategies of agricultural research in India as well as in other parts of the globe. In the past, the principal drive was to increase the yield potential of food crops and to maximize productivity. But, at present the drive for productivity is combined with a desire for sustainability. However, international emphasis on environmentally sustainable development with the use of renewable resources is likely to focus attention on the potential role of BNF in supplying nitrogen for agriculture. Biological nitrogen fixation has been the most effective system for sustaining production in low input traditional rice cultivation.

Cyanobacteria constitute the major component of rice field ecosystem as nitrogen budget is concerned. These organisms multiply rapidly in the flood water in neutral to alkaline soils and release nitrogen slowly for meeting the nutrient requirement by rice (Tirol *et al.,* 1982; Whitton *et al.,* 1988). The tropical conditions ensure increased incidence of cyanobacteria in the rice field soils which provide a suitable niche for their abundant growth. The luxuriant growth of cyanobacteria in rice fields is due to high humidity, temperature, light, water and nutrients (Roger and Reynaud, 1979). This could be the reason that cyanobacteria grow in higher abundance in rice field soils then in upland soils as reported in the widely different climatic conditions of the globe (Watanabe and Yamamoto, 1971; Singh, 1985; Whitton *et al.,* 1988). The role of cyanobacteria in the soil ecosystem is manifold, the most important consequences being the input of nitrogen and carbon (Roger and Kulasooriya, 1980). Besides, they liberate a variety of amino acids like aspartic acid, glutamic acid and alanine (Adhikary and Pattnaik, 1981). Being soluble in water, these substances form readily soluble source of nitrogen for the crop plants. The cyanobacteria affect the soil ecosystem in various ways like the formation of cynobacterial crusts on soil which checks soil erosion, increases the storage of rain water and reduces the water loss by evaporation (Van den Ancker *et al.,* 1985). They also enrich the soil with organic matter and are also capable of transforming soil characteristics including aggregation.

Phosphorus is the second major plant nutrient after nitrogen in terms of quantitative requirements for crop plants. Available phosphorus is estimated to be in insufficient amounts in most of Indian soils. According to one of the recent compilation based on about 9.6 million soil tests for available phosphorus in Indian soils, 49.3 per cent area spread in various states in the low category, 48.8 per cent is in medium category and only 1.9 per cent has high phosphorus status (Hassan, 1994). The problem of P management in soils is highly intricate as the applied phosphate through fertilizers is often and becomes unavailable to the crops. The P availability is mainly dependent on soil pH, soluble Fe, Al, Mn, and Ca ions, organic matter content and activities of soil micro-organisms (Airsang and Lakshman, 2007). In organic matter rich soils, P availability is gently enhanced through microbial activity, enzymatically as well as through acidic metabolites produced by micro-organisms.

Today, a large number of heterotrophic and autotrophic soil micro-organisms are known to solubilize inorganic phosphates including cyanabacteria (Roychoudry and Kaushik, 1989). Solubilization of Mussorie rock phosphate, a source of PO and raw material for fertilizer industry, by nitrogen fixing cyanobacteria *Tolypothrix tenuis, Scytonema cincinnatum* and *Hapalosiphon fontinalis* was observed *in vitro.* Cyanobacteria screened effluent showed pronounced increase in dissolved oxygen also. This increment is due to algal photosynthetic activity and also the solubility of Oxygen from the atmosphere under laboratory condition in which they are incubated. High dissolved Oxygen in batch cultures enhance the nutrient removal which subsequently reduce the Oxygen demands of the effluents (Kankal *et al.,* 1987).

Mechanism of Pi-Transformation in Soil

Blue green algae (BGA) have been shown to solubilize insoluble (Ca) (PO) (Bose *et al.,* 1971),FePO(Wolf *et al.,* 1985), AIPO (Dorich *et al.,*1985) and hydroxyapatite Ca(PO)OH (Cameron and Julian,1988) in soils, sediments or in pure cultures. There are mainly two hypotheses to explain how BGA solubilize such bound phosphates. One of the mechanisms in that BGA might synthesize a chelator of Ca^2z and drive the following dissolution reaction without changing the pH of the growth medium (Cameron and Julian, 1998; Roychoudhury and Kaushik, 1989).

$$Ca\,(OH)\,(PO) = 10Ca^2z2OH + 6PO^3z$$

Others (Bose *et al.,* 1971) were however of the opinion that HCO and other organic acids released by BGA during their growth could solubilize P from Ca sources through following the reaction:

Ca (PO)+2HCO = 2CaHPO+Ca (HCO)

A third group (Arora, 1969; Saha and Mandal, 1979) believed that the above mechanisms operate simultaneously. Once solubilised, the PO^3 is taken up the growing algal cells for their nutrition. After completing their growth cycle, when the cells undergo lysis the cell bound PO^3 increased in the growth medium and becomes available to plants on mineralisation. Observation of an initial decrease in the available P content in soils due to algal growth and an appreciable increase later during biomass decomposition (Saha and Mandal, 1979) supported this possible pathway. The solubilizing effect of BGA on bound PO^3 may also be used for the efficient utilization of low cost, low grade(in terms of P content) rock-phosphate fertilizers where PO^3 remains bound as hydroxyapatite, carbonapatite, chlorapatite, etc. These rock phosphates are generally used in acid soils as P fertilizers. It is possible that if rock phosphate is used with BGA inoculation in acid soils, its efficiency as a source of P may be increased.

Besides the conventional approach of screening natural isolates having mineral phosphate solubilizing phenotype, genetic engineering of non-mineral phosphate solubilizing cyanobacteria to become efficient phosphorus solubilizers is possible. So far, no mineral phosphorus solubilizing genes have been cloned in cyanobacteria, though Gyaneshwar *et al.* (1998), have cloned the genes responsible for conferring mineral phosphate solubilizing ability from *Synechocystis* PCC 6803 into *Escherichia coli* Cyanobacteria are reported to mobilize different insoluble forms of inorganic phosphate possibly by synthesizing chelator(s) and/or releasing organic acids. However, Dash and Mishra (1999) have implicated extracellular phosphatases in solubilization of fixed soil phosphates but a critical study is lacking. Reports suggesting an effective role of cyanobacteria at increasing phosphorus availability in saline soils do not suggest which cyanobacterial features are responsible for this property. Thus, the aim of the present investigation was to examine P-metabolism (uptake rate, cellular quota, and regeneration of inorganic phosphate) by *A. doliolum* cells under saline conditions.

Cyanobacterial cells store polyphosphate reserves at high levels during exponential growth, mediated by polyphosphate synthetase, which requires energy. These polyphosphate bodies are degraded during P-deficiency. However, activity of polyphosphate synthetase has been detected in P-deficient cells. The strategies adapted by cyanobacterial cells to deal with P deprivation involve accessing the stored intracellular polyphosphate bodies, the increase in the Pi uptake rate, and the induction of the synthesis of extracellular phosphatases for scavenging Pi moiety from variety of substrates present in the surroundings. Due to the conditions prevalent in lakes and paddy fields, it is generally the alkaline phosphatase activity that hydrolyses phosphate from a variety of organic phosphorus compounds. Phosphate uptake capacity is regulated by internal phosphorus pools, particularly polyphosphate. P-deficiency also induces extracellular alkaline phosphatase activity to alter the Pi availability. An inverse relation exists between cellular P levels and phosphatase activity. Thus, the kinetics of Pi uptake depends upon the both composition of the medium and nutritional state of the cells. The noticeable increase in the use of N- and P- fertilizers by intensive agricultural practices and the enhanced use of water for irrigation has resulted in soil salinization. Although salinity is one of the most deleterious factors for plant life and freshwater micro-organisms, there appears to be a few reports on the effect of elevated salt level on P metabolism in cyanobacteria.

While *Anabaena* and *Nostoc* sp showed poor utilization of phosphates, *Phormidium* screened effluent showed its level detectable limits. The capacity of cyanobacteria to remove large amount of phosphorous from waste had been documented earlier (Chan *et al.,* 1979, Tan and Wong, 1989). Filamentous cyanobacteria are not only able to retain these pollutants for a longer period, but can be suitably harvested easily (Kumar and Sharma, 1975).

Role of Cyanobacteria in Aquatic Environment

In India, much emphasis has been given to the quantity of produce to feed the ever increasing population. In the current year total food grain production has touched 209 million tonnes, of which rice alone accounts to 90 million tonnes. Apart from the quantity, the quality of grains is likely to improve by the use of biofertilizers. In a two year trail on rice variety PNR-381, the grain quality was improved in terms of nitrogen(1.53 per cent) and phosphorus content(0.27 per cent) over control (N 1.1 per cent and P 0.21 per cent), Similarly, the straw was also richer in nitrogen and phosphorus content than the grain or straw produced only by inorganic fertilizers.

As far as the total protein level is concerned, the fish farm effluent had brought down the total protein content of *Anabaena* and *Nostoc* species during the treatment period. Contrary to this in *Phormidium* sp, a initial level increase and subsequent reduction in protein level was noticed. The decrease in protein in proportion can be expected when nitrogen and phosphorus have been taken up by these algae during their active growth phase (Talbot and de la None, 1993). Similar observation of protein level reduction in *Oscillatoia* sp. was made while treating a paper mill effluent (Manoharan and Subramanian 1992). Also the environmental stress induced modification in the protein level while they grow in the effluent cannot also be ruled out (Kimpel and Key, 1985; Bhagwat and Apte, 1989). During the treatment period, the two strains *Anabaena* and *Phormidium* sp showed remarkable increase in chlorophyll-a and carotenoid pigments than the *Nostoc* sp which showed gradual reduction in the above pigment composition indicating their poor adaptation in the effluent. Cyanobacteria treated fish farm showed reduction in both BOD and COD levels. BOD reduction in waste waters using algal cultures had been reported (Pouliot *et al.,* 1989; Tripathi *et al.,* 1991). Similar observation of COD reduction by *Oscillatoria chlorina* screened effluent had also been documented earlier (Rana and Palria, 1987). It is clear from this study that these algae not only retrieve inorganic nutrients, but also helps in reducing BOD and COD level. Due to varied physiological adaptation of these algae in the effluent; there is a significant change on the biochemical composition of the organisms. The growth of these organisms in terms of chlorophyll-a, carotenoids and proteins showed considerable variations.

Algalization in the presence of fertilizers other than nitrogen: Phosphorus, lime and sometimes molybdenum application has been demonstrated frequently to have beneficial effect on the establishment and growth of nitrogen fixing algal flora. In field experiments, algalization in combination with lime, phosphorus and molybdenum application was more efficient than algalization alone (Jha *et al.,* 1965; Relwani, 1965; Relwani and Subrahmanyan, 1963; Sankaram *et al.,* 1966).

Nitrite at low concentration serves as nitrogen source for growth of cyanobacteria. In the fresh effluent the nitrite level was well below the detectable limits. But after cyanabacterial treatment, within 10 days its level has a marked increase in all the cases which may be due to the reduction of nitrate into nitrite which in turn is removed completely with in another 10 days. Similar intial level increase and subsequent depletions of nitrite paper mill effluent treatment the cyanobacterium *Oscillatoria pseudogeminata* has been made earlier (Manoharan and Subramanian, 1992).

Green house experiments of Watanabe *et al.* (1981) taking *Anabaena* and neem cake, ammonium sulphate, neem cake blended with urea and neem oil showed that neem cake increased the BGA population over the control at 25 and 35 days. It also reduced the depressing effect of ammonium-nitrogen and increased nitrogen fixation in non-nitrogen treated culture. Using *Aulosira* sp. and urea (50 kg N/ha) Saha *et al.* (1982) observed that the availability of nitrogen was more to the rice plants when the organism was present alone then in presence of urea. But Tirol *et al.* (1982) observed that the availability of N from the incorporated ammonium was higher than that of N from incorporated cyanobacteria. Exogenous nitrogen suppressed biological N-fixation and lower doses of nitrogen in presence of lime and molybdenum increased nitrogen fixing activity (Hassan *et al.,* 1984). Inoculation to field with five strains of cyanobacteria supplementing with muriate of potash (40 kg/ha), SSP (Kg/ha), urea, ammonium sulphate, monoammonium phosphate, ammonium nitrate and sodium nitrate (50 kg and 100 kg N/ha) showed that mono ammonium phosphate gave highest grain yield followed by ammonium sulphate, urea, ammonium nitrate and sodium nitrate both at 50 and 100 kg N/ha (Goyal, 1985).

Sharma (1982) had observed that *Anabaenopsis raciborskii,* isolated from an aquatic environment could assimilate urea, ammonia and nitrate up to 30, 50 and 300µg/ml respectively. Growth rate of the organism in presence of the combined nitrogen at these concentrations was higher than that of the value obtained in the culture with molecular nitrogen; rate of the growth was higher with ammonia and urea. Further, exponentially growing cultures taken from one source when inoculated in different sources of nitrogen showed lag period with decrease in their growth rates though there was no lag period when transferred to the same source of nitrogen. Anand and Karuppusamy (1987) have studied the interaction of six cyanobacterial species to 10, 20, 50, 100, 200 and 500µg/ml of urea potash, super phosphate, di-amino phosphate and NPK in laboratory culture. The result showed that all the fertilizers up to 100µg/ml stimulated the growth of the organisms. Lethal dose of urea varies from 200µg/ml to 500µg/ml; lethal dose of potash, SSP and DAP for all the organisms was invariably 500µg/ml. Kolte and Goyal (1989) studied the interaction of 16 species/strains of cyanobacteria isolated from the rice field soil of Vidarbha region of Maharashtra state with ammonium-nitrogen and observed that except *Anabaena khannae,* which exhibited a retardation of growth at 50µg N/ml, all other organisms were unaffected even up to 75µg N/ml. However, *Anabaena* sp., *Calothrix javanica,Calothrix membranacea, Cylindrospermum gorakhpurense, Nostoc muscorum* were more susceptible to ammonium-nitrogen in terms of nitrogen fixation. Prosperi *et al.* (1993) studied the effect of ammonium sulphate to 4 species of cyanobacteria *viz., Nostoc punctiforme, Anabaena variabilis, Calothrix marchica* and *Nodularia spumigena* and the influence of pH, light intensity and anaerobiosis on the short- term effect of ammonium of their nitrogenase activity. Ammonium completely inhibited nitrogenase activity in *Anabaena* regardless of culture pH on prolonged incubation experiments.

Singh and dash (1994), have studied the effect of urea and superphosphate on the biomass of 6 cyanobacterial species in the rice fields. They reported that the nitrogen fixing cyanibacteria grew well in non-N treated fields but on addition of N (urea) the biomass production was adversely affected and decreased gradually with increase of doses of the fertilizer. However, phosphorus (superphosphate) enhanced the biomass and ARA of *Aphanothece* sp. The increase in biomass at 10, 20 and 30 Kg P_2O_5/ha over control was 7.2, 41.2 and 74.6 per cent. At these doses, the ARA also increased over control by 15.2, 51.1 and 80 per cent. In field experiments Reddy and Roger (1988) treated the cultures of *A. variabilis T. tenuis. Aulosira fertilissima. Fischerella* sp.and *Nostoc* sp. with 20 kg/ha SSP and neem cake in five different types of soils. They observed that homocystous blue-greens grew well in all soils. Appearance of heterocystous BGA was higher in soils richer in organic matter. Considering the importance of use of algal biofertlizers for sustainable agriculture and nutrient management in the waterlogged rice fields, in the recent years research work in many laboratories has been focused in various directions. These are:

☆ Refinement of technique for the qualitative and quantitative examination of soil cyanobacteria (McCann and Cullimore, 1979).

☆ Definitive evaluation of the activity of these organisms and their importance to soil fertility (McCann and Cullimore, 1979).

☆ Evaluation of the relative chemical fertilizer sensitivities of different cyanobacterial types (Goyal and Marwaha, 1985; Kolte and Goyal, 1989). Many diazotrophic cyanobacterial species besides being able to reduce the elemental nitrogen have been shown to utilize various types of nitrogenous fertilizers for their growth. Selection of such strains for their subsequent fertilizer will be highly beneficial.

☆ Diazotrophic cyanobacteria also release biologically active compounds into the soil and that these compounds are assimilated by higher plants, significantly enhancing their growth (Whitton, 1965: Dadhich *et al.*, 1969: Venkataraman and Neelakantan, 1967: Adhikary and Pattnaik, 1981). In addition, cyanobacteria bind soil particles together due to adhesive properties of the copious mucilage excreted by them resulting in reduction of soil erosion and also prevents evaporative water loss from soil as desiccated algal crusts from a relatively permeable cover over the soil surface (Fogg *et al.*, 1973: De Winder *et al.*, 1990). Quantification of these extracellular substances under various agronomic practices has also been emphasized.

Conclusion

About 50 per cent of the cultivable lands in India are deficit in available form of soil P and only 20- 25 per cent of applied phosphates fertilizers are utilized by crop farming insoluble complexes with metal cat ions and clay mineral complexes. Added P is also readily fixed in most soils that its concentration in solution is always meager. Barber (1980) observed that the distance traversed by phosphate ions through soil in four days is between 0.80 to 0.006 cm. Thus, phosphate ions are practically immobile and hence its placement in root zone/moist layer is prerequisites for its efficient utilization. P in this root zone exists as an orthophosphate ion (H_2PO_4) and its availability to plant is dependent on its replenishment. In most lakes, the growth of phytoplankton and aquatic plants is restricted by the limited availability of phosphate and a usable form of nitrogen. However when these nutrients enter lakes in the effluent from sewage disposal plants and as farm land run off of animal wastes and excess fertilizer. Eventually, large amounts of algae die and decompose after which the aerobic micro-organisms of decay and deplete the lake's dissolved oxygen, causing die-offs of fishes and other aerobic organisms. However, there are evidences that soils of these lakes they liberate nutrients by feeding as bacteria, an indication of the complexity of nutrient mainly of limited phosphorus in soils. Importantly, human-made and catastrophic natural disturbances often have the same basic effects on beneficial micro-organism in soil, and many practices for restoring phosphate solubilizing organisms can be based as successional processes that occur naturally. Developing new techniques for reintroducing P solubilizers leads to sustainable agriculture and forestry and for restoring natural areas.

Acknowledgement

First author is indebted to UGC, New Delhi for awarding F.D.P (Faculty Development Programme) to pursue Ph.D. in Botany.

References

Adikary, S.P. and Pattnaik, H.1981. Effect of organic carbon sources on the liberation of extracellular amino acids by nitrogen–fixing blue-green alga, *Westiellopsis prolific* janet in light and dark. *Ind. J. Bot.* 4: 60–69.

Airsang, R.V. and Lakshman, H.C.2007. Role of AM fungi and Cyanobacteria in phosphate restoration in soil. M Phil Dissertation: AM– Quantification and Isolation of Spores from some Important Legumes of Dharwad. Periyar University Salem, 54: 54–57.

Anand, N. and Karuppusamy, A.1987. Growth of nitrogen fixing and non–nitrogen fixing blue–green alga in presence of some common fertilizers. *Phykos*, 26: 22–26.

Arora, S.K., 1969. The role of algae on the availability in paddy fields. ÉÉ *Riso.*, 18: 135–138.

Bose, P., Nagpal, U.S., Venkataraman, G.S., and Goyal, S.K., 1971. Solubilization of tricalcium phosphate by blue-green algae. *Curr.Sci.*, 40: 165.

Cameron, H.J. and Julian,G.R., 1988. Utilization of hydroxyapatite by cyanobacteria as their sole source of phophate and calcium. *Pl. Soil,* 109: 123–124.

Dadhich, K. S., Verma, A.K and Venkataraman, G.S. 1969. The effect of *Calothrix* inoculation in vegetable crops. *Plant and Soil*, 35: 377–379.

Dash, A.K. and Mishra, P.C. 1999. Role of cynobacteria in water pollution abetment. In: Cynobacteria and Algal metabolism and Environmental Biotechnology. (Ed) Tasneem Fatma. Narosa Publishing House, New Delhi.

Dorich, R.A., Nelson,D.W. and Sommers,L.E.,1985.Estimating algal–available phosphorus in suspended sediments by chemical extraction. *J. Environ. Qual.* 14: 400–405

De Winder, B.D., Stal, L.J. and Mur, L.R. 1990. *Crinalium spipsammum* sp.nov: a filamentous cyanobacterium with trichomes composed of elliptical cells and containing poly ß–(1, 4) glucan (cellulose). *J. Gen. Microbiol.* 136: 1645–1653.

Fogg, G.E., Stewart, W.D.P., Fay, P. and Walsby, A.E. 1973. The Blue–Green Algae. Academic Press, New York and London, pp.459.

Fuller,W.H., and Roger, R.N.,1952,Utilization of the phosphorus of algal cells as measured by the Neubauer technique. *Soil Sci.*, 74: 417–429

Goyal, S.K. 1993. Algal biofertlizer for vital soil and free nitrogen. In: Nitrogen Soils, Physiology, Biochemistry, Microbiology and Genetics. Eds. Abrol, Y.P., Tilak, K.V.B.R., Kumar, S.and Katyal, J.C.Ind. Natl.Sci. Academy, New Delhi, pp. 135–141.

Gyaneshwar, P., Kumar G.N.and Parekh, L. J.,1998.Cloning of mineral phosphate solubilizing genes from *Synechocystis* PCC 6803.*Curr.Sci.*, 74: 1097–1090

Kaushik B.D. 1989.Contribution of Cyanobacteria to Pi Nutrition in plants, in proceedings of mineral phosphate solubilization November–16, UAS Dharwad. Department of Agriculture in Microbiology UAS.Dharwad–5: pp. 32–35.

Mc Cann, A.E. and Cullimore, D.R. 1979. Influence of pesticides on the soil algal flora. *Residue Rev.*, 72: 1–31.

Manoharan,C. and Subramanian, G.1992. Interaction between paper mill effluent and the cyanobacterium *Oscillatoria pseudogeminata var, unigranulata*. *Pollution Research* 11(2): 73–84.

Relwani, L. L., and Subrahmanyan, R., 1963. Role of blue green algae, chemical nutrients and partial soil sterilization on paddy yield. *Curr, Sci.*, 32: 441–443.

Senthil, C.and Goyal, S.K.1996. Inorganic carbon assimilation in algae and its response to elevated levels of carbon dioxide. *Phykos* 35(1&2): 1–12.

Roychoudhary,P. and Kaushik,B.D.,1989. Solubilization of mussorie rock phosphate by cyanobacteria. *Curr.Sci.*, 58: 569–570.

Sankaram, A., Mudholkar,M.J. and Sahay, M.N., 1966. Algae in rice production. *Indian Fmg.*, 16: 37–38

Talbot, P. and J. de la Noue., 1993. Tertiary treatment of waste water with *Phormidium bohneri* (Schmidle) under various light and temperature conditions. *Water Research* 27(1): 153–159.

Venkataman, G.S. and Neelakantan, S. 1967. Effects of the cellular constituents of the nitrogen–fixing blue-green alga *Cylindrospermum muscicola* on the root growth of rice plants. *J. Gen. Appl. Microbiol.* 13: 53–62.

Whitton, B.A. 1965. Extracellular products of blue-green algae. *J. Gen. Microbiol.* 40: 1–11.

2015, Recent Trends in Microbiology, Mycology and Plant Pathology   *Pages* 115–133
Editor: **Dr. H.C. Lakshman**
Published by: **DAYA PUBLISHING HOUSE, NEW DELHI**

Chapter 9

Role of Micro-organism in Production of Single Cell Protein

Kiran P. Kolkar[1] and H.C. Lakshman[2]*

[1]*Department of Botany, Karnatak Science College,*
Dharwad – 580 001, Karnataka, India
[2]*Microbiology Lab, P.G. Department of Studies in Botany*
Karnatak University, Dharwad – 580 003, Karnataka, India
*E-mail: dr.hclakshman@gmail.com

Introduction

The explosive rate of population growth is the major problem faced by developing nations. In India, the rate of population growth is far greater than food production and supply from all sources available to the population. Conventional agricultural practices may be unable to provide optimum food to the ever-increasing mouths and this results in shortage of proteinaceous food supply. The Food and Agriculture Organization (FAO) predicted a widening of protein gap between developed and developing countries. At least 25 per cent of the world's population currently suffers from hunger, malnutrition and starvation.

The single-cell protein is a preparation containing dried cells which form a protein source for human beings and for cattle. Generally, single-cell proteins are produced from the culture of yeast, bacteria and algae. Germans first used the single-cell protein as food during the First World War. Since then, other countries have been taking considerable interest in the production of single-cell proteins.

New agricultural practices are making efforts to develop high protein cereals. The cultivation of soybeans and ground nuts is ever-expanding, protein may be extracted from liquid wastes by ulta filteration, and the use of micro-organisms as the producers of protein has gained experimental success. This area of study is called *Single Cell protein* or SCP production.

Prof. Scrimshaw (1960) of MIT, USA introduced the term *single cell protein* (*SCP*) for microbial protein that includes the proteins from unicellular (single-celled) micro-organisms like bacteria yeasts, algae and possibly protozoa. In addition, mushrooms which are fruiting bodies of certain basidiomycetous fungi are also rich sources of protein.

Best strains of micro-organisms are selected and cultured on a large scale in large-sized bioreactors. During the culture of microbes, they produce a large mass of cells which are dried into the fine granules. Such a preparation contains the proteins of a single species of micro-organisms. As the preparation contains proteins of a single species, it is called single-cell protein.

Advantages of Single-cell Protein

1. The micro-organisms such as algae and bacteria complete their life cycle within a short period. So, there is the possibility to get more biomass within a short duration.

2. These micro-organisms are most probably unicellular or a few-cellular organisms. Because of this reason, the genetic composition of the organisms can be easily changed to get a desired strain, by using genetic engineering methods.

3. Micro-organisms like *Chlorella, Scenedensmus,* etc., are having a large amount of proteins in their cells. These proteins are very similar to the proteins derived from the conventional protein sources.

4. The microbes are able to use different kinds of cheapest materials for their growth. They even use waste products released from industries for their growth. So the preparation of feed-stock for the culture of microbes is also easier.

5. The maintenance of micro-organisms in bioreactors is very easy. So, the microbes are easily grown in the bioreactors for a long time without any hazards.

6. They require a relatively small space for their culture.

7. During the large scale culture of microbes, they also produce some valuable secondary metabolites.

Micro-organisms – Source for SCP

Algae, fungi, yeast and bacteria are the micro-organisms from which SCP is produced. The microbes utilized for SCP production must be nontoxic and nonpathogenic to plants, animals and man. They should be fast growing, easy to separate from the medium and easy drying. Further, they must have good nutritional value and must be able to be cheaply produced on a commercially large scale.

Micro-organisms produce protein much more efficiently than any other animal. They can grow at remarkably rapid rates under optimum conditions; some microbes can double their mass every 0.5 to 1 hour. The protein-producing capabilities of 250 kg cow and 250 g of micro-organisms are often compared. The cow can produce 200 g of protein per day, whereas the microbes theoretically could produce 25 tonnes in the same time under growing conditions. However, the cow has its own significance because of its unique ability to convert grass into protein rich milk. After several years of research no rival method for that conversion process has been developed. Therefore, cow has been rightly described as a bioreactor.

Production of Bacterial and Actinomycetous Biomass

Bacteria are widely used as a source of single cell protein because of their short life cycle (20-30 minutes) and capacity to utilize a wide range of organic substrates as a source of energy. Actinomycetes also utilize these renewable sources as they have more or less same generation rate as bacteria.

Since the establishment of British Petroleum in 1960, a significant progress is made in the production of microbial products by using gaseous and liquid hydrocarbons and chemicals derived from them, for example, methanol, ethanol, etc. The Shell Research Limited, U.K. conducted research on pilot plant scale process for the production of bacterial SCP from methane by using *Mythylococcus capsulatus* or mixed culture of *Pseudomonas* sp., *Hypomicrobium* sp., *Acinetobacter* sp. and *Flavobacterium* (Litchfield, 1979). Several processes for the production of bacterial SCP from gaseous hydrocarbons have been developed at Kyowa Hakko Kogyo Company, Limited in Japan. *Brevibacterium ketoglutamicum* ATCC No.15587 was able to utilize methane-ethane, propane, n-butane, iso butane, propylene, butylenes or mixture of these hydrocarbons (Tanaka *et al.*, 1972). A pilot scale process for the production of *Achromobacter delvacvate* from diesel oil is developed at the Chinese Petroleum Corporation, Taiwan.

Streptomyces, sp. is capable of growing on methanol. *Thermonaspora fusca,* a thermophilic species, degrades 60-65 per cent paper mill fines resulting in 30 per cent protein product. Nowadays, cellulose, degrading thermophilic actinomycetes offer a great opportunity to yield SCP from cellulosic wastes.

Procedure for production

Roth (1972) has described the following steps for the production of bacterial biomass; (i) supply of a nutrient substrate; (ii) formulation of a suitable medium; (iii) multiplication of micro-organisms through fermentation (iv) separation of cellular substances from the left over medium; and (v) further treatment to kill and dry the bacterial biomass.

Imperial Chemical Industry (ICI) is a world's leader in biotechnology, as far as the production of a bacterial biomass, pruteen, from *Methylophilus methlotrophus* is concerned. Pruteen was produced on methanol but it served as high grade protein for animal feed. It contains 72 per cent protein, 86 per cent total lipid and an amino acid profile high in lysine and methionine. The conversion of methanol to SCP by *M. methylotrophus* is represented by the following equations;

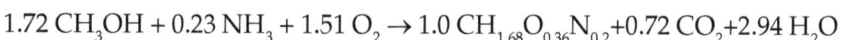

$$1.72\, CH_3OH + 0.23\, NH_3 + 1.51\, O_2 \rightarrow 1.0\, CH_{1.68}O_{0.36}N_{0.2} + 0.72\, CO_2 + 2.94\, H_2O$$

Consumption of Single-Cell Proteins

The nucleic acid contains two types of nitrogen bases, namely purines and pyrimidines. Of the two types of nitrogen bases, Pyrimidines are well metabolished in the human body. But purines are not easily metabolished because of the formation of uric acid at the end of purine catabolism. The human body does not have any mechanism to release it.

Biotechnologists have been using birds as intermedia agents to utilize the SCP as food. Birds have an effective mechanism to release uric acid during excretion. So they easily release the uric acid produced during the metabolism of purines (Figure 9.1). Hence purines do not cause any ill effects on birds. But in human beings uric acid effects the kidney and other organs also. The products of birds such as egg and flesh do not have any ill effects, while they have a larger proportion of proteins than the egg and flesh produced by other birds.

(If a man consumes the SCP, the end product of purine metabolism is deposited in his body. The end product is deposited in the form of allentonin in the muscles, and it changes the physiological conditions of his body).

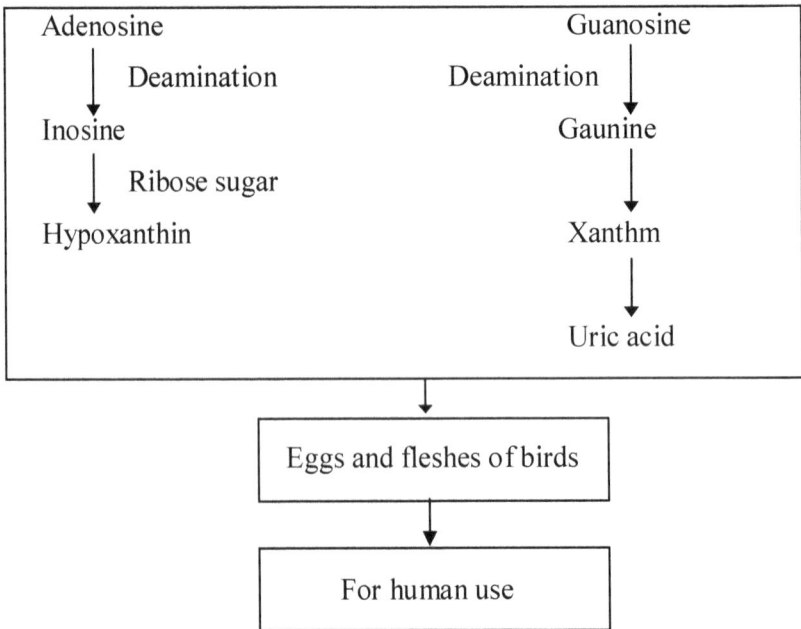

Figure 9.1: Showing Releasing Uric Acid during Metabolism of Purin in Birds.

Source of Single-Cell Proteins

Certain micro-organisms are likely to be beneficial owing to the presence of more proteins in their cells. Such microbes are directly used as a protein source in human diet. The important micro-organisms used as single-cell proteins are stated below:

Delbruck *et al.,* first established the culture of *Saccharomyces cerevisiae* for the production of single-cell protein. It was used along with soups and saugeses. It helped to solve the problem of food shortages in Germany during the First World War. Germans used the culture of *Candida arborea* and *C. utilis* during the Second World War.

The *British Petroleum* (BP) used hydrocarbons as raw materials for culturing micro-organisms in bioreactors. A British petroleum unit in Scotland has been purifying **n-Paraffins** to produce the biomass of *Candida.*

The petroleum refinery at Lavera, in France, has started to use crude oils as a substrate for the culture of Candida. The researchers harvested the biomass from the fermenters and removed the gas oils from the harvested microbes. Then they sterilized the microbes under stream pressure and the sterilized microbes were then dried into small granules. The dried granules wre marketed with the trade name "Toprina".

The KISR (Kuwait Institute of Scientific Research) in 982 cultured yeast cells for the production of single cell proteins by using methanol as substrate. Fungi, Torulospisis and bacteria like Methylomonas clara, Mythylophilus methylotrophus and Alcaligenes grow well by consuming methanol as a carbon source.

SCP from Algae

The use of algae as SCP producer is gaining interest because they grow well in ponds or tanks and need only the freely available CO_2 as carbon source and sunlight as an energy source for photosynthesis. Members of the genera *Chlorella and Senedesmus* have long been used as food in Japan, while *Spirulina* is widely used in Africa and Mexico. *Spirulina maxima* is commerclally produced in Mexico as a by-product of a large solar evaporator used for production of soda lime. It is used as animal feed and also suited for human use. It may be harvested by filtration or simply by skimming *Chlorella* is used as protein and vitamin supplement in some Japanese yoghurt, ice cream and breads. Algae can be used in ponds or lagoons to help in the removal of organic pollutants and the reaultant biomass is harvested, dried and the powder added to animal feed. Algal SCP contains about 60 per cent crude protein that increases their value as animal feed; the only disadvantage is its rich chlorophyll content that makes it unsuitable for human use.

SCP from Filamentous Fungi

Pekilo, a fungal product that was produced by fermentation of carbohydrates derived from sulphite liquor molasses, whey, waste fruits, woods, starch hydrolysates or agricultural hydrolysates, contains about 55 per cent protein and is rich in vitaminas.

The filamentous fungus, *Paecilomyces variotti,* is used in continuous fermentation processes to produce Pekilo, which is a good protein and vitamin supplement in the diet of farm animals and poultry birds. In UK, Rank-Hovis, McDoughall in conjunction with ICI (Marlow foods) is now commercially marketing the fungal protein, mycorprotein or Quorn, derived from the growth of *Fusarium graminearum* fungus on simple carbohydrates. Since they have high nucleic acid content (about 15 per cent

RNA, acting as the source of non-protein nitrogen), the process of continuous culture of the micro-organism involves the reduction of RNA that is achieved by addition of RNase (ribonuclease). Towersey, Logton and Cockram have patented this process in 1976. The process increases the actual percentage of protein in the reaction mixture.

Unlike almost all other forms of SCP, mycoprotein is produced for human consumption. The product is open to be sold in the UK since it is nontoxic and compatible with normal physiological function. In studies of its routine toxicity to man, one of the important physiological effects observed was a reduction in serum cholesterol.

SCP from Yeast

The commercial production of SCP from *Candida utilis* (Torula yeast), *C. intermedia, Kluyveromyces fragilis* and *Saccharomyces cerevisiae* (Baker's yeast) is currently practiced using the substrates like confectionary effluents, ethanol, sulphite liquor, whey, n-alkanes (C_{10}-C_{23} hydrocarbons), molasses, etc. The SCP produced from yeast has protein content upon 60 per cent. The first commercial production of SCP dates back to World War-1 when Torula yeast was produced in Germany and used in soups and sausages. Since then SCP produced from yeast is used both for human food and in animal food supplement.

Members of yeast also have high nucleic acid content which need to be reduced by treating them with R-Name; further, their slower growth rate may produce difficulties in production of SCP. However, there is minimum risk of bacterial contamination and their recovery by continuous centrifugation is easy.

Bacteria as Source of SCP

Several bacterial strains have been utlised as SCP producers. The Imperial Chemical Works (U.K.) produces proteen, the SCP product of *Methylophilus methylotrophus,* a bacterium that grows on C_1 compounds. This bacterium is grown on methanol derived from methane and the cell crop is harvested, centrifuged, dried and sold in pellet or granular form. Another bacterium *Brevibacterium sp.* is also exploited for its SCP-producing ability that uses, C_1-C_4 hydrocarbons as a substrate. Bacterial SCP has very high and crude protein (upto 80 per cent), which add to its nutritional value but at the same time it has high RNA content (upto 20 per cent), which is to be reduced; it is prone to contamination by pathogenic bacteria. Though their growth rate is highest and they use wide range in substrate, their recovery process is problematic.

SCP from Filentous Fungi

Pekilo, a fungal product that was produced by fermentation of carbohydrates derived from sulphite liquor molasses, whey, waste fruits, woods, starch-hydrolysates or agricultural hydrolysates, contains about 55 per cent protein and is rich in vitamins.

The filamentous fungus, *Paecilomyces variotti,* is used in continuous fermentation processes to produce Pekilo, which is a good protein and vitamin supplement in the diet of farm animals and poultry birds. In UK, Rank-Hovis, McDougall in conjunction with ICI (Marlow foods) is now commercially marketing the fungal protein,

```
┌──────────────────────────┐          ┌──────────────────────────┐
│  Carbohydrates as carbon │          │ Other constituents, water,│
│         sources          │          │    mineral salts, etc.    │
└──────────────────────────┘          └──────────────────────────┘
              ↘                          ↙
           ┌─────────────────────┐        ┌──────────┐
           │   Medium bending    │ ◄───── │   Air    │
           └─────────────────────┘        └──────────┘
                     ↓
           ┌─────────────────────┐
           │    Sterilisation    │
           └─────────────────────┘
                     ↓
┌──────────────┐   ┌─────────────────────┐
│   Inoculum   │──►│    Fermentation     │
└──────────────┘   └─────────────────────┘
                     ↓
           ┌─────────────────────┐
           │    RNA reduction    │
           └─────────────────────┘
                     ↓
         ┌──────────────────────────┐
         │  Purification and drying │
         └──────────────────────────┘
```

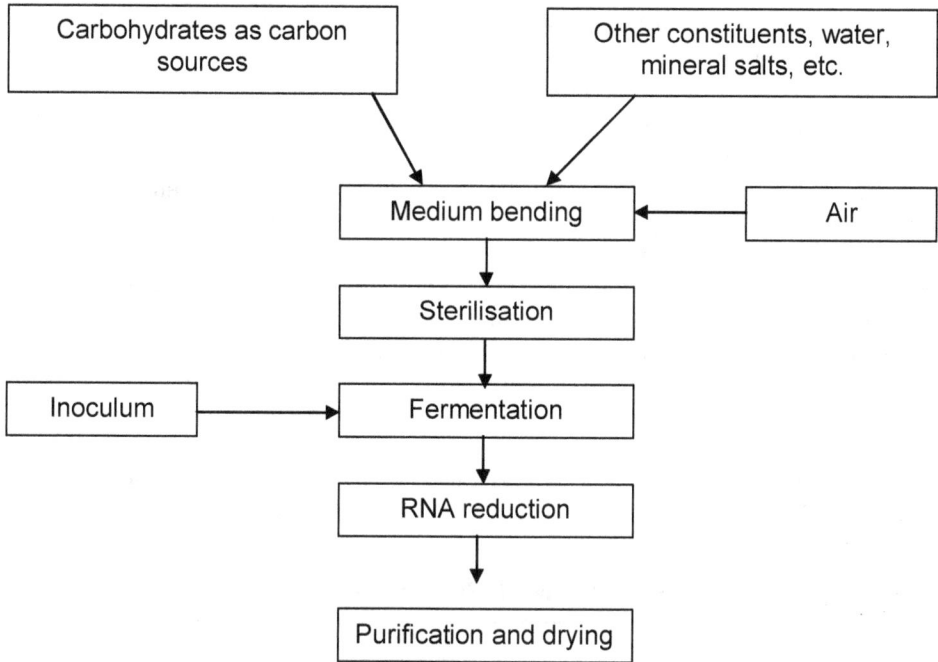

Flow Diagram of Mycoprotein Production.

mycoprotein or Quora, derived from the growth of *Fusarium graminearum* fungus on simple carbohydrates. The various stages for the production of mycoprotein from this fungal strain are shown in Figure. Since they have high nucleic acid conteitn (about 15 per cent RNA, acting as the source of non-protein nitrogen), the process of continuous culture of the micro-organism involves the reduction of RNA that is achieved by addition of RNase (ribonuclease). Towersey, Logton and Cockram have patented this process in 1976. The process increases the actual percentage of protein in the reaction mixture.

Unlike almost all other forms of SCP, mycoprotein is produced for human consumption. The production is open to be sold in the UK since it is nontoxic and compatible with normal physiological function. In studies of its routine toxicity to man, one of the important physiological effects observed was a reduction in serum cholesterol.

Mushroom as a Source of Protein

There are about 4000 species of mushrooms which mycologists have enumerated and named so far. Of this, nearly two thousands are known to be edible but about a dozen or two are regularly collected from natural environments and consumed, and at present, nearly a dozen types are cultivated. A packet of fresh mushroom can be purchased and used for breqakfast soup or for a variety of dishes that go with lunch or dinner. Cooking mushrooms is now familiar to majority of its lovers. They are stewed, baked, dish dried, or grilled. Many taste delicious when stuffed with a specy filling or when mixed in an omelette.

Mushrooms are becoming popular not only for their flavor and taste but also for their nutritive value. Like many other vegetables, mushrooms contain a high proportion of water apart from other constituents. They usually contain more protein (about 51 per cent) than many other comparable fruits or vegetables (on a dry weight basis) and as such can claim to be a "protective food". It is known that mushrooms are one of the best plant sources of nicotinic acid and riboflavin (vit. B_{12}), a good source of pathothenic acid, and a fair source of vitamins B,C,K, biotin and thiamin. It appears from the literature that its vitamins are well retained during the process of cooking, canning, drying and freezing. In some cases, linoleic acid has been found to be the main fatty acid component. Carbohydrate contents, however, vary between 28 to 76 per cent. Moreover, mushrooms being completely devoid of starch, constitute an ideal dish for diabetic patients. They also contain an appropriate amount of minerals, lipids and folic acid. Like many vegetables they can be stewed or backed, fried or sandwiched, can be used in soup or 'puree', in sauce, and also in toast tomatoes stuffed with mushrooms.

Some Cultivable Edible Mushrooms

Species	Local Name	Substrate	Country
Agaricus bisporus	Button mushroom	Horse, manure, wheat, rice straw	India, Europe, USA
Auricularia sp. *Coprinus* sp.	Woodear	Rice bran, saw dust straw	USA
Flammulina velutipes	Winter mushroom	Sawdust, straw	Europe, USA.
Grifola frondosa	Sitting hen mushroom or Limuo	Straw	Europe, USA
Hericium erinaceus	Monkey head or hedgehog mushroom	Wheat/rice straw	USA
Hypsizigus marmorens	Shimeji	Wheat/rice straw	Asia
Lentinula edodes	Shiitake or oak mushroom	Log, paper saw dust	USA, Asia, Europe
Fleurotus ostreatus	Oyster mushroom	Straw, rice bran, saw dust	India, China, Japan, USA, Europe
Pholiota nameko	Nameko or viscid mushroom	Wheat/rice straw	Asia, Europe
Termelia fuciformis	White jelly fungus or 'silver ear'	Wheat straw, rice bran	Asia, USA
Volvariella volvacea	Padi-straw mushroom or Chinese mushroom	Paddy/wheat/rice straw, cotton	Asia

Preparation of Culture Media

The pure culture of cultivated mushroom can be obtained on the following medial. These media are generally used as substrate for isolation, sub-culture, maintenance and preservation of mushroom cultures.

1. Potato Dextrose Agar (PDA) Medium
- ☆ Peeled potato   250g
- ☆ Dextrose   20g
- ☆ Agar agar   20 g
- ☆ Distilled water   1 L

Bil peeled and sliced potatoes in distilled water for 20-25 minutes till these become soft. Filter the extract with a muslin cloth. Add 20g dextrose and 20g agar powder to the filtrate over a hot plate by stirring. The final volume of the medium should be adjusted to 1L by adding required amount of distilled water. The medium (approx. 10ml) is taken in culture tubes and the tubes are plugged with non-absorbent cotton and sterilized in autoclave at 121°C (15 lb/sq. inch pressure) for 20-30 minutes. The autoclaved culture tubes should be kept in a slanting position for solidification.

2. Malt Extract Agar Medium
- ☆ Malt extract   25g
- ☆ Agar agar   20g
- ☆ Distilled water   1 L

3. Wheat Extract Agar Medium
- ☆ Wheat grains   100g
- ☆ Agar agar   20g
- ☆ Distilled water   1L

Wheat grains are boiled in 1L distilled water for one hour. Filter the extract after 25 hour. Agar is added to the supernatant by stirring with a glass rod over a hot plate. The final volume of medium is adjusted to 1L by adding required amount of distilled water. The medium is then autoclaved as above.

3. Compost Extract Agar Medium
- ☆ Ready synthetic compost   100g
- ☆ Agar agar   20g
- ☆ Distilled water   1 L

Boil ready synthetic compost in 1L water for 1 hour. Filter after 24 hour. Add 20g agar by stirring over a hot plate. The final volume of medium is adjusted to 1L by adding required amount of distilled water. The medium is then autoclaved as above.

4. Lambert's Agar Medium
- ☆ Dextrose   10g
- ☆ Magnesium sulphate   0.5g
- ☆ Agar agar   20g
- ☆ Distilled water   1 L

All the ingredients are dissolved in 1L distilled water and then autoclaved as above.

5. Yeast Extract Potato Dextrose Agar (YPDA)

- ☆ Yeast extract   1.0gm
- ☆ Peeled potato   250g
- ☆ Dextrose   20g
- ☆ Agar agar   20g
- ☆ Distilled water   1L

Preparation method is same as like potato dextrose agar medium.

6. Rice Bran Decoction Medium

- ☆ Rice bran   200gm
- ☆ Gelatin   20gm

Rice bran is boiled in water for 20 minutes and then filtered through muslin cloth. Added gelatin to both solution and then sterilized by autoclave.

Precautions

1. Proper tightening of cotton plugging of culture tubes should be done.
2. After sterilization, culture tubes should be carefully taken out and kept in a slanting position to increase the surface area of the medium.
3. The medium is slants should not touch the cotton plugs and at least 1 inch gap should be provided between the cotton plug and the end of slant medium to avoid contamination.

Preservation and Maintenances of Mushroom Cultures

After pure cultures has been established or obtained of mushrooms culture from any mushroom culture sources, the cultures are maintained and multiplied on culture tube slant or petriplates. Proper maintenance of pure cultures of cultivated mushroom is necessary to maintain vigour and productivity. For further investigation by researchers on mushroom cultivation, needed maintenances of mushroom cultures at time to time for vigour and productivity. Subculturing of stock cultures by periodic transfer on a suitable/different medium. There is no satisfactory techniques for examining and evaluating the qualities of cultures and spawn in the laboratory. If any strains of mushrooms cultivated continuously, it may eventually loose some of its desirable genetic straits. A tissue culture of such degenerated strain will rise to degenerated culture. Degeneration is edible fungi can be caused by lack of nutrients, oxygen, a change in pH of the substrate, accumulation of toxic metabolites or may be bacterial or viral infection. The strains of cultivated mushroom must be preserved and carefully tested from time to time for vigour and productivity. Maintenance of vigour and genetic characteristics of pure mycelium is main objective of stain preservation.

The preparation and maintenance of mushroom cultures require technical knowledge in different fields such as microbiology, mycology and taxonomy and it is not always possible for a small size mushroom farm to keep and maintain their own cultures. Different preservation techniques have been used to maintain the mushroom cultures and spawn.

General Method

Pure cultures are raised by using single or mass mushroom spores isolation and tissue culture techniques from freshly harvested mushroom primordial. Maintenance of pure culture or any obtained mushroom cultures by the periodical transfer technique such as sub-culturing. For sub-culturing, a small piece of medium with pure culture is cut under aseptic condition and put in sterilized petriplates or slant culture tubes. The pure mushroom culture tubes are open over flame, plug remove the mycelia bits or pieces are taken out with the help of sterilized scalpel/needle loop. Another new slant culture tube is open the same way and the mycelial bit is transferred and press in contact of medium and then plugged. The newly inoculated slants are incubated at 26°C or suitable temperature of mushroom strains. Sub-cultures are again used for sub-culture of mushroom strains. Sub-cultures are again used for sub-culture after two months for their maintenance. This preservation method of mushroom culture is generally applied *in-vitro* investigation or any other lab research work. There is partial loss of productivity and desired qualities because of degeneration and mutation during prolong vegetative propagation of stock cultures or from genetic remombination and selection in continuous field cultivation.

Compost Storage Method

In this method all mushrooms cultures can be best storage on compost for two years. Well prepared compost usually used for support of mycelium growth in the resting stages. For preparing compost substrate, 300gm dried compost is taken from well prepared pasteurized compost and ground to piece of 5mm size. The compost is moistened with one litre of tap water and then three times washed with hot water to remove undesirable gases and metabolites. This compost is filled in the culture tubes to one third of their capacity and plugged with non absorbent cotton and then two times at next one days interval at 121°C for two hours. Storage of mushroom cultures in such culture tubes containing compost medium at a temperature below 5°C.

Lyophilization

This method if a most important, economically and effective for preservation of sporulating fungi. Mushroom cultures can not be stored by this method, however mushroom spores collected from matured and healthy fruit bodies under aseptic condition and stored by this method. The fungal spores are frozen and at some time dried under low pressure in vacuum to remove water from spores. The fungal spores remain dormant and viable for longer time under this condition.

Freeze-Drying Method

In this method, mushroom cultures are fist grown on petriplates containing potato dextrose agar medium. These petriplates are transferred in heavy-walled borosilicate glass ampoules for freeze drying and storing in liquid nitrogen.

Cryogenic Freezing Method

This method involves very costly technology at its best and is not suitable for smaller laboratories. Above freeze-drying method of preservation causes freezing injury to biological system. There are compound that protect living organisms against damage due to freezing and thawing. The cryoprotective agents prevent the formation of cellular ice crystals, reduce electrolyte concentration and cells dehydration during cooling process. There are two types of cryoprotective agents applied in this cryogenic freezing method. Glycerol and dimethyl suflex are penetrating agents pass through the cell membrane and exercise their protective effect within the intracellular and extracellular environment. The non-penetrating agents do not cross the cell membrane. These agents are sucrose, lactose, glucose, mannitol, sorbotol, pyroliodone and dextran.

In this method mushroom cultures are fist grown in the slant culture tube containing potato dextrose agar medium. For optimum production of mushroom mycelium, slatns are flooded with 10 per cent glycerol or 5 per cent dimethyl sulfoxide (DMSO) and generally scraped to obtain a suspension for freezing. This prepared suspension of mushroom cultures are filled in heavy walled borosilicate glass ampoules and pre-cooled to 5°C for 30 minutes and then sealed with a semi-automatic sealer. These ampoules are freezing by dipping direct into the liquid nitrogen or by controlled freezing procedure. The best freezing result it obtained by slow cooling. After freezing of ampoules, they are transferred to storage in liquid nitrogen at -196°C or liquid nitrogen vapor storagen at -150°C to 180°C.

Wax Refrigerator Method

This is also popular maintaining method of mushroom cultures and can be stored for one year. In this method, mushroom cultures are fist grown in the slant culture tubes with cotton plugged dipping in hot wax solution for two minutes. These waxed mushroom culture tubes are kept in the maximum cooling refrigerator.

Mineral Oil Storage Method

The mineral oil of specific gravity 0.865 to 0.89 is used for this purpose. The mineral oil is sterilized for 30 minutes, cooled for some time and then poured in screw-capped tubes. Mushroom culture immersed under mineral oil and it can be stored in the room temperature or refrigerator.

Storage in Demineralized Water

In this method, mushroom culture are grown on potato dextrose agar medium and after full growth small bits of colonized agar medium are transferred aseptically into sterilized bottles containing demineralized water. The bottles containing demineralized water are sterilized in the autoclave at 15 lb for two hours and cooling at room temperature. The mushroom mycelial bits @ 2-3 pieces per bottle of 5mm are put in these bottle under aseptic condition.

Equipments Required for Mushroom Culture and Spawn Laboratory

The basic requirements of different equipments for establishing a spawn laboratory and mushroom cultivation are given here in detail;

1. Glass wares: For preparation and maintenance of mushroom cultures and spawn production.

2. Spirit lamp, inoculating needle: For sterilization and inoculation.

3. Boiling pans: For boiling of cereal grains.

4. Wooden trays: For spreading of boiled grains and mixing of gypsum and chalk.

5. pH meter: For checking the pH values of medium and spawn.

6. Hot air overn: For sterilization of glass wares.

7. Autoclave: For sterilization of culture and spawn media.

8. Laminar flow: For isolation and multiplication of cultures and spawn preparation.

9. B.O.D. Incubator: Needed to maintain and incubate mushroom cultures and spawn.

10. Refrigerator: Needed for short-term preservation of mushroom cultures and spawn.

11. Incubation chamber: For large scale spawn production, B.O.D. Incubator cannot be used as limited space is available in them and in such cases a chamber of 25x20x12 feet with entire surface area (all walls, floor, ceiling, doors) insolated with 2-3 inch thick insolation is required. Air-conditioners are employed for maintenance of temperature. Two air-conditioners of 1.5 tonne capacity each working alternatively is sufficient to maintain temperature in the incubation room at 26°C.

12. Cold storage room of 15'x10'x12' is required to store the spawn at 4-5°C. The wall, roof and floor as well as door is given heavy insulation (3-4") and air-conditioners (2 of 1.5 tonne capacity) are provided to maintain the temperature in the room.

13. Other items include office tables and chairs, working tables, stools, almirah, racks for keeping glass wares and chemicals, troughs, sieves, inoculatin needle, scalpels, cotton, saline bottles polypropylene bags, culture tubes, petriplates, oven, paraffin wax, aluminium foil, etc.

14. Chemicals are required for media preparation. Formaldehyde, calcium carbonate, calcium sulphate, mercuric chloride, spirit and other chemicals are required in bulk.

Spirulina as a Single Cell Protein

Many pilot plants for the production of *Spirulina* powder have been established in Japan, United States and European countries. Sosa Texcoco is the first Mexican

Company to set up the first pilot plant in 1973 which produced about 150 tonnes *Spirulina* powder per annum; but the yield was increased to 1000 tonnes in 1982. This company exported powder to the United States and prepared lozenges and capsules from the powders by adding vitamin A and C. The Mexican company also supplied *Spirulina* powder to government institutions which were in charge of improving the nutritional situation of the population: Institutions used *Spirulina* powder to make biscuits and confectionery with a high protein content (Sasson, 1984).

Mass Cultivation of *Spirulina* SCP

At present two types of farms for mass production of *Spirulina* SCP are under operation. A third type (*i.e.,* enclosed system using transparent tube, biocoil or photobioreacter) is under development (Henrikson, 1990).

(*i*) **Semi-natural lake system**: Sosa Texcoco Lake (Mexico) and Lake Chand (Africa) offer an ideal environment for the natural growth of Spirulina. The product is expensive but of low quality due to contamination and pollution by uncontrolled natural factors. SCP of these lakes are good for fish and animal feed. Researches are in progress to refine the powder and make the products of good quality.

(*ii*) **Artificially built cultivation system**: The climatic conditions of most of the developing countries are such that favour mass outdoor production of *Spirulina.*

Requirement for Growth of Spirulina

1. **Algal tanks**: Generally, circular or rectangular cemented tanks are constructed. The circular tanks are more preferred over the rectangular one because of ease in handling. Size may be according to convenience and yield needed. Depth should be about 25cm. Open tanks are suitable for tropical and subtropical regions.

2. **Light**: Low light intensity is required at the beginning to avoid photolysis. Spirulina exposed to high light intensity is lysed.

3. **Temperature**: Temperature for optimum growth should be between 35-40°C.

4. **pH**: Spirulina grows at high pH ranging from 8.5 to 10.5. Initially, culture should be maintained at pH 8.5 which automatically is elevated to 10.5.

5. **Agitation**: Agitation of culture is very necessary to get good quality and better yield. The culture is agitated by brush, paddle power, pipe pumps, wind power, rotators, etc.

6. **Harvesting**: The filaments of Spirulina float on surface of water forming thick mat. Therefore, it can be harvested by fine mesh steel screens, nylon or cotton cloths, etc.

7. **Drying**: As it has a thin wall, sun drying is the most suitable and economical. Various trials done at CFTRI, Mysore and MCRC, Madras with sun drying have given good results.

8. Yield: An average yield of 8-12 g Spirulina powder/m^2/day has been obtained in India and other countries. This is equivalent to 20 tonnes/ha/annum. In warmer climate, the yield can increased to about 20 g/m^2/day.

Avoiding Contamination

Although there is the least chance for contamination, yet regular monitoring of algal culture is necessary. Because the microbial load is likely to affect the quality and safety of the product. At MCRC and CFTRI the cultures of Spirulina were found either within or very close to safety limits of Indian Standard Institute (ISI) for baby food, to about 5x10^5 propagules per gram (Venkataraman and Becker, 1985). Dried Spirulina powder is packed in aluminium bags or sealed in bottles and sent to market.

Uses of *Spirulina* as a Single Cell Protein

(*i*) **As protein supplemented food**: Since Spirulina is a rich source of protein (60-72 per cent) vitamins, amino acids, minerals, crude fibres, etc,. It is used as supplemented food in diets of under-nourished poor children in developing countries (Jeegi Bai and Seshadri, 1986; Sachan, 1991). The United Nations, Mexican National Institute of Nutrition, French Petroleum Institute and National Institute of Nutrition, Hyderabad have formulated four algal recipes as a weaning substitute for infants. In India, the Village Health and Energy System Projects are operated at CVS (Wardha). At MCRC, the product are distributed to the local under-nourished children. It has been found that 1g of *Spirulina* tablets contains as much nutrition as one kg assorted vegetable.

(*ii*) **As health food**: Spirulina is very popular as health food. Most of Sosa Texococo products are exported to USA, Europe and France where it is sold in health and food stores. It is the part of the diet of the US Olympic team. Jaggers take spirulina tablets for instant energy. Since it provides all the essential nutrients without excess calories and fats, it is taken by those who want to control obesity. The MCRC has for the first time launched the project as health and baby food, and multivitamin powder and tablet under trade name 'Multin' and 'Multinal'.

(*iii*) **In the therapeutic and natural medicine**: *Spirulina* possesses many medicinal properties. Therefore, it is used as social and preventive medicine also. It has been recommended by medicinal experts for reducing body weight, cholesterol and pre-menstrual stress and for better health. It lowers sugar level in blood of diabetics due to the presence of gamma-linolenic acid and prevents the accumulation f cholesterol in human body (Nayak *et al.,* 1988). It is a good source of β-carotene (a precursor of vitamin A) and, therefore, helps in monitoring healthy eyes and skin. B-carotene is known as the best anticancer substance (Schwartz *et al.,* 1986). In 1989, UN National Cancer Research Institute announced that substances from blue-green algae are active against AIDS and cancer virus. In Vietnam its tablets are used to increase lactation in nourishing mothers.

In Cosmetics

Spirulina contains high quality of proteins and vitamin A and B. These play a key role in maintaining healthy hair. Many herbal cosmeticians are making efforts to develop a variety of beauty products. Phycocyanin pigment has helped in formulating biolipstics and herbal face cream in Japan. These products can replace the present coaltar-dye based cosmetics which are known as carcinogenic.

SCP Derived from Fossil Carbon Sources

The substrates having high commercial value as energy sources or derivatives of such chemicals, *e.g.* gaseous hydrocarbons, liquid hydrocarbon, gas oil, n-alkenes, methanol and ethanol have wide commercial importance in SCP production.

Methane (C_1-hydrocarbon) as an SCP source has been extensively studied but there are technical difficulties, like risk of explosion when more than 12 per cent O_2 (v/v) is used and requirement of efficient cooling during the process which hampered the production. In contrast, methanol offers great interest. In UK, ICI constructed a large scale fermentation plant for producing the methanol-utilising bacterium – *Metholophilus methylotrophus.* Hoechst (West Germany) and Mitsubishi (Japan) worked on a process – they used yeast as the SCP source instead of bacteria. Methanol as a carbon source for SCP has many advantages over n-paraffins, methane and even polysaccharides, since methanol is nontoxic, easily soluble in water, and its composition is independent of seasonal functions. C_5 to C_8 n-alkanes are liquid at room temperature and they are usually toxic to cells due to their solvent action. Generally C_9-C_{18} n-alkane4s are used as SCP substrates. British Petroleum was producing SCP at an industrial scale using gas oil in combination with ammonia and mineral nutrients with the help of *Candida* yeast. Ethanol is also acceptable as a substrate for SCP production for human use. Bacteria, yeast and filamentous fungi exploit ethanol for SCP production. In USA, Amco Foods produces SCP by growing *Torula* yeast.

SCP from Wastes

Agricultural, forestry and industrial wastes like straw, bagasse, sawdust, sulphite liquor, effluents from different industries, starch and cellulose hydrolysate, whey, molasses, animal manures and sewage, significantly contribute to pollution of water bodies. The vitalization of such materials for SCP production offers two benefits – reduction in level of enrivonmental pollution and production of edible protein. Most of the organic wastes are available at low cost in most countries thus ensuring independence in supply. Molasses acted upon by Baker's and Torula yeast, whey by *Kluyyeromyces* fragilis in combination with *Lactobacillus bulgaricus,* cellulosic material by fungus *Chaetomium cellulolyticum*, starch hydrolysate by *Fusarium graminearum*, and industrial effluent by *Candida utilis* and *Paecilomyces varioti* can produce SCP of human interest.

Evaluations of SCP

A tremendous amount of attention has been paid to the problems related to the safety, nutritional value and acceptability of SCP.

1. The chemical assessment of SCP in terms of protein, nucleic acid, lipid and vitamin content must be done.

2. The nature of raw material used in SCP process must be analyzed, *e.g.* the possibility of carcinogenic hydrocarbons in gas oil or n-alkenes, the heavy metals or other contaminations in the mineral salts and the mycotoxin production by certain fungi.

3. The process organism must e non-pathogenic and non-toxigenic.

4. Rigorous sanitation and quality control procedures must be maintained throughout the process to avoid spoilage or contamination by pathogenic or toxigenic microbes.

5. The physical characteristics like texture, odour, taste, colour, particle size, density, storage, etc., must be determined.

6. Toxicological testing of final product including short-term acute toxicity must be done with different laboratory animal species.

7. For the evaluation of nutritional value of the product, long-term testing should be done on target species.

Thus, if SCP is to be used directly as food for human beings, the skills of the food technologist will be greatly challenged.

Economic Importance of SCP

The process of SCP production, the micro-organisms involved in the process and the product itself offer several advantages which includes the following;

1. SCP is rich in high-quality protein, with poor quantity of lipids/fats which make it desirable for use as human food supplement.

2. The SCP may be used in animal feed to at least partially replace the currently used protein-rich soybean meal and fish-proteins and even cereals that can be diverted for human consumption.

3. The SCP may be a good source of vitamins particularly B-complex which adds to its nutritional value.

4. SCP can be stored and shipped over long distance without spoilage.

5. The process of production can be continued throughout the year independent of the seasonal fluctuations.

6. The micro-organisms involved in the production process have remarkably fast-growing ability and they can produce large quantity of SCP in a relatively small space.

7. The microbes can easily be modified genetically by gene transfer technology to have plenty of quality protein.

8. The substrates used in the process are of low cost and the effluents produced are easily degraded by microbes to reduce pollution.

9. Mushroom cultivation provides a great advantage in converting industrial, urban and wood wastes into a product that is directly edible by human beings.

10. The production of SCP and mushroom cultivation have evolved with new technical solutions to fill the gap between demand and supply of quality protein.

Uses of Single-Cell Proteins

1. Single cell proteins are used as active ingredients of human diet. Thus, they increase the nutritive value of human diet.

2. The Germans used SCPs during food shortages during the world wars. They believed that SCP forms a good source of protein during the shortage of foods.

3. SCP is used as feed for cattle and birds.

4. During the production of biomass, the microbes produce a number of organic acids, amino acids, fats, methanol, starch and sugars. The fats and oils are extracted and used in soap industry. Methanol is an alcohol which is used in soap industry. Methanol is an alcohol which is used in fermentation and distillation laboratories.

5. The starch and sugars are used in the production of some chemicals and polymers.

6. A mutant strain of *Sporotrichum pulverullentum* utilizes lignin found in the wood biomass. So, it releases cellulose freely from the lignin. This method is used in the preparation of cellulose from wastes and the wood biomass. This reaction is carried out by Phanero chaets and *Chrysiosporum* which degrade lignin in the presence of hydrogen peroxide (H_2O_2). The crystalline cellulose is used as a raw material in the paper industry.

7. Pure cellulose is produced from starch and sugars by the action of *Acetobacter xylinum*. It is used in the manufacture of *Cellophane* and the other cellulosic polymers.

8. The biomass of *Methylophilus methylotrophus* is also used in the manufacture of glutamic acid.

Conclusion

A few micro-organisms play certain important roles in the manufacture of some food items. This is specially true in the preparation of milk products like cultured butter milk, Bulgarian milk, acidophilus milk, yogurt, kerif and kumiss. Micro-organisms are also involved in the manufacture of some fermented food items like pickles, saverkraut sousage etc. Thus microbes actively participate in the manufacture of fermented food items for the improvement of the fashion of society to a certain extent.

References

Ahindra Nag, 2008. Text book of Agriculture Biotechnology PH Learning Private Ltd, New Delhi. 264 pp.

Bowrges, H. Sotomayor, A. Mendoza, E and Clavez, A. 1971. Utilization of the alga spirulina as a protein source. *Nutr. Rep. Internet.* 4(1): 31.

Dabah, R 1970. Protein from micro-organisms. *Food Tech* 24(1): 95.

Debey R.C. 2002. Single cell protein and microprotein. In: *A Textbook of Biotechnology.* S. Chand and Company Ltd., New Delhi, pp. 300–327.

Deshmukh, A.M. 1998. *Selection Topics in Biotechnology.* Pama Publication, Karad, 151pp.

Dubey, R.L. and Maheshwari D.K. 2002. *A Textbook of Microbiology.* S. Chand and Company Ltd., Ramnagar, New Delhi, 684 pp.

Heritage, J Evans, *E.G.*V and Kathington, R.A. 1996. *Introductory Microbiology.* Cambridge University Press, Great Britain, 238 pp.

Johri, C.N and Goel, R. 1998. Fungi in biotechnology: Historical to current prescription. In: *Fungi in Biotechnology.* Ed Anil Prakash. CBS Publisher, New Delhi, pp. 1–15.

Joshi, N.K. 2007. *Biotechnology in Agriculture.* Aavishkar Publisher Distributor, Jaipur, India, 454 pp.

Lohar, P.S. 2005. *Biotechnology.* MJP Publisher, Chennai, India, pp. 183–194.

Nehra, S.L. 2013. *Applied Microbiology.* Pointer Publisher, Jaipur, 264 pp.

Patel, A.H. 1996. *Industrial Microbiology.* MacMillan India Limited, New Delhi, 178–187 pp.

Pelezar, M.J. Chan, E.C.S and Krieg, N.R. 2006. *Microbiology.* Tata McGraw--Hill Publishing Company Ltd., New Delhi, 920pp.

Ram, R.C., 2007. *Mushrooms and their Cultivation Techniques.* Aavishkar Publisher, Distributors, Jaipur, India, 168pp.

Rangaswwmy G. and Bagyaraj, D.J. 1996. *Agricultural Microbiology.* Second edition, Prentice-Hall of India. Private limited New Delhi, 438 pp.

Shekhawat, M.S and Vikarant, 2011. *Plant Biotechnology.* MJP Publisher, Chennai India, pp. 530.

Smith, J.E. 2009. Single cell protein derived from high energy source. *Biotechnology,* Cambridge University Press, 178–185 pp.

2015, **Recent Trends in Microbiology, Mycology and Plant Pathology**   *Pages* **135–142**
Editor: **Dr. H.C. Lakshman**
Published by: **DAYA PUBLISHING HOUSE, NEW DELHI**

Chapter 10

Micro-organisms as Pollution Indicators in Selected Lentic Habitats of Dharwad, Karnataka State, India

Doris M. Singh

Department of Botany, Karnatak University's,
Karnatak Science College, Dharwad, Karnataka, India
E-mail: limnokcd@gmail.com

ABSTRACT

Water quality assessments from different ponds of Dharwad are presented. The results show that the samples have physico-chemical properties well within the permissible limits. But increased turbidity and objectionable microbial load, which indicate water pollution, can serve as an added tool for the rapid evaluation of water quality. Principal component analysis when subjected to the data matrix, revealed the status of water quality in the three selected lentic water bodies.

Introduction

Water is one of the most important natural resource required for all living organisms. Water harbors' micro-organisms from soil, sewage, air, organic matter, dead plants and animals etc. Though the problem if water pollution is worldwide, it is depressing that it draws the attention of scientists only when it becomes hazardous for human health

In India, studies on the problems of water pollution have geared up earlier but analysis of water quality is being attended to only during the last few decades, when the crisis became alarming. In Karnataka water quality are not adequate (Ayed, 2002), while the water bodies in the villages around Dharwad taluk are untouched. The microbial statuses of these ponds are not known. Hence the present work was undertaken to study the present status of the microbial load in three ponds of Dharwad, which are sensitive indicators of water quality. The continued anthropogenic activity in and around the ponds of Dharwad has not only caused water quality deterioration, which may lead to serious health hazard to people and cattle.

Materials and Methods

Sampling Method (APHA, 1995)

Grab samples were collected from selected localities of Dharwad taluk. The description of the sampling sites are given in Figure 10.1.

Sampling Sites

Three lentic water bodies *i.e.*, Banadur Tank, Murkatti tank and Mavinkoppa tank were selected. The location of the three water bodies are shown in Figure 10.1.

Microbiological Quality

In the microbiological studies, six microbial indicators were estimated. Aerobic microbial count; E.coli; Total Coliforms, *Pseudomonas aeruginosa* and yeast and moulds. (APHA, 1975; Trivedy and Goel, 1986).

Physico-chemical Parameters

The physico-chemical parameters of water samples were estimated according to APHA (1995).

Statistical Analysis

Principal component analysis (PCA) is a way of identifying patterns in data, and expressing the data in such a way as to highlight their similarities and differences. Since patterns in data can be hard to find in data of high dimension, where the luxury of graphical representation is not available, PCA is a powerful tool for analyzing data. The other main advantage of PCA is that once you have found these patterns in the data, and you compress the data, *i.e.* by reducing the number of dimensions, without much loss of information. (Lindsay, 2002).

PCA (Pearson, 1901 and Jolliffe, 2002), a type of multivariate analysis has been used by authors such Petr (2007), Bartolomeo. *et al.,* 2004 and Zeng. *et al.,* 2005, to study relationships among the environmental variables and patterns in sample clustering. PCA can be a useful tool to obtain view on the water quality in any urban or other geographical territory, when analyzing large data sets without *a priori* knowledge about them. Petr (2007). In the present paper the xl stat (2013) software was used to obtain a PCA bi-plot, in determining the water quality and microbiological indicators.

The analyzed data matrix were transformed to Log_{10} forms and subjected to Pearson Correlation matrix and PCA bi-plot (xl STAT, 2013).

In the Present study the water quality parameters are studied to assess the suitability of tank water for human usage. The permissible values of the environmental parameters for drinking water recommended by WHO (1984), ISI and ICMR have been quoted (Table 10.3).

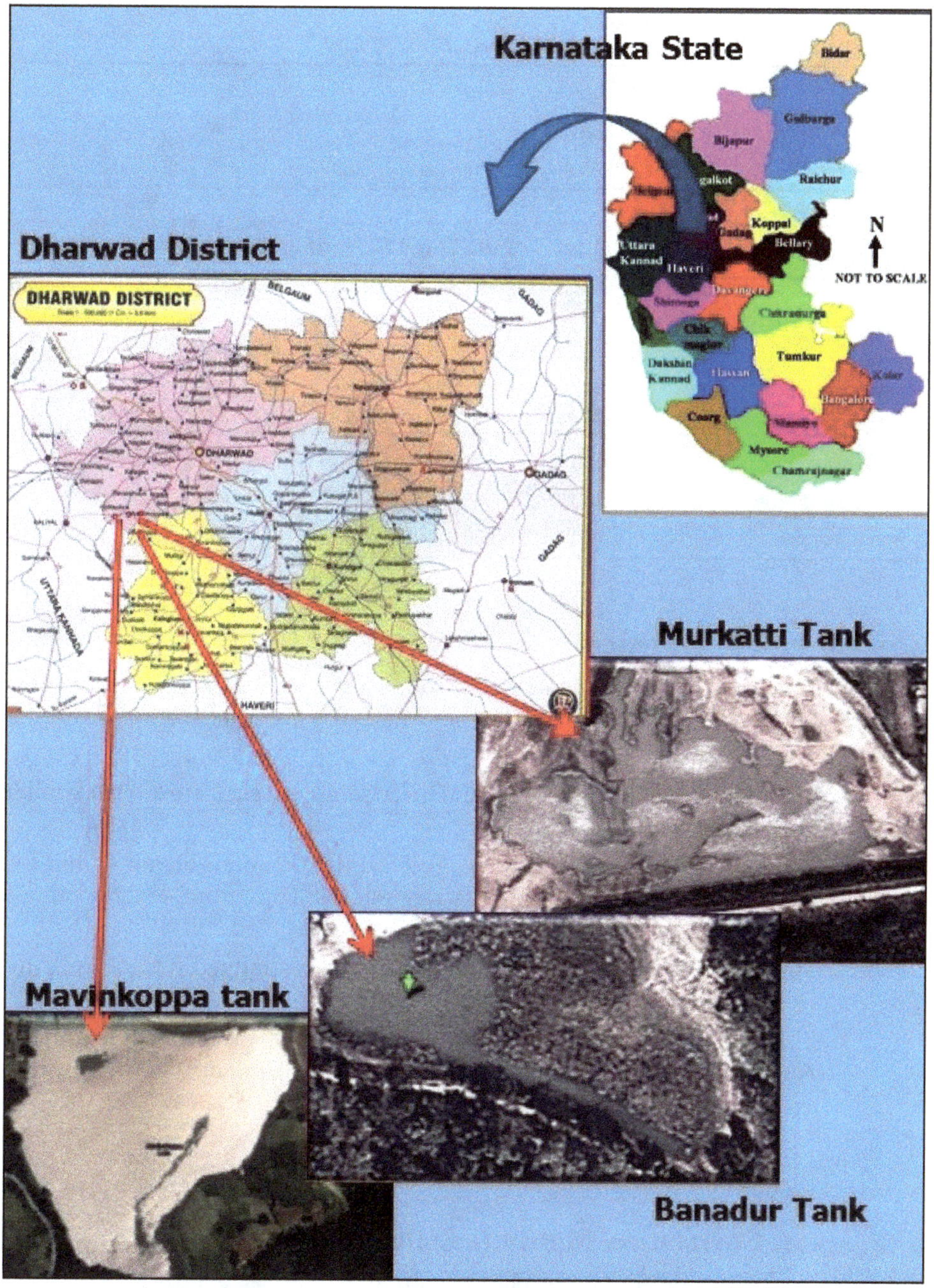

Figure 10.1: Map Showing the Location of Sampling Sites.

Table 10.1: Physico-chemical Characteristics of Water Samples in the 3 Sampling Sites of Dharwad Taluk

Sites (Tanks)	Total Alkalinity	Total Hardness	Total Dissolves Solids	Acidity	Turbidity	pH
	ppm	ppm	ppm	mg/l	(ntu)	
Bandur	16.2	164	125	5	21	8.23
Murkatti	18	225	254	5	17	8.88
Mavinkoppa	20	45	297	6.3	18	8.25

Table 10.2: Microbiological Quality of the Three Sampling Sites of Dharwad Taluk

Sites (Tanks)	Total Coliforms	E. coli	Total Plate Count	Yeast and Moulds	Aerobic Microbial Count	Pseudomonas aeroginosa
	MPN/ 100 ml	MPN/ 100 ml	Colony Count	Colony Count	Colony Count	MPN/ 100 ml
Bandur	100	160	300	100	85	300
Murkatti	250	200	350	150	90	245
Mavinkoppa	75	100	365	60	60	1

Results and Discussion

The laboratory results obtained from the analysis of water samples of three different ponds are shown in Tables 10.1 and 10.2. The analyzed data showed the following results:

1. In the Present study the pH in the three sampling sites are almost similar ranging from 7.9 to 8.4

2. Total alkalinity, Total Hardness, Total dissolved solids are all within the permissible limits, total alkalinity ranged from 16-20 ppm, 45-225 ppm for Total Hardness and 125-297 ppm for Total dissolved solids.

3. The turbidity values in the three sampling sites ranged from 21 – 3 ntu. Hence not fit for consumption or for domestic use

4. It was found to be that in all the water samples, there was contamination of *E. coli* and Total Coliforms, Murkatti tank was contaminated by *Pseudomonas aeroginosa*, which may due to high turbidity

When the analyzed data were subjected to PCA Bi-plot, the following results were revealed:

a. Pearson Correlation Matrix (as shown in Table 10.4)

Notable correlations in the correlation matrix:

Table 10.3: Drinking Water Standards (Thakor *et al.,* 2011; EPA, 2009; WHO, 1996; BIS 1993 and ICMR, 1975) (All parameters except pH are in mg/l)

Water Quality Parameters	Standards	Recommended Agency
Total Alkalinity	120	ICMR
Total Hardness	300	ICMR/BIS
Total Dissolved Salts	500	ICR/BIS
Turbidity	1	WHO
pH	6.5-8.5	ICMR/BIS
Coliform	0	EPA
E. coli	0	EPA

☆ Turbidity shows positive correlation with the presence of *Pseudomonas aeroginosa*

☆ Aerobic microbial count, *E.coli,* Yeast and moulds and Total Coliforms high in Murkatti tank

The above correlations are supported by the *Variables (axes F1 and F2:100 per cent)* as shown in Figure 10.2.

Table 10.4: Pearson Correlation matrix (xL STAT. 2013)

Variables	Col	E.co	TPC	YM	AMC	Ps	TA	TH	TDS	AC	TUB	pH
Col	1	0.875	0.168	0.947	0.726	0.464	−0.162	0.839	0.148	−0.610	−0.592	0.987
E. co	0.875	1	−0.331	0.984	0.968	0.836	−0.620	0.998	−0.350	−0.918	−0.127	0.787
TPC	0.168	−0.331	1	−0.157	−0.556	−0.795	0.945	−0.395	1.000	0.679	−0.894	0.323
YM	0.947	0.984	−0.157	1	0.908	0.724	−0.471	0.969	−0.178	−0.832	−0.302	0.884
AMC	0.726	0.968	−0.556	0.908	1	0.946	−0.796	0.983	−0.573	−0.988	0.125	0.607
Ps	0.464	0.836	−0.795	0.724	0.946	1	−0.949	0.871	−0.807	−0.985	0.439	0.318
TA	−0.162	−0.620	0.945	−0.471	−0.796	−0.949	1	−0.673	0.952	0.881	−0.699	−0.003
TH	0.839	0.998	−0.395	0.969	0.983	0.871	−0.673	1	−0.414	−0.943	−0.059	0.742
TDS	0.148	−0.350	1.000	−0.178	−0.573	−0.807	0.952	−0.414	1	0.693	−0.885	0.303
AC	−0.610	−0.918	0.679	−0.832	−0.988	−0.985	0.881	−0.943	0.693	1	−0.277	−0.476
TUB	−0.592	−0.127	−0.894	−0.302	0.125	0.439	−0.699	−0.059	−0.885	−0.277	1	−0.713
pH	0.987	0.787	0.323	0.884	0.607	0.318	−0.003	0.742	0.303	−0.476	−0.713	1

The correlations between variables and factors (shown in Table 10.5), reveals the following results

☆ In the F1 factor, except TPC, TA, TDS and AC, the remaining variables show positive increase, where as the exceptional factors show negative factors

☆ In the F2 factor, all the variables increases negatively

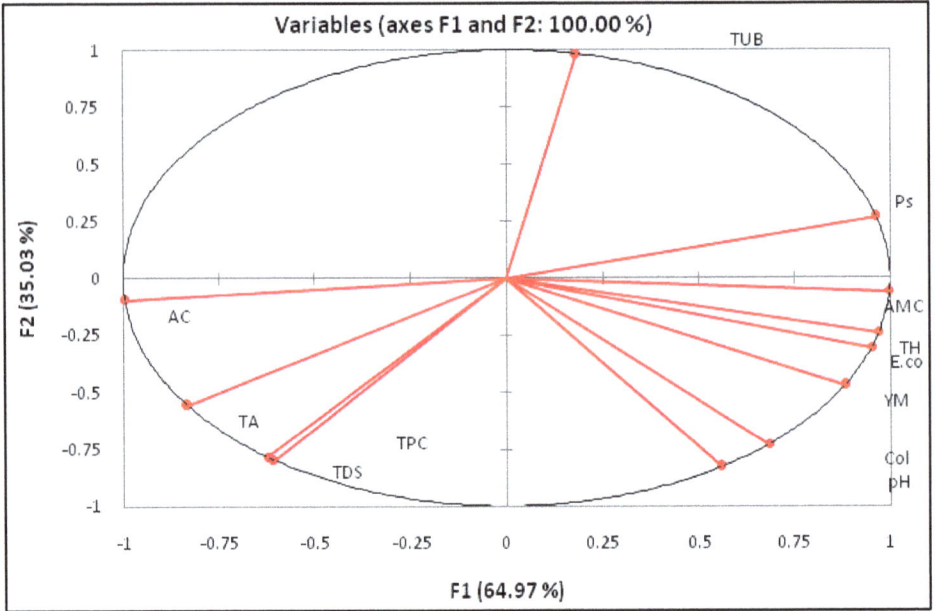

Figure 10.2: Variables (Axes F1 and F2:100 per cent) (xL STAT, 2013).

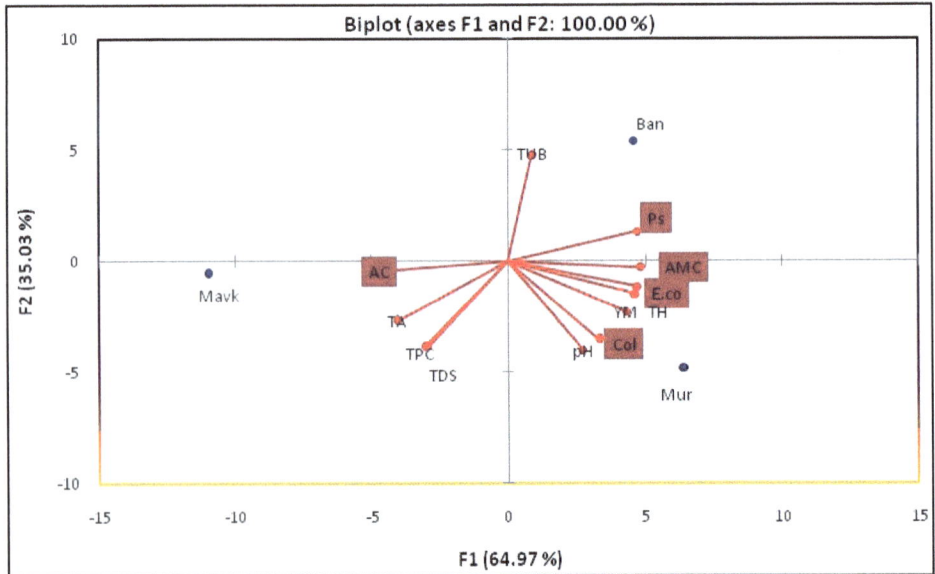

Figure 10.3: PCA bi-plot (xL STAT, 2013).

Abbreviations: Mavk: Mavinakoppa tank; Ban: Banadur tank; Mur: Murkatti tank; TA: Total Alkalinity; TH: Total Hardness; TDS: Total dissolved solids; AC: Acidity; TUB: Turbidity; pH: pH; Col: Total coliforms; E.co: *Escheridia coli***; TPC: Total plate count; YM: Yeast and moulds; AMC: Aerobic Microbial count; Ps:** *Pseudomonas aeruginosa.*

Table 10.5: Correlations between Variables and Factors (xL STAT. 2013)

	F1	F2
Col	0.684	−0.730
E.co	0.952	−0.307
TPC	−0.605	−0.797
YM	0.882	−0.472
AMC	0.998	−0.060
Ps	0.964	0.266
TA	−0.831	−0.556
TH	0.971	−0.241
TDS	−0.621	−0.784
AC	−0.995	−0.096
TUB	0.183	0.983
pH	0.559	−0.829

Based on the previous interpretations (Correlations and Variable axis plot), the interpretations of the *PCA Bi-plot* (as shown in Figure 10.3) are given below:

☆ In Mavinkoppa tank, Total plate count is high, clearly indicates, it's non-suitability for drinking but may be used for domestic purposes, which is due to low counts of *E.coli, Pseudomonas aeroginosa* and total Coliforms.

☆ In Banadur Tank, the occurrence of *Pseudomonas aeroginosa* is high, which is influenced by high turbidity, clearly indicates that it is not suitable as recreational pool.

☆ In Murkutti tank, Aerobic microbial count, *E.coli*, Total Coliforms and (yeast and moulds) count is high, clearly indicates non-suitability for drinking and domestic purposes.

Acknowledgment

The Author would like to express her thanks to her students Mrs. Ashwini K. for helping in the laboratory analysis of the water samples.

References

Bartolomeo, A., D., Poletti, L., Sanchini, G., Sebastiani, B., Morozzi, G.: Relationship among parameters of lake polluted sediments by multivariate statistical analysis. *Chemosphere 55 (10), 2004, 1323-1329*.

BIS. Analysis of water and waste water. Bureau of Indian Standards, New Delhi. (1993).

Drinking water contaminants. 2009. http://water.epa.gov/drink/contaminants/index.cfm. Information accessed September 2013.

ICMR Manual of standards of quality for drinking water supplies. ICMR, New Delhi. (1975).

Jolliffe, I., T. Principal component analysis (2nd edn.). *Springer-Verlag, New York, 2002.*

Lindsay, I S. (2002). A tutorial on Principal Components Analysis. *http: www.ce.yildiz.edu.tr/personal/songul/file/./principal_components.pdf.* (Pdf accessed September 5, 2013)

Pearson, K. (1901). "On Lines and Planes of Closest Fit to Systems of Points in Space". *Philosophical Magazine* 2 (11): 559–572.

Petr, P. (2007). " Urban water quality evaluation using multivariate analysis". *Acta Montanistica Slovaca* 2 (12): 150-158.

Robert G Wetzel. 1983. Limnology. 2nd Edn. University of Michigen

Thakor, F.J., Bhoi, D.K., Dabhi, H.R., Pandya, S.N. and Chauhan, Nikitraj B. 2011. Current world environment. Vol. 6(2), 225-231

Trivedi, R.K. and P.K. Goel: Chemical and Biological methods for water pollution studies 209 pp. Enviromedia publications, Karad (1986).

WHO. 1996. Guidelines for drinking-water quality [electronic resource]: incorporating first addendum. Vol. 1, Recommendations. – 3rd ed. (Pdf accessed September, 2013).

Zeng, X., Rasmussen, T.,C.: Multivariate statistical characterization of water quality in lake Lanier, Georgia, USA. *J. Environ. Qual. 34 (6), 2005, 1980-1991.*

2015, Recent Trends in Microbiology, Mycology and Plant Pathology   *Pages* **143–149**
Editor: **Dr. H.C. Lakshman**
Published by: **DAYA PUBLISHING HOUSE, NEW DELHI**

Chapter 11

Short-Term Water Stress Induced Accumulation of Proline in *Triticum aestivum* L. Varieties Inoculated with AM Fungus *Rhizophagus fasciculatus*

V.S. Bheemareddy[1]*, S.B. Gadi[1] and H.C. Lakshman[2]

[1]*Department of Botany, J.S.S. Banashankari Arts, Commerce and S.K. Gubbi Science College, Dharwad – 580 004, Karnataka, India*
[2]*Post Graduate Department of Botany, Karnatak University, Dharwad – 580 003*
*E-mail: venupra1964@gmail.com

ABSTRACT

Arbuscular Mycorrhizal fungi are known to enhance the adoption ability of host plants under water stress conditions and help the host plants to cope up with situations of drought. Mycorrhizas were involved in protection against drought stress through improved nutritional status and osmotic adjustments. These fungi help the host plants to increase uptake of nutrients and tolerance to abiotic and biotic stresses. *Triticum aestivum* L. varieties DWR-162, DWR-195, DWR-225 and NI 5439 cultivated in North Karnataka of India, were selected for the study. *Rhizophagus fasciculatus* is the efficient arbuscular mycorrhizal fungus selected or the inoculation. Plants were grown in open field conditions with and without mycorrhizal inoculation under well watered and water stress conditions. Leaves were harvested after 60 and 90 days after sowing and are used for the determination of proline content. Inoculated plants accumulated more proline than control plants grown under well watered as well as water stress conditions.

Plants accumulated relatively more proline at 60 days after sowing stage compared to 90 days after sowing stage. DWR-225 and DWR-162 have shown more accumulation of proline to tolerate water stress by improving osmotic adjustment showing that plants belonging to these varieties are more tolerant to water stress

Keywords: *Arbuscular mycorrhizal fungi, Triticum aestivum L. varieties, Rhizophagus fasciculatus, Proline and water stress.*

Introduction

Drought is a worldwide problem seriously affecting the productivity of many crop plants. In nature plants are frequently exposed to adverse environmental conditions that have a negative effect on plant survival, development and productivity. Drought affects the quantity and quality of the grains produced (Shao *et al.*,2005). Water stress is considered to be the main environmental factor limiting growth and development of crop plants (Krammer and Boyer, 1997). Plants respond to water stress at morphological, anatomical, cellular and physiological levels with modifications that allow the plants to avoid stress and increase their tolerance (Bray, 1997). Accumulation of plant metabolites have been reported by the earlier workers. Barnet and Naylar (1966), reported that accumulation of proline was occured when the plants are subjected to water stress. Many plants accumulate proline as non toxic and protective osmolyte to maintain osmotic balance under water stress conditions. Zhu (2002), reported the accumulation of organic solutes such as carbohydrates and proline under stress conditions.

Contributions of AM (Arbuscular mycorrhizal) fungi to agriculture are well known. Mycorrhizas were involved in protection against drought stress through improved nutritional status and osmotic adjustments. AM fungi are known to enhance the adoption ability of host plants under water stress conditions and help the host plants to cope up with situations of drought. AM fungi can influence the host plants against water stress (Schellembaum *et al.*, 1998). AM fungi help the host plants to increase uptake of nutrients and tolerance to abiotic and biotic stresses (Ruiz-Lozano *et al.*, 1995). Mycorrhizal plants enhance the photosynthesis and assimilation of carbohydrates more than those in non mycorrhizal plants (Ghorbanli *et al.*, 2004). The accumulation of metabolic substances may suggest that AM colonization could improve osmotic adjustment originating not only from proline but also from carbohydrates and proteins resulting in the enhancement of water stress tolerance.

Rhizophagus fasciculatus is one of the common arbuscular mycorrhizal fungus present in the rhizospheric soil samples of agricultural fields of North Karnataka. The AMF *Rhizophagus fascicularis* is used for the inoculation. *Triticum aestivum* L. varieties commonly cultivated in North Karnataka of India, DWR-162, DWR-195, DWR-225 and NI 5439 were selected for the study.

The objective of this study was to evaluate the relative proline accumulation in mycorrhizal and non-mycorrhizal *Triticum aestivum* L. varieties in response to water stress applied at 60 and 90 DAS (Days After Sowing) stages. Proline quantification was studied under well watered as well as water stress conditions in mycorrhizal and non-mycorrhizal plants at 60 and 90 DAS stages.

Materials and Methods

Grains of four *Triticum aestivum* L. varieties commonly cultivated in Karnataka state of India DWR-162, DWR-195, DWR-225 and NI 5439 were procured from Wheat Research station, University of Agricultural Science, Dharwad, India. Grains were surface sterilized by placing them in 2 per cent sodium hypo chloride and were washed thoroughly with distill water to remove the traces of sodium hypo chloride.

Isolation of AM Fungal Spores, their Observation and Identification

Rhizospheric soil samples were collected from each wheat growing field for the recovery of AMF spores. AM fungal spores were collected by decanting and wet sieving method (Gerdamann and Nicolson, 1963).

Establishment of Monospecific Culture by using Funnel Technique

AM spores recovered by using wet sieving and decanting method, were collected by using needle and were stored in watch glass at 4°C. Spores were examined daily till the day of inoculation and spores with changed morphology were discarded. Before the inoculation water is added to the watch glass containing spores. The selected spores were surface sterilized using 200 ppm streptomycin Sulphate solution for 3-5 minutes and later washed with distil water and again surface sterilized with 2 per cent chloramines T solution for 3-5 minutes. After surface sterilization, the spores were serially washed with distil water for 4 to 5 times. Funnels used in this experiment were plugged with absorbent cotton and filled with 1:1 sand and soil mixture was added to the funnel. Sand and soil mixtures were added to the top of the funnel at the ratio of 1:1. Top of the funnel was then wrapped with aluminum foil and sterilized in the autoclave for 45 minutes at 120°C and 15 lbs pressure. This process of sterilization was repeated for three consecutive days. The surface sanitized spores were added to the funnel by making a small slit at the centre of aluminum foils. The spores were then placed 2-3 cm below the soil and covered with soil. Over this, 3-4 *Sorghum* seeds were sown. The funnels were then placed in conical flasks containing sterile distilled water. Soil in the funnel maintained wet by capillary action of water. The entire set up was placed in glass house for 45-50 days. After 45-50 days, the plants along with soil were transferred to 15 cm diameter pot containing sterile sand soil mix (3:1 W/W). About 40-50 sterilized *sorghum* seeds were planted and watered regularly. Hoagland solution was given as and when required. After 60 days, *Sorghum* roots were checked for the colonization. After confirming the colonization shoot portion is chopped off at the soil level, the roots were then chopped to fine pieces and mixed well with the soil. This mixture is taken as pure AMF inoculum.

Plants were grown in open field conditions under natural photoperiods with following treatments.

Treatment 1

Plants grown in polybags without mycorrhizal inoculation (AM) watered on every alternate day till harvest.

Treatment 2

Plants grown in polybags without mycorrhizal (AMF) inoculation watered on alternate day. It was just before 60th and 90th day plants were subjected to water stress by withholding water for 10 days. After each stress period the plants were rewatered till harvest.

Treatment 3

Plants grown in polybags with AMF *Rhizophagus fasciculatus* inoculation. Plants were watered regularly on alternate days till harvest.

Treatment 4

Plants grown in polybags with AMF *Rhizophagus fasciculatus* inoculation. Plants were watered regularly on alternate days. Just before 60th and 90th day they were subjected to water stress by withholding water for 10 days. After each stress period the plants were rewatered till harvest.

Hoagland's solution minus P was applied once in 15 days to all the plants grown in 4 treatments. Plants were maintained in triplicates and arranged in complete randomized design.

Harvest

Leaves harvested after 60 and 90 DAS. Leaves are used for the determination of proline content.

Estimation of Proline

0.5g of plant tissue from leaf was extracted by homogenizing in 10mL of 3 per cent aqueous Sulphosalicylic acid. The homogenate was filtered through Whatman's filter paper No. 2.

2 mL of acid Ninhydrin was added. Samples were placed in boiling water bath for 1 hour. Reaction was terminated by placing samples in ice bath. 4 mL of Toluene was added to the reaction mixture and stirred well for 20-30 minutes.

Warm the sample to room temperature to separate Toluene layer. Red colour intensity was measured at 520 nm with UV- visible spectrophotometer (Systronics double spectrophotometer 2203). Standard curve was prepared with series of pure proline taken at different concentrations. Amount of proline present in plant sample was measured by using standard curve.

Amount of proline in the samples was measured by using following equation.

$$\text{Moles of proline/g fresh tissue} = \frac{\text{g proline/mL} \times \text{mL of Toluene}}{115.5} \times \frac{5}{\text{g sample}}$$

Where 115.5 is the molecular weight of proline.

Results and Discussion

Experimental results revealed that there is more proline accumulation in plants subjected to water stress. It was observed that inoculated plants accumulated more proline than control plants grown under well watered as well as water stress conditions. Proline accumulation was increased considerably in Triticum *aestivum* L. varieties due to water stress and mycorrhizal inoculation (Figure 11.1). Mycorrhizal inoculation resulted in more accumulation of proline both in well watered and stress

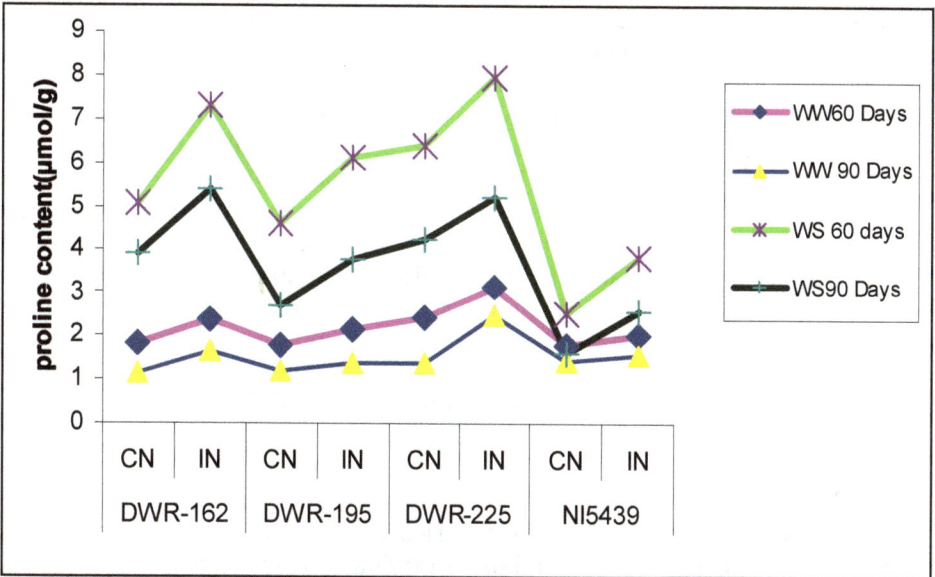

WW: Well watered; WS: Water stress; CN: Control; IN: Inoculated

Figure 11.1: Proline Content in *Triticum aestivum* L. Varieties at 60 and 90 DAS under Well Watered and Water Stress Conditions with and without AM Fungus (*Rhizophagus fasciculatus*) Inoculation.

induced plants. Water stress resulted in more proline accumulation. It was observed that plants accumulated relatively more proline at 60 DAS stage compared to 90 DAS.

Under water stress treatment, at 60 DAS stage plants belong to DWR 225 variety have shown highest proline content, moderate accumulation was observed in plants belong to DWR 162 variety and least was noted in DWR-195 and NI 5439 varieties.

It has been shown that mycorrhizal colonization and drought interact in modifying free amino acid and sugar pools in *Rosa* roots (Auge *et al.*,1992). Proline is a non protein amino acid formed in tissues under water stress conditions(Barnett and Naylor, 1966; Wu *et al.*, 2008). The proline together with sugar is readily metabolized upon recovery from the drought in white clover leaves. Proline serves as a sink for energy to regulate redox potentials, such as hydroxy radical scavenger, as a solute that protects macromolecules against denaturation (Shao *et al.*,2005). In plants, proline is an important organic compound that participates in the osmotic adjustment. It has been shown that plants produce more proline when they are subjected to more stress. Greater accumulation of amino acids might also indicate plants more capably osmotically adjusted to water stress (Auge, 2001).

Mathur and Vyas, (2000), have shown that mycorrhizal colonization and drought interact in modifying free amino acids and sugar pools and proline contents in *Ziziphus mauritiana* Lam. plants under water stress conditions.In present investigation similar results were observed confirming the accumulation of proline due to water stress and mycorrhizal inoculation. Accumulation of amino acid proline is one of the most frequently reported modifications induced by water stress in plants. As soil dries out the soil water potential becomes more negative and plants must decrease their water potentials to maintain water flow from soil in to the roots. To achieve such an effect, plants develop osmotic adjustment by active accumulation of organic ions or solutes as suggested by Morghan, (1984); Hoekstra *et al.* (2001). The increase showed in free amino acids is due high protein hydrolases. Sircelj *et al.* (2005), explained that under water stress the free amino acid proline is accumulated in apple trees to retain water through osmotic adjustment. Proline might confer drought stress tolerance to wheat plants by increasing anti oxidant system rather as an osmotic adjustment. It has also been shown that mycorrhizal colonization and drought stress interact in modifying free amino acids and sugar pools in roots (Auge, 2001). In dry or saline environments osmotic pressure increases to protect cellular constituents. Their protective effects also extend to temperature extremes and other stresses.

Conclusion

The present study showed that AM fungi led to enhancement of drought tolerance in experimental plants subjected to short term water stress at critical stages of plant growth followed by recovery. Increased accumulation of proline in presence of AM fungus *Rhizophagus fascicularis*(Thaxt.) was observed during water stress. This helps to improve osmotic adjustment of plants. Among four *Triticum aestivum* L.varieties, DWR-225 and DWR-162 have shown more accumulation of proline to tolerate water stress by improving osmotic adjustment. The results conclude that plants belonging to these two varieties are more drought tolerant.

References

Auge, R.M, Foster, J.G., Loescher, W.H. and Stodola, A.W.J. 1992. Symplastic molality of free amino–acids and sugars in *Rosa* roots with regard to VA mycorrhizae and drought. *Symbiosis.* 12: 1–17.

Auge, R.M. 2001. Water relation drought and vesicular arbuscular mycorrhizal Symbiosis. *Mycorrhiza.* 11: 3–42.

Barnett,N.M., Naylor,A.W. 1966. Amino acids and protein metabolism in Barmuda grass during water stress. *Plant Physiology.*41: 1222–1230.

Bray, A.E. 1997. Plants responses to water deficit. *Trends Plant Science.* 2: 48–54.

Gerdemann, J. W. and Nicolson, T. H. 1963. Spores of mycorrhizal *Endogone* species extracted from soil by wet sieving and decanting. *Trans Brit. Mycol. Soc.,* 46: 235.

Ghorbanli,M., Ebrahimzadeh,H. and Sharifi,M. 2004. Effects of NaCl and mycorrhizal fungi on antioxidative enzymes in soybean. *Biologia Plantarum.* 48: 575–581.

Hoekstra, F.A., Golovina, A. and Buitink, J. 2001. Mechanism of plant desiccation tolerance. *Trends in Plant Science.* 6: 431–438.

Kramer, P.J. and Boyer, J.S. 1997. Water relations of plants and soils. Academic Press, New York, pp. 278.

Mathur, N. and Vyas, A. 2000. Influence of Arbuscular mycorrhizae on biomass production, nutrient uptake and physiological changes in *Ziziphus mauritiana* Lam. Under water stress. *J Arid. Enviorn.* 45: 191–195.

Morghan, J.M. 1984. Osmoregulation and water stress in higher plants. *Annual Review of Plant Physiology.* 33: 299–319.

Ruiz–Lazano, J.M., Azcon, R. and Gomez, M. 1995. Effects of arbuscular mycorrhizal Glomus species on drought tolerance: Physiological and nutritional plant responses. *Appl. Enviorn Microbiol.* 61: 456–460.

Schellenbaum, L., Muller, J., Boller, T., Wiemken, A., Schuepp, H. 1998. Effects of drought on non–mycorrhizal and mycorrhizal maize: Changes in the pools of non–structural carbohydrates, in the activities of invertase and trehalase and in the pools of amino acids and imino acids. *New Phytol.* 138: 59–66.

Shao, H.B., Liang, Z. S., Shao, M. A. 2005. Changes of antioxidative enzymes and ABA content under soil water deficits among the ten wheat (*Triticum aestivum* L.) genotypes at maturation stage. Colloids and surfaces B: *Biointerfaces.* 45: 7–13.

Sircelj, H., Tausz, M., Grill, D. and Batic, F. 2005. Biochemical responses in leaves of two apple tree cultivars subject to progressing drought. *Journal of Plant Physiology.* 162: 1308–1318.

Wu, Q.S., Xia, R.X., Zou, Y.N. 2008. Improved soil structure and citrus growth after inoculation with three Arbuscular–mycorrhizal fungi under drought stress. *Euro Journal of Soil Biology.* 44: 122–128.

Zhu, J.K. 2002. Salt and drought stress signal transduction in plants. *Annual Review of Plant Biology.* 53: 247–273.

SECTION II
MYCOLOGY

2015, Recent Trends in Microbiology, Mycology and Plant Pathology   *Pages* 153–176
Editor: **Dr. H.C. Lakshman**
Published by: **DAYA PUBLISHING HOUSE, NEW DELHI**

Chapter 12

Antioxidants from Mushrooms may Antagonize Convulsions

Imtiyaz Ahmad Sheikh*, Keerti Dehariya,
Vinita Singh, Poonam Dehariya and Deepak Vyas

Laboratory of Microbial Technology and Plant Pathology,
Department of Botany, Dr. Hari Singh Gour University Sagar, M.P., India
*E-mail: simi.myco@gmail.com

ABSTRACT

Botanicals have a long history in their use for different human diseases. The medicinal properties so far reported in them included, antitumor, immunomodulatory, antiallergic, anticancer, antidiabetic, hepatoprotective, nephroprotective, anticonvulsant, antioxidant, etc. Mushrooms form a major group of botanicals used against various ailments. Antioxidant properties concerning mushrooms are due to the presence of huge quantities of bioactive principles such as phenolic compounds, flavonoids, saponins and triterpenoids. Reports regarding the anticonvulsant activities of mushrooms although scarce are mainly due to flavonoids and saponins. Various animal models have demonstrated antiepileptic activity of mushrooms namely *Armillarea mellea*, *Amanita muscaria, Ganoderma lucidum* etc. Epilepsy is the second most common neurological disorder characterized by unprovoked seizures due to an imbalance in the excitatory and inhibitory nerve impulses. Different types of seizures arise due to different etiological factors making it difficult to control with a single synthetic drug. Almost all the known synthetic antiepileptic drugs (AEDs) are associated with side effects; hence, patients with epilepsy are not wholly satisfied with AED treatment. One of the known mechanisms of the epileptogenesis is the generation of reactive oxygen and reactive nitrogen free radicals in brain. Hence, naturally occurring antioxidants also prove to be anticonvulsive in nature. On the other hand, various antioxidants and a few bioactive principles isolated from mushrooms have been reported as potent modulators of neurotransmitters of brain. Although mushrooms are less investigated in this regard but isolated bioactive compounds and antioxidants from mushrooms could be of immense value in developing alternative antiepileptic therapy.

Keywords: *Epilepsy, Antioxidants, Antiepileptic drugs, Mushrooms, Neurotransmitters, GABA, Glutamate.*

Introduction

Since the beginning of human civilization, ethnobotanicals such as plants and mushrooms have been valued for both culinary and medicinal properties (Wasser and Weis, 1999; Mahady *et al.,* 2001). At present, herbal therapies are being tried by the patients in developing as well as developed countries for controling diseases or adverse effects from synthetic drugs or for general health maintenance. Among the botanicals, various plant species and macrofungal species have been significantly investigated for their pharmacological properties. Some of the noteworthy among plants are *Rauwolfia serpentina, Withania somnifera, Aloe vera, Crocus sativus, Passiflora edulis, Artemisia abrotanum, Sutherlandia frutescens, Taxus wallichiana, Ipomoea stans, Magnolia grandiflora* and so on, whereas macrofungi also form an unbound list (Kumar *et al.,* 1997; Hawksworth, 1991).

Estimates of fungal diversity suggest that we have described less than 5 per cent (Hawksworth, 2001), and in some estimates less than 2 per cent (O'Brien *et al.,* 2005), of the fungi present on our planet. Macromycetes arranged in the phylum Basidiomycota and some of them in the Ascomycota are known as the higher fungi (Moradali *et al.,* 2007, Sicoli *et al.,* 2005). Mushrooms are a special group of macroscopic fungi. As described by Chang and Miles (1992), mushrooms are macrofungi with a distinctive fruiting body, epigeous or hypogeous, large enough to be seen with the naked eye and can be picked by hand'. Of the 1.5 million estimated fungi, 14,000 species produce fruiting bodies of sufficient size and suitable structure to be considered macro fungi, which can be called "mushrooms". Of these, about 7000 species are considered to possess varying degrees of edibility, and more than 3000 species are regarded as prime edible mushrooms. To date, only 200 of them are experimentally grown, 100 economically cultivated, approximately 60 commercially cultivated, about 10 have reached an industrial scale of production in many countries (Chang and Mshigeni, 2004). Fresh mushrooms can be acquired from grocery stores and markets, including straw mushrooms (*Volvariella volvacea*), oyster mushrooms (*Pleurotus Ostreatus*), shiitakes (*Lentinula edodes*) and enokitake (*Flammulina velutipes*). Other forms of mushroom include milk mushrooms, morels, chanterelles, truffles, black trumpets and porcini mushrooms (*Boletus edulis,* also known as "king boletes") (Ghorai *et al.,* 2009).

Mushrooms have been studied for nutritional and medical purposes through different scientific approaches. Novel and sophisticated cultivation technologies have been implemented to enhance the mushroom yield while various productivity factors are also taken into consideration. It has been demonstrated that seasonal and substrate variation, humidity, temperature, photoperiodism and other factors influence the productivity of mushroom nutraceuticals (Chaubey *et al.,* 2010; Dehariya *et al.,* 2011). Various potential anti-tumoral and immunomodulatory substances, mainly polysaccharides, have been isolated from mushrooms (Zhang *et al.,* 2007). In China, the dietary supplements and nutraceuticals made from mushroom extracts are used, along with various combinations of their herbal preparations (Barros *et al,.* 2008a; Carbonero *et al,.* 2008).

There are some prominent and bestselling edible and medicinal mushrooms such as: *Grifola frondosa, Coriolus versicolor, Lentinula edodes, Cordyceps sinensis, Schizophyllum commune, Hericium erinaceus, Pleurotus* species and *Ganoderma lucidum* (Chang and Mshigeni, 2004; Smith *et al.,* 2002, Dehariya, 2010). Among these cultivated mushrooms for food and medicine, *Ganoderma* is cultivated only for medicine. Recent scientific studies demonstrate that *Ganoderma* mushrooms possess anti-cancer, anti-allergy, anti-oxidation, anti-hypertension, cholesterol reduction, anti-ageing, anti-microbial activities and inhibiting platelet aggregation properties (Chang and Mshigeni, 2004; Zhang *et al.,* 2007). Extracts from *Ganoderma lucidum* and *Morchella esculenta* have been shown to possess the antimicrobial, antioxidant and antidiabetic activity (Wagey *et al.,* 2011). Triterpenes in *Ganoderma lucidum* suppress growth and invasive behaviour of cancer cells, whereas the polysaccharides stimulate the immune system resulting in the production of cytokines and activation of anti-cancer activities of immune cells (Sliva, 2006). Different antioxidant assays like hydroxyl radical, DPPH (1,1-Diphenyl-2-picrylhydrazyl) radical, superoxide radical, nitric oxide (NO) scavenging, ferrous ion chelating ability and β-carotene/linoleic acid assay have revealed that mushrooms such as *Pleurotus squarrosulus* extracts possess significant antioxidant activity. It was also confirmed that this antioxidant potential is due to the presence of higher phenolic, total flavonoid, β-carotene and lycopene content (Jaita and Krishnendu, 2010). Hence, the use of herbal remedies and dietary supplements including mushrooms is widespread throughout the world, and may find their potential use in convulsions (Epilepsy) as demonstrated scientifically using different animal models. This account highlights; primarily the understanding of epilepsy and antiepileptic drugs; secondarily the pharmacological properties of mushrooms, antioxidant activity in particular, role of antioxidants in epilepsy and, finally the potential of mushrooms towards the formation of antiepileptic medications on account of the antioxidants present in them.

Epilepsy

Epilepsy is the chronic disorder of the central nervous system manifested by recurrent unprovoked seizures. A seizure is a paroxysmal event due to changes in motor activity, autonomic function, consciousness or sensation that results from abnormal, excessive, hyper-synchronous discharges from an aggregate of central nervous system (CNS) neurons (Daniel, 2001). The tendency to have recurrent attacks is known as epilepsy but a single attack does not constitute epilepsy (Dhillon and Sander, 2003). It has been shown to affect several brain activities and promote long-term changes in multiple neural systems. This disorder, if untreated, can lead to impaired intellectual function or death and is typically accompanied by psychopathological consequences such as lose of self esteem. Epilepsy is the second most common neurological disorder after stroke. It shows a prevalence rate in 1-2 per cent of the world population (Angeli *et al.,* 2006). As an estimate, approximately 7 million people are affected in India and 50 million worldwide, 40 per cent of them are women. In developed countries, where drugs are easily available, epilepsy responds to treatment in up to 70 per cent of the patients. However, in developing countries, 75 per cent of people with epilepsy do not receive the treatment (Choi and Grind, 2006). There is an increased mortality of people with epilepsy; most studies have given overall standardized mortality ratios between two and three times higher than that of the general population (Dhillon and Sander, 2003).

Types of Epilepsy

The agreed clinical classification recognizes two categories, namely; partial seizures and generalized seizures, although there are some overlaps and many varieties of each. Partial seizures are those in which the discharge begins locally and often remains localized. These may produce relatively simple symptoms without loss of consciousness, such as involuntary muscle contractions, abnormal experiences or autonomic discharge, or they may cause more complex effects on behavior, often psychomotor epilepsy (Rang *et al.,* 2003). Partial seizures are subdivided further into simple partial and complex partial seizures.

Generalized seizures involve the whole brain, including the reticular system, this producing abnormal electrical activity throughout both hemispheres. Immediate loss of consciousness is characteristic of generalized seizures (Webster 1989, Sudarsky 1990, Bienvenu *et al.,* 2002; Rang *et al.,* 2003). The main categories are generalized tonic-clonic seizures (grand mal) and absence seizures (petit mal). These consist of initial strong contraction of the whole muscles, causing a rigid extensor spasm respiration stops and defaecation, micturition and salivation often occurs. This is followed by a series of violent, synchronous jerks, which gradually dies out in 2-4 minutes (Sudarsky, 1990).

Etiology: Causes of Convulsions

The accurate observations in the traditional medicine emphasize the concept that many causes of seizures and epilepsy result from a dynamic interplay among endogenous factors, epileptic factors and precipitating factors. The cause of convulsions must be clearly understood through some precise observations. The type of seizure depends on the site of the focus in the brain. Epileptic attack can be caused by biochemical insults to the brain, such as during hypoglycemia, anoxia, hypocalcaemia, hyperventilation, water intoxication and sudden withdrawal of certain drugs such as barbiturates or alcohol (Bienvenu *et al.,* 2002). Epilepsy can also be caused by previous active pathology, such as birth trauma to the brain, during or following meningitis, trauma to the skull and brain later in life, cerebral abscesses, cerebral infarction, cerebral haemorrhage or subarachnoid haemorrhage (Biller 1997, Bienvenu *et al.,* 2002).

The major amino acid neurotransmitters in the brain are GABA, an inhibitory transmitter and glutamic acid, an excitatory transmitter. Studies show that the blockade of postsynaptic gamma-amino butyric acid (GABA) receptors or an inhibition of

GABA synthesis is the principal origin of brain discharge (Delgado *et al.,* 1970, Muhizi 2002). While it is evident that a reduction in GABAergic activity is associated with seizures, it is true for glutamic acid also (Rang *et al.,* 2003). An epileptic attack can also be triggered by a sensory stimulus, which is specific for individuals. To date there is no single unifying explanation as to how these diverse factors cause seizures. Hence, it is difficult to determine the exact cause of seizures, even when it has been possible to investigate the physiological events which participate in the genesis of the epilepsy (Sudarsky, 1990).

Involvement of Free Radicals in the Seizure Mechanism

Free radicals, in addition to contributing to neuronal injury in cerebral ischemia and hemorrhage, may be involved in neuronal degeneration in schizophrenia, tardive dyskinesia, normal ageing, and Parkinson's and Alzheimer's diseases (Sonavane, *et al.,* 2002)

Concerning epileptic seizures, excitatory amino acid receptor activation by glutamate or N-methyl-D-aspartic acid (NMDA) has been known to accompany generation of reactive oxygen species (ROS) and reactive nitrogen species (RNS) (Raza, *et al.,* 2001; Vyawahare, *et al.,* 2007; Amin, *et al.,* 2008). Infact, free OH^- is detectable after pentylene tetrazole(PTZ) induced seizure and kindling (Rauca *et al.,* 1991). ROS and RNS are related in their metabolic pathway in that, $ONOO^-$ formed from NO and O_2^- is a potent oxidant that may exert injurious effects in the brain.

RNS, especially NO, is produced in several epilepsy animal models. NO is thought to function in the brain as a neuromodulator of cerebral blood flow and to play a role in learning and memory. However, the presence of excess NO may cause neuronal cell injury. NO may be associated with convulsive seizures inn that excess synthesis and release of NO occurs with the stimulation of NMDA receptor (Garthwaite *et al.,* 1991; Dawson *et al.,* 1991).

Head injuries or haemorrhagic cortical infarction result in extravasation of blood with breakdown of red blood cells and haemoglobin. Iron liberated from haemoglobin and haemoglobin itself is known to generate Reactive Oxygen Species (ROS) (O'Brien, *et al.,* 1973). Transient formation of ROS is found after the injection of iron salt into the rat cerebral cortex (Willmore and Hiramatsu, 1983). ROS, especially OH^-, are responsible for the induction of peroxidation of unsaturated fatty acids that are components of neuronal membranes. Such damage to neuronal memebranes may result in depolarization. On the other hand, ROS accelerate production of neurotoxic gaunidino compounds, endogenous compounds known to be convulsants in the brain (Hiramatsu *et al.,* 1984). Such reactions may be followed by excitatory and inhibitory neurotransmitter changes, especially increased release of aspartic acid and decreased release of gamma amino butyric acid (Janjua *et al.,* 1990). Such transmitter changes may be related directly to epileptogenicity. Considering possible known mechanisms, a broad list of synthetic antiepileptic agents have been developed from time to time antagonizing the provocation of convulsive attacks. They have proven best in the history of epilepsy management on a timely basis and research of novel therapeutic agents is underway till the most compatible and effective drug is approved as a drug of choice.

Choice of Antiepileptics in Management of Convulsive Disorders

Each system of medicine is the art and science of diagnosing the cause of disease, treating disease and maintaining health in the broadest sense of physical, spiritual, social and psychological well-being. Each culture has found solutions to the preventive, promotive and curative aspects of health that resonate in harmony with the worldview of that culture.

1857 was chosen as the start of the first edition history because it was, of course, the year of the first appearance of bromide therapy, so effective that within a few years it was used worldwide for the control of epilepsy. In the theatre of epilepsy history, the introduction of Phenytoin played the starring role but other developments were important supporting actors. Similarly, drugs have been developed to modulate GABA function; the inhibitors of GABA transaminases, which metabolize GABA, have been shown to be effective anticonvulsants. These are derivatives of valproic acid that do not only inhibit the metabolism of GABA, but also act as antagonist of GABA autoreceptor and thereby enhance the release of the neurotransmitter (Leonard, 2003). Drugs have also been developed to modulate glutamic acid function. Reduction of excitatory glutamatergic neurotransmission is potentially important; AMPA receptor blockade probably contributes to the antiepileptic effect. topiramate and NMDA receptor blockade contribute to the antiepileptic effect of drugs such as lamotrigine (Porter and Meldrum, 2001).

Although several antiepileptic drugs (AEDs) are available to treat epilepsy, the treatment of epilepsy is still far from adequate. The current therapy of epilepsy with modern antiepileptic drugs is associated with side effects (Table 12.1), dose related and chronic toxicity, teratogenic effects and approximately 30 per cent of the patients continue to have seizures with current antiepileptic drug therapy (Sonavane, *et al.,* 2002; Raza, *et al.,* 2001). In many cases even multi-drug therapy is not effective and neurosurgical procedures may be indispensable (Saberi *et al.,* 2008). Consequently a real need exists to develop new anticonvulsant compounds to cover seizures which are so far resistant to presently available drugs (Amin *et al.,* 2008).

Natural Products as Anticonvulsants

Natural products from folk remedies have contributed significantly in the discovery of modern drugs and can be an alternative source for development of drugs with novel structures and better safety and efficacy (Raza *et al.,* 2003). To provide a live evidence for any activity of a herbal component, it is essential to carry out the activity in living animal models. Consequently, a huge number of studies have been undertaken to evaluate different types of pharmacological properties of herbal extracts or the isolated compounds thereof. Some of the commonest activities which these herbal extracts exhibit are antitumor, hepatoprotective, immunomodulation, antidiabetic, antimicrobial, antioxidant, anti-inflammatory, antiallergic activities and so on. One of the rarest activities which are demonstrated for plant active compounds is anticonvulsant activity, although recently it has also gained momentum in many countries of the world. In such types of studies the main goal is to isolate a particular active substance which could antagonize the provocation of epileptic episodes. However, most of the studies only reveal the potential of plant extract to cover seizures

Table 12.1: Antiepileptic Drugs with Side Effects

Seizure Type	Drugs	Side Effects
Generalized Tonic-Clonic, Simple Partial and Complex Partial Seizures	Carbamazepine, phenobarbital, phenytoin, valproate, Lamotrigine	Idiosyncratic effects, diplopia, drowsiness, headache, nausea, orofacial dyskinesia, arrhythmias, drowsiness, ataxia, dizziness, weight gain
Absence Seizures	Ethosuximide and valproate	Lupus erythematosus and psychoses, nausea, vomiting, headache, lethargy, drowsiness
Tonic Seizures, Atonic Seizures and Atypical Absence Seizures	Phenobarbital, phenytoin, valproate, clonazepam, Gabapentin	Ataxia, nystagmus, drowsiness, gingival hyperplasia, hirsutism, diplopia, folate deficiency, orofacial dyskinesia, asterixis, fatigue, depression, poor memory, impotence, hypocalcaemia, osteomalacia, folate deficiency
Myoclonic Seizures	Valproate or clonazepam	High relapse rate, dyspepsia, hair loss, anorexia, drowsiness, nausea, vomiting
Infantile Myoclonic Epilepsy	Clonazepam	Severe brain damage, fatigue, fatigue, drowsiness, ataxia
Febrile Convulsions	Diazepam, Phenobarbital, Vigabatrin	Hepatotoxicity, neurological abnormalities, fatigue, depression, poor memory, impotence, hypocalcaemia, osteomalacia, folate deficiency
Status Epilepticus	Diazepam clonazepam, phenytoin, Topiramate	Dizziness, drowsiness, nervousness, fatigue, weight loss

and active principles remain to be isolated. Apart from the plant extracts assessed as a whole, a large number of bioactive principles have been isolated from the mushrooms which have shown direct impact on the reduction and delay of the epileptic episodes. These form the frequently known antioxidants from mushrooms, and through different mechanisms they were found to be at par when compared with the standard anticonvulsants. Many natural phenolic antioxidants have been found in plants including vegetables, teas and herbal medicines. Vitamin E (tocopherol), vitamin C and glutathione are the most well known radical scavenging antioxidants in animals. Pretreatment with a free radical scavenger or antioxidant, such as α-tocopherol, prevents the development of iron induced epileptiform activity in rats, decrease the formation of peroxides at the iron injection site, hastens the resolution of brain edema, and also prevents the development of cavitation and gliosis (Cui, *et al.,* 2005). Some of the naturally occurring antioxidants and bioactive principles have proved to be effective against seizures as detailed in Table 12.2.

Pharmacological Properties of Mushrooms

Medicinal mushrooms have become even more widely used as traditional medicinal ingredients for the treatment of various diseases and related health problems largely due to the increased ability to produce the mushrooms by artificial methods. As a result of large numbers of scientific studies on medicinal mushrooms especially in Japan, China and Korea, over the past three decades, many of the traditional uses have been confirmed and new applications developed (Table 12.3). Data given in the above table clearly indicates the potential of mushrooms in various pharmacological activities.

Wild Mushrooms as a Source of Antioxidants

Mushrooms have become attractive as functional foods and as a source of physiologically beneficial medicine (Cheung, 2008). The antioxidants found in mushrooms are mainly phenolic compounds (phenolic acids and flavonoids), followed by tocopherols, ascorbic acid and carotenoids. These molecules were quantified in terms of different species mainly from Finland, India, Korea, Poland, Portugal, Taiwan and Turkey (Table 12.3). The values are available in literature, but expressed in different basis (dry weight, fresh weight and extract). *Helvella crispa* from India revealed the highest content of phenolic compounds expressed per g of extract (34.65 mg/g), while *Sparassis crispa* from Korea revealed the highest value expressed in a dry weight basis (0.76 mg/g). *Auricularia fuscosuccinea* (white) from Taiwan (32.46 mg/g of extract), *Agaricus silvaticus* (3.23×10^{-3} mg/g of dry weight) and *Ramaria Botrytis* (2.50×10^{-4} mg/g of fresh weight) from Portugal, were the richest species in tocopherols. *Auricularia fuscosuccinea* (brown) from Taiwan (11.24 mg/g of extract), *Suillus collinitus* from Portugal (3.79 mg/g of dry weight) and *Agaricus bisporus* from Poland (0.22 mg/g of fresh weight) revealed the highest levels of ascorbic acid. *Lactarius deliciosus* from Portugal revealed the highest contents in β-carotene (0.09 mg/g of extract).

A few studies concerning the analysis of the phenolic components and other antioxidant active principles of wild mushrooms can be found in the literature

Table 12.2: Natural Antioxidants as Anticonvulsants

Bioactive principle	Source	Activity	Reference
Condensed tannins: epigallacatechin and epigallocatechin-3-O-gallate, procyanidine	Entire plant kingdom	Hypertension and to prevent stroke, scavenge O_2^- and OH^- radicals, prevented the occurrence of iron-induced epileptiform discharges.	Elmastas *et al.*, 2007; Ferreira, *et al.*, 2007
Adenosine	Peripheral and central nervous systems	ROS scavenging activities, suppressed and delayed occurrence of iron induced epileptiform discharges.	Turkoglu *et al.*, 2007
Fermented papaya	Yeast fermentation product of *Carica papaya*	Potent OH- scavenger and significantly inhibits thiobarbituric acid reactive substances formation in iron induced seizure focus of rats	Barros *et al.*, 2008
4-hydroxybenzyl alcohol and vanillyl alcohol	*Gastrodia elata*	Antioxidant at cellular and molecular level in the brain, anti-convulsive in ferric chloride-induced epileptic seizures	Murugkar *et al.*, 2005
Flavonoids:			
Rutin	Plants such as buckwheat, apples and black tea	Dose dependent anticonvulsant activity against pentylenetetrazole induced minimal clonic and generalized tonic clonic seizures in rats.	Kuntic *et al.*, 2007 Nassiri-Asl *et al.*, 2008
Apigenin	Flowers of *Matricaria chamomilla*	Reduced the latency in the onset of picrotoxin induced convulsions in rats	Avallone *et al.*, 2000
Goodyerin	*Goodyera schlechtendaliana*	Prolonged the latency of onset of seizure and reduced the duration of seizures and exhibited complete protection against induced convulsions in rats	X.M. Du, *et al.*, 2002
Wogonin	*Scuttellaria baicalensis*	Decreased the seizure response induced by PTZ in male mice, decreased the intensity of electrogenic seizures induced.	Park, *et al.*, 2007
Hispidulin	*Artemisia* and *Salvia* species	Markedly reduced the number of animals suffering from seizures through its interaction with benzodiazepine binding site	Kavvadias, *et al.*, 2004
Alkaloids:			
Sanjoinine A	*Zizyphi spinosi* semen	Significantly decreased seizure score, increased the latency of seizure onset against NMDA elicited convulsions in mice	Ma *et al.*, 2008

Contd...

Table 12.1–Contd...

Bioactive principle	Source	Activity	Reference
Nantenine	*Nandina domestica*	Reduced extensor/flexor ratio and mortality and showed an inhibition of 30, 60 and 90 per cent tonic phase occurrence against MES and PTZ induced seizures in mice	Ribeiro *et al.*, 2005
Piplartine	*Piper tuberculatum*	Decreased the latency to death against PTZ induced seizures in mice	Felipe *et al.*, 2007
Terpenes:			
Betulin	Marcgraviaceae Family	Antgonised the BCL induced myoclonic jerks due to its direct binding to the GABAA receptor GABA site	Muceniece *et al.*, 2008
Safranal	*Crocus sativus*	Exerted its anticonvulsant behaviour through GABAA-benzodiazepine receptor complex and little role of opoid receptors may also be involved	Hosseinzadeh *et al.*, 2007
Ursolic acid	*Nepeta sibthopii*	Increased the latency period and decreased the number of clonic-tonic PTZ induced convulsions	Taviano *et al.*, 2007
Others:			
Vanillyl alcohol	*Gastrodia elata*	Anticonvulsant effect of vanillyl alcohol resulted mainly from its free radical scavenging activities	Hsieh *et al.*, 2000
Barakol	*Cassia siamea*	Prolonged the latency of clonic convulsion induced by picrotoxin in mice	Sukma *et al.*, 2002
Thymoquinone	*Nigella sativa*	Reduced the duration of myoclonic seizures induced by PTZ administration in mice	Hosseinzadeh *et al.*, 2004

Table 12.3: Some of the Pharmacologically Active Mushrooms and their Medicinal Properties (Wasser and Weis, 1999a)

Activity	Mushroom Species													
	Aa	Ab	Abi	Am	Ff	Fv	Gf	Gl	He	Io	Le	Po	Sc	Tv
Antifungal	−	−	−	+	−	+	+	−	−	−	−	−	−	−
Antiinflammatory	−	−	−	−	−	+	−	+	−	−	+	−	+	−
Antitumour	+	+	+	−	+	+	+	+	+	+	+	+	+	+
Cardiovascular disorders	+	−	−	+	−	−	−	+	−	−	−	−	−	−
Hypercholesterolemia, hyperlipidemia	+	−	−	−	−	−	−	−	−	−	+	+	−	−
Antidiabetic	−	−	−	−	−	+	+	+	−	−	+	−	−	−
Immunomodulating	−	−	+	−	−	−	+	+	+	+	+	−	+	+
Kidney tonic	−	−	+	−	−	−	−	+	−	−	+	−	+	−
Nerve tonic	−	−	−	−	−	−	−	+	+	−	−	+	−	−
Hepatoprotective	−	−	−	−	−	−	+	+	−	+	+	−	+	+
Chronic bronchitis	+	−	−	−	−	−	+	+	+	−	−	−	−	−
Sexual potentiator	−	−	−	−	−	−	−	+	−	−	+	−	−	−
Antiviral (*e.g.* anti-HIV)	−	−	−	−	−	+	+	+	−	−	+	−	−	+
Antibacterial and Antiparasitic	−	−	−	−	+	−	+	+	−	−	+	−	−	+
Blood pressure regulation	+	−	−	+	−	−	+	+	−	−	+	−	−	−

+: Non-commercially developed mushroom product; −: Unknown data.

Aa: *Auricularia auricula-judas* (Bull.) Wettst.; Ab: *Agaricus blazei* Murr.; Abi: *Agaricus bisporus* (J.Lge) Imbach; Am: *Armillarea mellea* (Vahl.:Fr.) P. Karst.; Ff: *Fomes fomentarius* (L.:Fr.) Fr.; Fv: *Flammulina velutipes* (Curt.:Fr.)P. Karst; Gf: *Grifola frondosa* (dicks.:Fr) S.F. Gray; Gl: *Ganoderma lucidum* (Curt.: Fr.) P. Karst; He: *Hericium erinaceus* (bull.:Fr.) Pers.; Io: *Inonotus obliquus* (Pers.:Fr.) Bond et Sing.; Le: *Lentinus edodes* (Berk.) Sing.; Po: *Pleurotus ostreatus* (Jacq.:Fr.) Kumm.; Sc: *Schizophyllum commune* Fr.; Tv: *Trametes versicolor* (L.:Fr.) Lloyd.

(Puttaraju *et al.,* 2006, Ribeiro *et al.,* 2006, Mattila *et al.,* 2001). Their individual profiles were obtained by high-performance liquid chromatography coupled to photodiode array detector (HPLC-DAD) (Ribeiro *et al.,* 2006, Kim *et al.,* 2008, Valentao *et al.,* 2005, Barros *et al.,* 2008, Ribeiro *et al.,* 2007), or to an ultraviolet detector (Jayakumar *et al.,* 2008), or by gas chromatography-mass spectrometry selected ion monitoring (GC-MS SIM) (Mattila *et al.,* 2001). Studies also reveal the antioxidant activities correlated to secondary metabolites using more sophisticated antioxidant assays.

Robaszkiewicz *et al.* (2010), observed a strong correlation (5 > 0.9) between the concentration of total phenolics and reducing power/scavenging effects in both aqueous and methanolic extracts, while this correlation was moderate for flavonoids.

Gao *et al.* (2012) studied the antioxidant activities of hot-water extract from cultured mycelia of *Armillaria mellea* assessed in different *in vitro* systems. The extract showed strong inhibitory effect on microsomal liposome oxidation with an inhibition rate over 85 per cent at 1.0–2.0 mg/mL, scavenged superoxide radical in a concentration dependent fashion with IC_{50} value of 1.03 mg/mL, scavenged more than 77.2 per cent of 1,1-Diphenyl-2-picrylhydracyl (DPPH) radical at the concentrations of 4–8 mg/mL. The extract at 1.25–5.0 mg/mL also protected supercoiled DNA strand in plasmid pBR322 against scission induced by Fenton-mediated hydroxyl radical.

On the other hand, there are reports isolating a number of bioactive principles from mushrooms which could potentially antagonize convulsions on account of their free radical scavenging activity. Table 12.4 presents a few studies of active compounds isolated from several mushroom species.

Apart from antifungal, antibacterial, antitumour, antiallergic, antidiabetic and hepatoprotective activities, mushrooms have been evaluated for their antioxidant and anticonvulsant activities also, though a deficit seems to be there in the literature regarding anticonvulsant activity. Mushrooms primarily contain a large number of secondary metabolites which could be potential anticonvulsants; secondarily their free radical scavenging activity could be of immense value to combat convulsions. A considerable number of studies reveal that natural antioxidants through various mechanisms antagonize convulsions due to their bioactive components like flavonoids, saponins, triterpenoids, and steroids and so on. These bioactive principles may either inhibit the reactive substances formed during iron induced seizure models or may directly bind to $GABA_A$ receptor GABA site thereby inhibiting the formation of epileptic discharges. Thus there is a possibility that mushrooms serve as potential anticonvulsants.

An account is given here in support of the anticonvulsant potential of some of the mushrooms:

1. The mushroom, *Armillarea mellea* has been shown to have anticonvulsant properties. The fermentation extracts of *A. mellea* raised the seizure threshold in PTZ-induced seizures in mice (New Drug Group, 1977). *A. mellea* polysaccharides also have therapeutic effect on vertigo induced by machinery rotation (Yu *et al.,* 2006). N6-(5-hydroxy-2-pyridyl)-methyl-adenosine (AMG-1) from the mycelia of *A. mellea*, is 1000 times stronger

Table 12.4: Antioxidant Principles Detected in Wild Mushrooms

Bioactive Principle	Mushroom Species	Reference
Benzoic acid	*Agaricus blazei, Sparassis crispa, Phellinus linteus*	Kim *et al.* (2008)
p-Hydroxybenzoic acid	*Agaricus bisporus* (white), *Agaricus bisporus* (brown), *Lentinus edodes, Amanita rubescens, Russula cyanoxantha, Tricholoma equestre, Amanita rubescens, Suillus granulates, Sparassis crispa, Phellinus linteus, Ionotus obliquus*	Mattila *et al.* (2001)
Gallic acid	*Termitomyces heimii, Termitomyces mummiformis, Lactarius deliciosus, Pleurotus sajor-caju, Hydnum repandum, Lentinus squarrulosus, Sparassis crispa, Morchella conica, Russula brevepis, Geastrum arinarius, Cantharellus cibarius, Lactarius sangifluus, Macrolepiota procera, Cantherallus clavatus, Auricularia polytricha, Pleurotus djamor, Lentinus edodes, Morchella conica, Termitomyces microcarpus, Helvella crispa, Termitomyces shimperi*	Puttaraju *et al.* (2006)
Homogentisic acid	*Pleurotus ostreatus, Flammulina velutipes, Ionotus obliquus, Termitomyces heimii, Pleurotus sajorcaju, Hydnum repandum, Lentinus edodes, Morchella conica, Russula brevepis, Lactarius sangifluus, Macrolepiota procera, Cantherallus clavatus, Auricularia polytricha, Pleurotus djamor*	Kim *et al.* (2008)
Vanillic acid	*Flammulina velutipes, Sparassis crispa, Phellinus linteus, Ganoderma lucidum*	Puttaraju *et al.* (2006)
5-Sulfosalicylic acid	*Ionotus obliquus*	Ribiero *et al.* (2007)
Vanillin	*Cantharellus cibarius*	Kim *et al.* (2008)
p-Coumaric acid	*Termitomyces heimii, Boletus edulis, Sparassis crispa, Geastrum arinarius, Cantharellus cibarius, Lactarius* sp., *Macrolepiota procera, Pleurotus djamor, Lentinus sajor caju*	Valentao *et al.* (2005)
o-Coumaric acid	*Agaricus arvensis, Agaricus silvicola, Lepista nuda Fistulina hepatica*	Puttaraju *et al.* (2006)
Caffeic acid	*Sparassis crispa, Cantharellus cibarius, Ionotus, Flammulina velutipes, Agaricus blazei, Sparassis crispa, Ganoderma lucidum, Lonotus obliquus*	Kim *et al.* (2008)
Rutin	*Cantharellus cibarius, Pleurotus ostreatus*	Ribiero *et al.* (2007)
Kaempferol	*Sparassis crispa, Ganoderma lucidum, Ionotus*	Kim *et al.* (2008)
Tannic acid	*Termitomyces heimii, Termitomyces mummiformis, Boletus edulis, Lactarius deliciosus, Pleurotus sajor-caju, Hydnum repandum, Lentinus squarrulosus, Morchella conica, Russula brevepis, Geastrum arinarius, Cantharellus cibarius, Lactarius sangifluus, Macrolepiota procera, Cantherallus clavatus, Auricularia polytricha, Pleurotus djamor, Termitomyces tylerance, Morchella anguiticeps, Termitomyces Microcarpus, Helvella crispa, Termitomyces shimperi*	Valentao *et al.* (2005), Jayakumar *et al.* (2008), Kim *et al.* (2008), Puttaraju *et al.* (2006)

Contd...

Table 12.4–Contd...

Bioactive Principle	Mushroom Species	Reference
α-tocopherol	*Auricularia mesenterica, Auricularia polytricha, Auricularia fuscosuccinea* (brown), *Auricularia fuscosuccinea* (white), *Tremella fuciformis Grifola frondosa, Morchella esculenta, Termitomyces albuminosus.Agaricus bisporus, Polyporus squamosus, Pleurotus ostreatus, Lepista nuda, Russula delica, Boletus badius, Verpa conica, Hypsizigus marmoreus*	Mau *et al.* (2001) Elmastas *et al.* (2007)
	Agaricus blazei, Agrocybe cylindracea, Boletus edulis	Lee *et al.* (2007)
	Agaricus arvensis, Agaricus bisporus, Agaricus silvaticus, Agaricus silvicola	Tsai *et al.* (2007)
β-tocopherol	*Boletus edulis, Calocybe gambosa, Cantharelus cibarius, Craterellus cornucopioides, Marasmius oreades, Hypholoma fasciculare, Lepista nuda, Lycoperdon molle, Lycoperdon perlatum, Pleurotus ostreatus, Agaricus arvensis, Agaricus bisporus, Agaricus romagnesii*	Barros *et al.* (2008)
δ-tocopherol	*Agaricus silvaticus, Agaricus silvicola, Boletus edulis, Calocybe gambosa, Cantharelus cibarius, Craterellus cornucopioides, Marasmius oreades, Hypholoma fasciculare, Lepista nuda, Ramaria botrytis, Tricholoma acerbum*	Barros *et al.* (2008)
γ-tocopherol	*Auricularia polytricha, Auricularia fuscosuccinea* (brown), *Tremella fuciformis, Hericium erinaceus, Grifola frondosa, Morchella esculenta, Termitomyces albuminosus, Hypsizigus marmoreus, Agaricus blazei, Agrocybe cylindracea, Boletus edulis , Hypsizigus marmoreus*	Mau *et al.* (2001) Mau *et al.* (2004)
	Ganoderma lucidum, Ganoderma tsugae, Grifola frondosa, Trichloma giganteum, Lentinula edodes, Pleurotus cystidiosus, Pleurotus ostreatus, Grifola frondosa, Morchella esculenta, Termitomyces albuminosus Agaricus blazei, Agrocybe cylindracea, Boletus edulis	Lee *et al.* (2007) Tsai *et al.* (2007) Lee *et al.* (2008) Yang *et al.* (2003) Tsai *et al.* (2007)

than adenosine in cerebral protecting activity (Watanabe *et al.*, 1990). It has been shown that AMG-1 acts on the presynapse (may be the A1 receptor) to attenuate the release of neurotransmitters. This compound abolished the neurogenic twitch responses induced by electrical field-stimulation, while the responsiveness of rat vas deferens to exogenous acetylcholine was decreased, showing both presynapse and post-synapse depression (Xiong and Huang, 1998).

2. Muscimol, isolated from *Amanita muscaria,* is a potent agonist at GABA-inhibitory synapses in mammalian brain. Given systemically at 7 µmol per kilogram, it blocks topical penicillin seizures and delays the onset of generalized metrazol convulsions in rats. When applied topically to cortex, muscimol blocks focal penicillin, bicuculline, and picrotoxin discharges in a dose-response relationship. It has no effect against topical strychnine. Muscimol offers a potential new approach to the treatment of epilepsy (Robert and Collins, 1980).

3. Similarly, extracts from *Ganoderma lucidum, Hericium erinaceus* and *Cordyceps sinensis* have been demonstrated to have central depressant and anticonvulsant activities (Chu *et al.*, 2007; Chen, 2008; Hazekawa *et al.*, 2010; Liu *et al.*, 2010). Studies revealing the anticonvulsant potential of mushrooms are far from adequate but the significant presence of antioxidants and other bioactive compounds responsible for anticonvulsant activity suggest that mushrooms have anticonvulsant potential.

Concluding Remarks

The Research and Development thrust in the pharmaceutical sector is focused on development of new innovations/indigenous plant based drugs through investigation of leads from the traditional system of medicine. Herbal medicine has gained the momentum and it is evident from the fact that certain herbal remedies peaked at par with synthetic drugs. Despite the widespread use of mushrooms as functional food and medicine by patients with different diseases, mushrooms potentially yield new treatment options for patients whose seizures are uncontrolled despite available antiepileptic drugs (AEDs), and may also represent inexpensive, culturally acceptable treatments for the millions of people around the world with untreated epilepsy. However, there is a striking lack of controlled evidence to support their use in epilepsy, which could be ascertained by future investigations on mushrooms. Due to easy cultivability, better cultural acceptability, better compatibility with human body, wide biological activities, higher safety margin and lesser costs than the synthetic drugs, there is great demand of mushroom medicines in the developed as well as developing countries. Medicinal mushrooms and modern medicine techniques must be coupled in order to bring out high quality mushroom products with rapid onset of action and good bioavailability. The possible mechanism of actions shown in this review can be exploited further for the identification of particular fraction and or active constituent which can provide more extensive results. The biotechnological approach is a useful alternative for the production of bioactive components of mushrooms. With strain improvement and optimization of culture conditions, the production of secondary metabolites by culturing could be greatly enhanced. The review explored various active principles as antioxidants, responsible for anticonvulsant activity too. These reports could be a better target for the development of alternatives to synthetic antiepileptic drugs. On the other hand, compounds that originated from mushrooms are offering interesting opportunities for the evaluation of their novel bioactivities in anticonvulsant, immunostimulating properties, among others. It is necessary that the relationship of structure-activity and mechanism of such biological action be further investigated.

Acknowledgements

IA Sheikh was supported by Central Research Fellowship (PLAN GRANT XII) of University Grants Commission, New Delhi, India. The authors are thankful to the Head Department of Botany, Dr. H. S. Gour Central University Sagar, and the anonymous reviewers for critical comments and suggestions during preparation of the review. Authors are very grateful to Dr. Ashok Shukla (Dr. DS Kothari PDF Fellow) for careful reading of the first draft of this review.

References

Acharya K, Yonzone P, Rai M and Rupa A. 2005. Antioxidant and nitric oxide synthase activation properties of *Ganoderma applanatum*. *Indian J. Exp. Biol.* 43: 926–929.

Amin KM, Rahman DEA and Al–Eryani YA. 2008. Synthesis and preliminary evaluation of some substituted coumarins as anticonvulsant agents. *Bioorg Med Chem.* 16: 5377–5388.

Atigari DV, Gundamaraju R, Sabbithi S, Chaitanya K and Ramesh D. 2012. Evaluation of Antiepileptic Activity of Methanolic Extract of *Celastrus paniculatus* Willd Whole Plant in Rodents. *Int.J.Pharm.Phytopharmacol.Res.*, 2(1): 20–25.

Avallone R, Zanoli P, Puia G, Kleinschnitz M, Schreier P and Baraldi M. 2000. Pharmacological profile of apigenin, a flavonoid isolated from *Matricaria chamomilla*. *Biochem Pharmacol.* 59: 1387–1394.

Azikiwe CCA, Siminialayi IM, Brambaifa N, Amazu LU, Enye JC and Ezeani MC. 2012. Anticonvulsant activity of the fractionated extract of *Crinum jagus* bulbs in experimental animals. *Asian Pacific Journal of Tropical Disease*, S446–S452.

Barros L, Baptista P, and Ferreira. 2007. ICFR Effect of *Lactarius piperatus* fruiting body maturity stage on antioxidant activity measured by several biochemical assays. *Food Chem. Toxicol.*, 45: 1731–1737.

Barros L, Correia DM, Ferreira ICFR, Baptista P and Santos–Buelga C. 2008. Optimization of the determination of tocopherols in *Agaricus* sp. edible mushrooms by a normal phase liquid chromatographic method. *Food Chem.*, 110: 1046–1050.

Barros L, Falcão S, Baptista P, Freire C, Vilas–Boas M and Ferreira ICFR. 2008. Antioxidant activity of *Agaricus* sp. mushrooms by chemical, biochemical and electrochemical assays. *Food Chem.*, 111: 61–66.

Bienvenu E, Amabeoku GJ, Eagles PK, Scott G and Springfield EP. 2002. Anticonvulsant activity of aqueous extract of *Leonotis leonurus*. *Phytomedicine*, 9: 217–223.

Biller, J. 1997. *Practical neurology*. Lippincott–Raven, Philadelphia, pp. 55–436.

Carbonero ER, Gorin PAJ, Iacomini M, Sassaki GL, and Smiderle FR. 2008. Characterization of a heterogalactan: Some nutritional values of the edible mushroom Flammulina velutipes. *Food Chemistry*, 108: 329–333.

Chang R. 1996. Functional properties of edible mushrooms. *Nutr. Rev.,* 54: S91–93.

Chang S. 1999. Global impact of edible and medicinal mushrooms on human welfare in the 21st century: Nongreen revolution. *Int. J. Med. Mushrooms,* 1: 1–7.

Chang ST and Mshigeni KE. 2004. Mushroom and their human health: their growing significance as potent dietary supplements. The University of Namibia, Windhoek, Namibia. pp. 1–79.].

Chaubey A, Dehariya P And Vyas D. 2010. Seasonal Productivity and Morphological Variation in *Pleurotus djamor. Indian J.Sci.Res.* 1(1): 47–50.

Chen QX, Miao JK, Li C, Li XW, Wu XM and Zhang XP. 2013. Anticonvulsant Activity of Acute and Chronic Treatment with a–Asarone from *Acorus gramineus* in Seizure Models. *Biol Pharm Bull.* 36(1): 23–30.

Cheung LM and Cheung PCK. 200.5 Mushroom extracts with antioxidant activity against lipid peroxidation. *Food Chem.* 89: 403–409.

Cheung LM, Cheung PCK and Ooi VEC. 2003. Antioxidant activity and total phenolics of edible mushroom extracts. *Food Chem.,* 81: 249–255.

Cheung PCK 2008 *Mushrooms as Functional Foods*. A John Wiley and Sons Publications, *Ed.* 2008.

Choi Y, Lee SM, Chun J, Lee HB and Lee J. 2006. Influence of heat treatment on the antioxidant activities and polyphenolic compounds of Shiitake (*Lentinus edodes*) mushroom. *Food Chem.,* 99: 381–387.

Choi YH, Yan GH, Chal O, Choi YH, Zhang X, Lim JM, Kim JH, Lee MS, Han EH and Kim HT. 2006. Inhibitory effects of *Agaricus blazei* on mast cell–mediated anaphylaxis like reactions. *Biol Pharm Bull.* 29: 1366–71.

Chu Q, Wang LE, Cui XY, Fu HZ, Lin ZB, Lin SQ, Zhang YH. 2007. Extract of Ganoderma lucidum potentiates pentobarbital–induced sleep via a GABAergic mechanism. *Pharmacology, Biochemistry and Behavior* 86 693–698.

Cui Y, Kim DS and Park KC. 2005. Antioxidant effect of *Inonotus obliquus. J. Ethnopharmacol.,* 96: 79–85.

Daniel HL. 2001. Diseases of the central nervous system. In: Braunwald E, Fauci AS, Kasper DL, Hauser SL, Longo DL and Jameson JL ed. *Harrison's Principles of Internal Medicine*. McGraw–Hill, New Delhi; 2354–2369.

Dawson VL, Dawson TM, London ED, Bredt DS and Snyder SH. 1991. Nitric oxide mediates glutamate neurotoxicity in primary cortical cell cultures. *Proc. Natl. Acad. Sci., USA,* 88: 6368–6371.

Dehariya P, Chaubey A and Vyas D. 2011. Effect of proteinaceous substrate supplementation on yield of Pleurotus sajor–caju. *Indian Phytopathology*, 64(3): 291–295.

Delgado, JN and Isaacson, EI. 1970. Anticonvulsants. In Burger, A. (ed.), Medicinal Chemistry, third edition. John Wiley and sons, Chichester, pp.1386–1400.

Dhanabal SP, Paramakrishnan N, Manimaran S and Suresh B. 2007. Anticonvulsant potential of essential oil of *Artemisia abrotanum. Curr Trends Biotech Pharm.* 1(1): 112–116.

Dhillon S and Sander JW. 2003. Epilepsy. In: R. Walker, C. Edwards ed. Clinical Pharmacy and therapeutics. Churchill Livingstone, Scotland; 465–481.

Dore CMPG, Azevedo TCG, de Souza MCR, Rego LA, de Dantas JCM, Silva FRF, Rocha HAO, Basela IG and Leite EL. 2007. Antiinflammatory, antioxidant and cytotoxic actions of beta–glucan–rich extract from *Geastrum saecatum* mushroom. *Int. Immunopharmacol.,* 7: 1160–1169.

Du XM, Sun NY, Takizawa N, Guo YT and Shoyama Y. 2002. Sedative and anticonvulsant activities of goodyerin, a flavonol glycoside from *Goodyera schlechtendaliana. Phytotherapy Res.* 16: 261–263.

Elmastas M, Isildak O, Turkekul I, Temur N. 2007. Determination of antioxidant activity and antioxidant compounds in wild edible mushrooms. *J. Food Comp. Anal.,* 20: 337–345.

Felipe FCB, Filho JTS, Souza LEO, Silveira JA, Uchoa DEA, Silveira ER, Pessoa ODL, Muceniece GSB, Saleniece RK, Rumaks J, Krigere L, Dzirkale Z, Mezhapuke R, Zharkova O and Klusa V. 2008. Betulin binds to Z–aminobutyric acid receptors and exerts anticonvulsant action in mice. *Pharmacol Biochem Beh.* 90: 712–716

Fernández SP, Wasowski C, Loscalzo LM, Granger RE, Johnston GAR, Paladini AC and Marder M. 2006. Central nervous system depressant action of flavonoid glycosides. *Eur J Pharmacol.* 539: 168–176.

Ferreira ICFR, Baptista P, Vilas–Boas M, Barros L. 2007. Free radical scavenging capacity and reducing power of wild edible mushrooms from northeast Portugal: Individual cap and stipe activity. *Food Chem.,* 100: 1511–1516.

Filipa SR, Pereira E, Barros L, Sousa MJ, Martins A and ICFR Ferreira. 2011. Biomolecule Profiles in Inedible Wild Mushrooms with Antioxidant Value. *Molecules 16*(6), 4328–4338.

Gao LW and Wang JW. 2012. Antioxidant Potential And DNA Damage Protecting Activity of Aqueous Extract from *Armillaria Mellea. Journal of Food Biochemistry,* 36: 139–148.

Garthwaite J. 1991. Glutamate, nitric oxide and cell–cell signaling in the nervous system. *Trends Neurosci* 14: 60–67.

Ghorai S, Banik S P, Verma D, Chowdhury S, Mukherjee S and Khowala S. 2009. Fungal biotechnology in food and feed processing. *Food Research International.* doi: 10.1016/j.foodres.2009.02.019.

Grind B, Hetland G and Johnson E. 2006. Effects on gene expression and viral load of medicinal extract from *Agaricus blazei* in patients with chronic hepatitis C infection. *Int Immunopharmacol.* 6: 1311–4.

Hawksworth DL. 1991. The fungal dimension of biodiversity: magnitude, significance, and conservation. *Mycol Res.*, 95: 641–655.

Hawksworth DL. 2001. The magnitude of fungal diversity: the 1.5 million species estimate revisited. *Mycological Research* 105: 1422–1432.

Hazekawa M, Kataoka A, Hayakuwa K, Uchimasu T, Furuta R, Irie K, Akitake Y, Yoshida M, Fujioka T, Egashira N, Oishi R, Kenji M, Mishima K, Uchida T, Iwasaki K and Fujiwara M. 2010. Neuroprotective effect of repeated treatment with *Hericium erinaceum* in mice subjected to middle cerebral artery occlusion. *Journal of Health Science*, 56(3): 296–393.

Hiramatsu M, Mori A and Kohno M. 1984. Formation of peroxyl radicals after $FeCl_3$ injection into rat isocortex. *Neurosciences,* 10: 281–284.

Hosseinzadeh H and Parvaedeh S. 2004. Anticonvulsant effects of thymoquinone, the major constituent of *Nigella sativa* seeds, in mice. *Phytomedicine.* 11: 56–64.

Hosseinzadeh H and Sadeghina HR. 2007. Protective effect of safranal on pentylenetetrazol–induced seizures in the rat: Involvement of GABAergic and opoid systems. *Phytomedicine.* 14: 256–262.

Hosseinzadeh H and Talebzadeh F. 2005. Anticonvulsant evaluation of safranal and crocin from *Crocus sativus* in mice. *Fitoterapia.* 76: 722–724.

Hsieh CL, Chang CH, Chiang SY, Li TC, Tang NY, Pon CZ, Hsieh CT and Lin JG. 2000. Anticonvulsive and free radical scavenging activities of vanillyl alcohol in ferric chloride–induced epileptic seizures in Sprague–Dawley rats. *Life Sci.* 67: 1185–1195.

Hu SH, Liang ZC, Chia YC, Lien JL, Chen KS, Lee MY and Wang JC. 2006. Antihyperlipidemic and antioxidant effects of extracts from *Pleurotus citrinopileatus. J. Agric. Food Chem.*, 54: 2103–2110.

Jaita P, Sourav G, Saifa TK and Krishnendu A,. 2010 *In vitro* free radical scavenging activity of wild edible mushroom, *Pleurotus squarrosulus. Indian Journal of Experimental Biology*, 48(12): 1210–1218.

Janjua NA, Mori A and Hiramatsu M. 1990. Increased aspartic acid release from the iron–induced epileptogenic focus. *Epilepsy Res,* 102: 24–27.

Jaworska G, Bernas E, Cichon Z and Possinger P. 2008. Establishing the optimal period of storage for frozen *Agaricus bisporus*, depending on the preliminary processing applied. *Int. J. Refrigeration,* 31: 1042–1050.

Jayakumar T, Thomas PA and Geraldine P. 2008. In–vitro antioxidant activities of an ethanolic extract of the oyster mushroom, *Pleurotus ostreatus. Innovat. Food Sci. Emerg. Technol., In Press, Corrected Proof.*

Jesberger JA and Richardson JS. 1991. Oxygen free radicals and brain dysfunction. *Int J Neuro Sci.*, 57; 1–17.

Kavyadias DP, Sand KA, Youdim MZ, Qaiser C, Rice–Evans R, Baur E, Sigel WD, Rausch P, Ma Y, Yun SR, Nam SY, Kim YB, Hono JT, Kim Y, Cuoi H, Lee K and Oh KW. 2008. Protective effects of Sanjoinine A against N–methyl– D–aspartate–induced seizure. *Biol Pharm Bull.* 31(9): 1749–1754.

Kim MY, Seguin P, Ahn JK, Kim JJ, Chun SC, Kim EH, Seo SH, Kang EY, Park SL, Ro HM and Chung IM. 2008. Phenolic compound concentration and antioxidant activities of edible and medicinal mushrooms from Korea. *J. Agric. Food Chem.,* 56: 7265–7270.

Kitzberger CSG, Smania A, Pedrosa RC and Ferreira SRS. 2007. Antioxidant and antimicrobial activities of shiitake (*Lentinula edodes*) extracts obtained by organic solvents and supercritical fluids. *J. Food Eng.,* 80: 631–638.

Kumar S, Shukla YN, Lavania UC, Sharma A and Singh AK. 1997. Medicinal and Aromatic Plants: Prospects for India. *J. Med. Arom. Pl. Sc.* 19 (2): 361–365.

Kuntic V, Pejic N, Ivkovic Z, Ilic K, Micic S and Vukojevic V. 2007. Isocratic RP–HPLC method for rutin determination in solid oral dosage forms. *J Pharmaceutical Biomed Anal.* 43: 718–721.

Lee BC, Baea JT, Pyoa HB, Choeb TB, Kimc SW, Hwangc HJ and Yun JW. 2003. Biological activities of the polysaccharides produced from submerged culture of the edible Basidiomycete *Grifola frondosa. Enzyme Microb. Technol.,* 32: 574–581.

Lee IK, Kim YS, Jang YW, Jung JY and Yun BS. 2007. Antioxidant properties of various extracts from *Hypsizigus marmoreus. Food Chem.,* 104: 1–9.

Lee YL, Jian SY, Lian PY and Mau JL. 2008. Antioxidant properties of extracts from a white mutant of the mushroom *Hypsizigus marmoreus. J. Food Comp. Anal.,* 21: 116–124.

Leonard, BE. 2003. Fundamentals of Psychopharmacology. Third edition, John Wiley and Sons, Chichester, pp. 295–318.

Lindequist U, Niedermeyer THJ and Julich WD. 2005. The pharmacological potential of mushrooms. *eCAM,* 2: 285–299.

Liu Z, Li P, Zhao D, Tang H and Guo J. 2010. Protective effect of extract of *Cordyceps sinensis* in middle cerebral artery occlusion–induced focal cerebral ischemia in rats. *Behavioral and Brain Functions,* 6: 61.

Lo KM and Cheung PCK. 2005. Antioxidant activity of extracts from the fruiting bodies of *Agrocybe aegerita* var. alba. *Food Chem.,* 89: 533–539.

Mahady GB, Fong HHS and Farnsworth NR. 2001. Botanical Dietary Supplements: Quality, Safety and Efficacy. Swets and Zeilinger.

Mandegary A, Sharififar F, Abdar M, Arab–Nozari M. 2012. Anticonvulsant activity and toxicity of essential oil and methanolic extract of *Zhumeria majdae* Rech, a unique Iranian plant in mice. *Neurochem Res,* 37(12): 2725–30

Mattila P, Konko K, Eurola M, Pihlava JM, Astola J, Vahteristo L, Hietaniemi V, Kumpulainen J, Valtonen M and Piironen V. 2001. Contents of vitamins, mineral elements, and some phenolic compounds in cultivated mushrooms. *J. Agric. Food Chem.,* 49: 2343–2348.

Mau JL, Chang CN, Huang SJ and Chen CC. 2004. Antioxidant properties of methanolic extracts from *Grifola frondosa, Morchella esculenta* and *Termitomyces albuminosus* mycelia. *Food Chem.*, 87: 111–118.

Mau JL, Chao GR, Wu KT. 2001. Antioxidant properties of methanolic extracts from several ear mushrooms. *J. Agric. Food Chem.*, 49: 5461–5467.

Menoli RCRN, Mantovani MS, Ribeiro LR, Speit G and Jordão BQ. 2001. Antimutagenic effects of the mushroom *Agaricus blazei* Murrill extracts on V79 cells. *Mutat Res.* 496: 5–13.

Moradali MF, Mostafavi H, Ghods S and Hedjaroude GA. 2007. Immunomodulating and anticancer agents in the realm of macromycetes fungi (macrofungi). *International Immunopharmacology* 7, 701–724.

Moshi M.J., Kagashe GAB and Mbwanbo ZH. 2005. Plants used to treat epilepsy by Tanzanian traditional healers. *J Ethnopharmaco.* 97: 327–336.

Muceniece R, Saleniece K, Rumaks J, Krigere L, Dzirkale Z, Mezhapuke R, Zharkova O and Klusa V. 2008. Betulin binds to Z–aminobutyric acid receptors and exerts anticonvulsant action in mice. *Pharmacol Biochem Beh.* 90: 712–716.

Murcia MA, Martinez–Tome M, Jimenez AM, Vera AM, Honrubia M and Parras P. 2002. Antioxidant activity of edible fungi (truffles and mushrooms): losses during industrial processing. *J. Food Prot.*, 65: 1614–1622.

Murugkar AD and Subbulakshmi G. 2005. Nutritional value of edible wild mushrooms collected from the Khasi hills of Meghalaya. *Food Chem.*, 89: 599–603.

Nassiri–Asl M, Shariati–Rad S and Zamansoltani F. 2008. Anticonvulsive effects of intracerebroventriculsar administration of rutin in rats. *Prog Neuro–Pharmacol Biol Psy.* 32: 989–993.

Nuran CY, Turkoglu S, Yildirim N and Ince OK. 2012. Antioxidant Properties of Wild Edible Mushroom *Pleurotus eryngii* collected from Tunceli Province of Turkey. *Digest Journal of Nanomaterials and Biostructures*, Vol. 7, p. 1647–1654.

O'Brien HE, Parrent JL, Jackson JA, Moncalvo JM and R Vilgalys. 2005. Fungal community analysis by largescale sequencing of environmental samples. *Applied and Environmental Microbiology* 71(9): 5544–5550.

O'Brien PJ,. 1969. Intracellular mechanisms for the decomposition of a lipid peroxide by metal ions, heme compounds, and nucleophiles. *Can J Biochem,* 47: 485–492.

Ojewole JAO. 2008. Anticonvulsant property of *Sutherlandia frutescens* R. BR. (variety Incana E. MEY.) [Fabaceae] shoot aqueous extract. *Brain Res Bull.* 75: 126–132.

Oliveira OM, Vellosa JC, Fernandes AS, Buffa–Filho W, Hakime–Silva RA, Furlan M and Brunetti IL. 2007. Antioxidant activity of *Agaricus blazei. Fitoterapia,* 78: 263–264.

Park HG, Yoon SY, Choi JY, Lee GS, Choi JH, Shin CY, Son KH, Lee YS, Kim WK, Ryu JH, Ko KH and Cheong JH. 2007. Anticonvulsant effect of wogonin isolated from *Scutellaria baicalensis. Eur J Pharmacol.* 574: 112–119.

Pooja S, Irchhaiya R, Bhawna S, Gayatri S and Kumar S. 2011. Anticonvulsant and muscle relaxant activity of the ethanolic extract of stems of *Dendrophthoe falcata* (Linn. F.) in mice. *Indian J Pharmacol,* 43(6): 710–713.

Porter, RJ and Meldrum, BS. 2001. Antiseizure drugs. In: Katzung, B.G (ed.), Basic and Clinical Pharmacology, eighth edition. McGraw–Hill, London, pp.395–416.

Puttaraju NG, Venkateshaiah SU, Dharmesh SM, Urs SM and Somasundaram R. 2006. Antioxidant activity of indigenous edible mushrooms. *J. Agric. Food Chem.,* 54: 9764–9772.

Rang HP, Dale MM, Ritter JM and Moore PK. 2006. Antiepileptic Drugs. In: Rang HP, Dale MM, Ritter JM and Moore PK ed. Pharmacology. International Print–O–Pac Limited, Noida; 550–561.

Rauca C, Zerbe R and Jante H. 1999. Formation of free hydroxyl radicals after pentylenetetrazol–induced seizure and kindling. *Brain Res* 847: 347–351.

Raza M, Shaheen F, Choudhary MI, Sombati S, Rafiq A, Suria A, Rahman A and DeLorenzo RJ. 2001. Anticonvulsant activities of ethanolic extract and aqueous fraction isolated from Delphinium denudatum. *J Ethnopharmacol.* 78: 73–78.

Raza M, Shaheen F, Choudhary MI, Sombati S, Rafiq A, Suria A, Rahman A and DeLorenzo RJ. 2003. Anticonvulsant effect of FS–1 subfraction isolated from roots of Delphinium denudatum on hippocampal pyramidal neurons. *Phytotherapy Res.* 17: 38–43.

Ribeiro B, Valentao P, Baptista P, Seabra RM, Andrade PB. 2007. Phenolic compounds, organic acids profiles and antioxidative properties of beefsteak fungus (*Fistulina hepatica*). *Food Chem. Toxicol.,* 45: 1805–1813.

Ribeiro RA and Leite JR. 2003. Nantenine alkaloid presents anticonvulsant effect on two classical animal models. *Phytomed.* 10: 563–568.

Robaszkiewicz A, Bartosz G, Lawrynowicz M and Soszyriski M. 2010. The Role of Polyphenols, 57–Carotene, and Lycopene in the Antioxidative Action of the Extracts of Dried, Edible Mushrooms. *Journal of Nutrition and Metabolism,* Volume 2010, 9 pages.

Robert C, Collins MD,. 1980. Anticonvulsant effects of muscimol. *Neurology,* vol. 30 no. 6, pp 575.

Saberi M, Rezvanizadeh A and Bakhtiarian A. 2008. The antiepileptic activity of *Vitex agnus castus* extract on amygdala kindled seizures in male rats. *Neurosci Lett.* 441: 193–196.

Sarikurkcu C, Tepe B, Yamac M. 2008. Evaluation of the antioxidant activity of four edible mushrooms from the Central Anatolia. *Bioresour. Technol.,* 99: 6651–6655.

Schreier P. 2004. The flavone, hispidulin, a benzodiazepine receptor ligand with positive allosteric properties, traverses the blood–brain barrier and exhibits anticonvulsive effects. *Bri J Pharmacol.* 142: 811–820.

Sicoli G, Rana GL, Marino R, Sisto D, Lerario P and Luisi N. 2005. Forest Fungi as Bioindicators of a Healthful Environment and as Producers of Bioactive

Metabolites Useful For Therapeutic Purposes. 1st European Cost E39 Working Group 2 Workshop: "Forest Products, Forest Environment and Human Health: Tradition, Reality, and Perspectives" Christos Gallis (editor) – Firenze, Italy 20th–22nd.

Sliva, D. 2003. *Ganoderma lucidum* (Reishi) in cáncer treatment. *Integrative Cancer Therapies* 2(4): 358–364

.Soares AA, de Souza CGM, Daniel FM, Ferrari GP, da Costa SMG, Peralta RM. 2009. Antioxidant activity and total phenolic content of *Agaricus brasiliensis (Agaricus blazei* Murril) in two stages of maturity. *Food Chem.,* 112: 775–781.

Sonavane GS, Palekar RC, Kasture VS and Kasture SB. 2002. Anticonvulsant and Behavioural Actions of Myristica fragrans seeds. *Ind J Pharmacol.* 34: 332–338.

Song YS, Kim SH, Sa JH, Jin C, Lim CJ, Park EH. 2003. Antiangiogenic, antioxidant and xanthine oxidase inhibition activities of the mushroom *Phellinus linteus. J. Ethnopharmacol.,* 88: 113–116.

Sudarsky, L. 1990. Pathophysiology of the nervous system. Little Brown and Company, Boston, pp. 139–181.

Sukma M, Chaichantipyuth C, Murakami Y, Tohda M, Matsumoto K and Watanabe H. 2002. CNS inhibitory effects of barakol, a constituent of *Cassia siamia* Lamk. *J Ethnopharmacol.* 83: 87–94.

Taviano MF, Miceli N, Monforte MT, Tzakou O and Galati EM. 2007. Ursolic acid plays a role in Nepeta sibthorpii Bentham CNS depressing effects. *Phytotherapy Res.* 21: 382–385.

Tsai SY, Tsai HL and Mau JL. 2007. Antioxidant properties of *Agaricus blazei, Agrocybe cylindracea,* and *Boletus edulis. Lwt–Food Sci. Technol.,* 40: 1392–1402.

Turkoglu A, Duru ME, Mercan N, Kivrak I, Gezer K. 2007. Antioxidant and antimicrobial activities of *Laetiporus sulphureus* (Bull.) Murrill. *Food Chem.,* 101: 267–273.

Tyagi A and Delanty N. 2003. Herbal Remedies, Dietary Supplements, and Seizures. *Epilepsia,* 44(2): 228–235, Blackwell Publishing, Inc. © 2003 International League against Epilepsy.

Valentão P, Andrade PB, Rangel J, Ribeiro B, Silva BM, Baptista P, Seabra RM. 2005. Effect of the conservation procedure on the contents of phenolic compounds and organic acids in Chanterelle (*Cantharellus cibarius*) mushroom. *J. Agric. Food Chem.,* 53: 4925–4931.

Valentao P, Lopes G, Valente M, Barbosa P, Andrade PB, Silva BM, Baptista P, Seabra RM. 2005. Quantitation of nine organic acids in wild mushrooms. *J. Agric. Food Chem.,* 53: 3626– 3630.

Venkateswarlu G, Edukondalu K, Chennalakshmi BGV, Sambasivarao P, Raveendra G and Reddy VR. 2012. Evaluation of Anti Epileptic Activity of Leaf Extract of *Cynodon dactylon* (L.) Pers. in Validated Animal Models. *Current Pharma Research* 571–579.

Vyawahare NS, Khandelwal AR, Batra VR and Nikam AP. 2007. Herbal Anticonvulsants. *J Herbal Med Toxicol.* 1(1): 9–14.

Wagay JA, Javed J and Vyas D. 2011. Antihyperglycemic and Insulin enhancing Potential of Morchella esculenta. *Research Journal of Agricultural Sciences* 2(3): 445–790.

Wasser SP. 2002. Review of Medicinal Mushrooms Advances: Good News from Old Allies. Herbal Gram. 2002; 56: 28–33 American Botanical Council.

Wasser SP and Weis AL. 1999. Medicinal properties of substances occurring in higher Basidiomycetes mushrooms. Current perspectives (Review). *Int J Med Mushr.* 1: 31–62.

Watanabe N, Obuchi T, Tamai M, Araki H, Omura S, Yang JS, Yu DQ, Liang XT and Huan JH. 1990. A novel N6–substituted adenosine isolated from mi huan jun (*Armillaria mellea*) as a cerebral–protecting compound. *Planta Med.* 56: 48–52.

Webster RA and Jordan CC. 1989. Neurotransmitters, Drugs and Disease. Blackwell Scientific Publications, London, pp. 301–344.

Willmore LJ, Hiramatsu M, Kochi H and Mori A. 1983. Formation of superoxide radicals after $FeCl_3$ injection into rat isocortex. *Brain Res,* 277: 393–396.

Xiong J and Huang JH 1998. The A1– and non A1–effects of N6–(5–hydroxy– 2–pyridyl)–methyl–adenosine on rat vas deferens. *Acta Pharmaceut. Sin.* 33: 175–179 (in Chinese).

Yang JH, Lin HC and Mau JL. 2002. Antioxidant properties of several commercial mushrooms. *Food Chem.,* 77: 229–235.

Yu L, Shen YS, Miao HC. 2006. Study on the anti–vertigo function of polysaccharides of *Gastrodia Elata* and polysaccharides of *Armillaria mellea. Chin. J. Information, TCM* 13: 29–36 (in Chinese).

Zanoli P, Avallone R and Baraldi M. 2000. Behavioral characterization of the flavonoids apigenin and chrysin. *Fitoterapia.* 71: S117–S123.

Zhang M, Cui SW, Cheung PCK and Wang Q. 2007. Antitumor polysaccharides from mushrooms: a review on their isolation process, structural characteristics and antitumor activity. *Trends Food Sci Technol.* 39: 14–9.

2015, Recent Trends in Microbiology, Mycology and Plant Pathology   *Pages* **177–196**
Editor: **Dr. H.C. Lakshman**
Published by: **DAYA PUBLISHING HOUSE, NEW DELHI**

Chapter 13

Fungal Metabolites and their Importance

A. Channabasava, H.C. Lakshman and T.C. Taranath*

P.G. Department of Studies in Botany,
Karnatak University, Pavate Nagar, Dharwad – 580 003, Karnataka, India
*E-mail: achannabasava@gmail.com

Introduction

What are Fungi?

About 80000 to 120000 species of fungi have been described to date, although the total number of species is estimated at around 1.5million (Hawksworth, 2001; Kirk *et al.*, 2001). This would render fungi one of the least-explored biodiversity resources of our planet. It is notoriously difficult to delimit fungi as a group against other eukaryotes, and debates over the inclusion or exclusion of certain groups have been going on for well over a century. In recent years, the main arguments have been between taxonomists striving towards a phylogenetic definition based especially on the similarity of relevant DNA sequences, and others who take a biological approach to the subject and regard fungi as organisms sharing all or many key ecological or physiological characteristics the 'union of fungi' (Barr, 1992). Being interested mainly in the way fungi function in nature and in the laboratory, we take the latter approach and include several groups in this book which are now known to have arisen independently of the monophyletic 'true fungi' (Eumycota) and have been placed outside them in recent classification schemes (Figure 13.1). The most important of these 'pseudofungi' are the Oomycota. Based on their lifestyle, fungi may be circumscribed by the following set of characteristics (modified from Ainsworth, 1973).

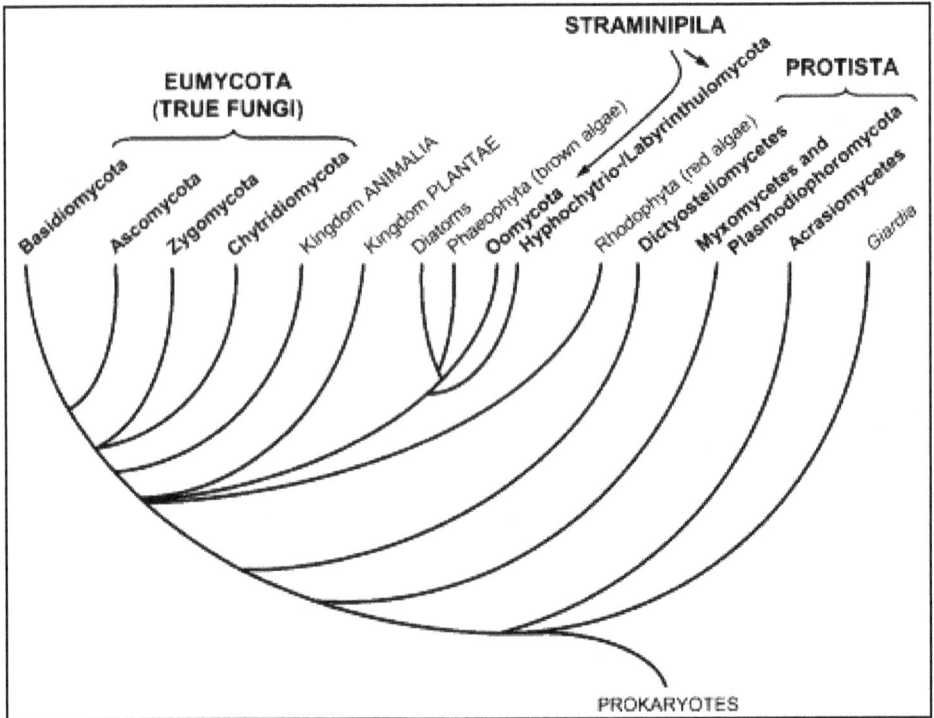

Figure 13.1: The Phylogenetic Relationship of Fungi and Fungus like Organisms Studied by Mycologists with other Groups of Eukaryotes. The analysis is based on comparison of 18s rDNA sequence. Modified and redrawn from Bruns *et al.* (1991) and Berbee and Taylor (1999).

1. *Nutrition*: Heterotrophic (lacking photosynthesis), feeding by absorption rather than ingestion.

2. *Vegetative state*: On or in the substratum, typically as a non-motile mycelium of hyphae showing internal protoplasmic streaming. Motile reproductive states may occur.

3. *Cell wall*: Typically present, usually based on glucans and chitin, rarely on glucans and cellulose (Oomycota).

4. *Nuclear status*: Eukaryotic, uni- or multi-nucleate, the thallus being homo- or heterokaryotic, haploid, dikaryotic or diploid, the latter usually of short duration (but exceptions are known from several taxonomic groups).

5. *Life cycle:* Simple or more usually, complex.

6. *Reproduction*: The following reproductive events may occur: sexual (*i.e.* nuclear fusion and meiosis) and/or parasexual (*i.e.* involving nuclear fusion followed by gradual de-diploidization) and/or asexual (*i.e.* purely mitotic nuclear division).

7. *Propagules*: These are typically microscopically small spores produced in high numbers. Motile spores are confined to certain groups.

8. *Sporocarps*: Microscopic or macroscopic and showing characteristic shapes but only limited tissue differentiation.

9. *Habitat*: Ubiquitous in terrestrial and fresh water habitats, less so in the marine environment.

10. *Ecology*: Important ecological roles as saprotrophs, mutualistic symbionts, parasites, or hyperparasites.

11. *Distribution*: Cosmopolitan.

Fungi are the most important biotechnologically useful organisms (Kurtzman, 1983). Fungi have been used to elucidate the complex biochemistry and genetic principles of eukaryotes. They are important not only for antibiotics, but also used in commercial production of flavoring foods, production of biochemicals such as organic acids and enzymes (Blain, 1975; Eveleigh, 1981). The fungal kingdom offers enormous biodiversity, with around 70,000 known species, and an estimated 1.5 million species in total. Most of these are filamentous fungi, which differ from the yeasts not only in their more complex morphology and development (*e.g.* asexual and sexual structures), but also in their greater metabolic complexity. In particular, they are known for production of enzymes and secondary metabolites. Of the 12,000 antibiotics known in 1995, about 22 per cent could be produced by filamentous fungi. Ascomycetes and fungi imperfecti are the most frequent producers of secondary metabolites accounting for 6400 compounds (Berdy, 1995). *Aspergillus, Penicillium* and *Fusarium* species produce 950, 900 and 350 compounds respectively from the group representing ascomycetes. Basidiomycetes or mushrooms produce 2000 active compounds.

What is Need of Fungal Metabolites?

There is a need to search new ecological niches for potential sources of natural bioactive agents for different pharmaceutical, agriculture, and industrial applications; these should be renewable, eco-friendly and easily obtainable (Liu *et al.*, 2001). The most prominent producers of natural products can be found within different groups of organisms including plants, animals, marine macro-organisms (sponge, corals and algae), and micro-organisms (bacteria, actinomycetes, and fungi). Secondary metabolites have been the most successful source of potential drug leads (Rey-Ladino *et al.*, 2011; Cragg, and Newman, 2005; Butler, 2004; Haefner, 2003). Nevertheless, natural products continue to provide unique structural diversity in comparison to standard combinatorial chemistry, which presents opportunities for discovering mainly novel low molecular weight lead compounds. Since less than 10 per cent of the world's biodiversity has been evaluated for potential biological activity, many more useful natural lead compounds await discovery with the challenge being how to access this natural chemical diversity (Cragg, and Newman, 2005). The present chapter includes the some important secondary metabolites or bioactive compounds derived from fungi and their importance. Fungi are remarkable organisms that readily produce a wide range of natural products often called secondary metabolites. However, interest in these compounds is considerable, as many natural products are of medical, industrial and/or agricultural importance. Some natural products are deleterious (*e.g.*, mycotoxins), while others are beneficial (*e.g.*, antibiotics) to humankind. Although it has long been noted that biosynthesis of natural products is usually associated with cell differentiation or development, and in fact most secondary metabolites are produced by organisms that exhibit filamentous growth and have a relatively complex morphology, until recently the mechanism of this connection was not clear.

Fungi as Producers of Biologically Active Metabolites

Fungi are among the most important groups of eukaryotic organisms that are well known for producing many novel metabolites which are directly used as drugs or function as lead structures for synthetic modifications (Kock *et al.*, 2001; Bode *et al.*, 2002, Donadio *et al.*, 2002, Chin *et al.*, 2006, Gunatilaka 2006, Mitchell *et al.*, 2008, Stadler and Keller 2008). The success of several medicinal drugs from microbial origin such as the antibiotic penicillin from *Penicillium* sp., the immunosuppressant cyclosporine from *Tolypocladium inflatum* and *Cylindrocarpon lucidum*, the antifungal agent griseofulvin from *Penicillium griseofulvum* fungus, the cholesterol biosynthesis inhibitor lovastatin from *Aspergillus terreus* fungus, and β-lactam antibiotics from various fungal taxa, has shifted the focus of drug discovery from plants to microorganisms. Suryanarayanan *et al.* (2009) discussed many fungal secondary metabolites with various chemical structures and their wide ranging biological activities and this reflects the high synthetic capability of fungi (Suryanarayanan and Hawksworth, 2005). About 1500 fungal metabolites have been reported to show antitumor and antibiotic activity (Peláez 2005) and some have been approved as drugs. These include micafungin, an anti-fungal metabolite from *Coleophoma empetri* (Frattarelli *et al.*, 2004), mycophenolate, a product of *Penicillium brevicompactum*, which is used for preventing renal transplant rejection (Curran and Keating 2005), rosuvastatin from *Penicillium citrinum* and *P. brevicompactum*, which used for treating dyslipidemias (Scott *et al.*, 2004) and cefditoren pivoxil, a broad spectrum antibiotic derived from *Cephalosporium* sp. (Darkes and Plosker 2002). Others include derivatives of fumagillin; an antibiotic produced by *Aspergillus fumigates* (Chun *et al.*, 2005), and illudin S, a sesquiterpenoid from *Omphalotus illudens* (McMorris *et al.*, 1996) which exhibits anticancer activities. Also, fungal metabolites are important in agriculture applications (Anke and Thines, 2007).

Soil fungi have been the most studied of fungi, and typical soil genera such as *Acremonium, Aspergillus, Fusarium* and *Penicillium* have shown ability to synthesis a diverse range of bioactive compounds. More than 30 per cent of isolated metabolites from fungi are from Aspergillus and Penicillium (Bérdy 2005). *Penicillia* are some of the most widespread hyphomycetes among microscopic fungi. Their natural habitats are soils of northern latitudes. They are known as producers of various types of biologically active compounds such as ergot alkaloids, diketopiperazines, uinolines, quinazolines and polyketides (Cole and Cox, 1981). Ergot alkaloids possess various kinds of biological activity. They can act peripherally leading to uterine vasoconstriction and contraction; neurohormonally, by blocking the action of adrenaline and serotonin; and can also affect the central nervous system, by reducing

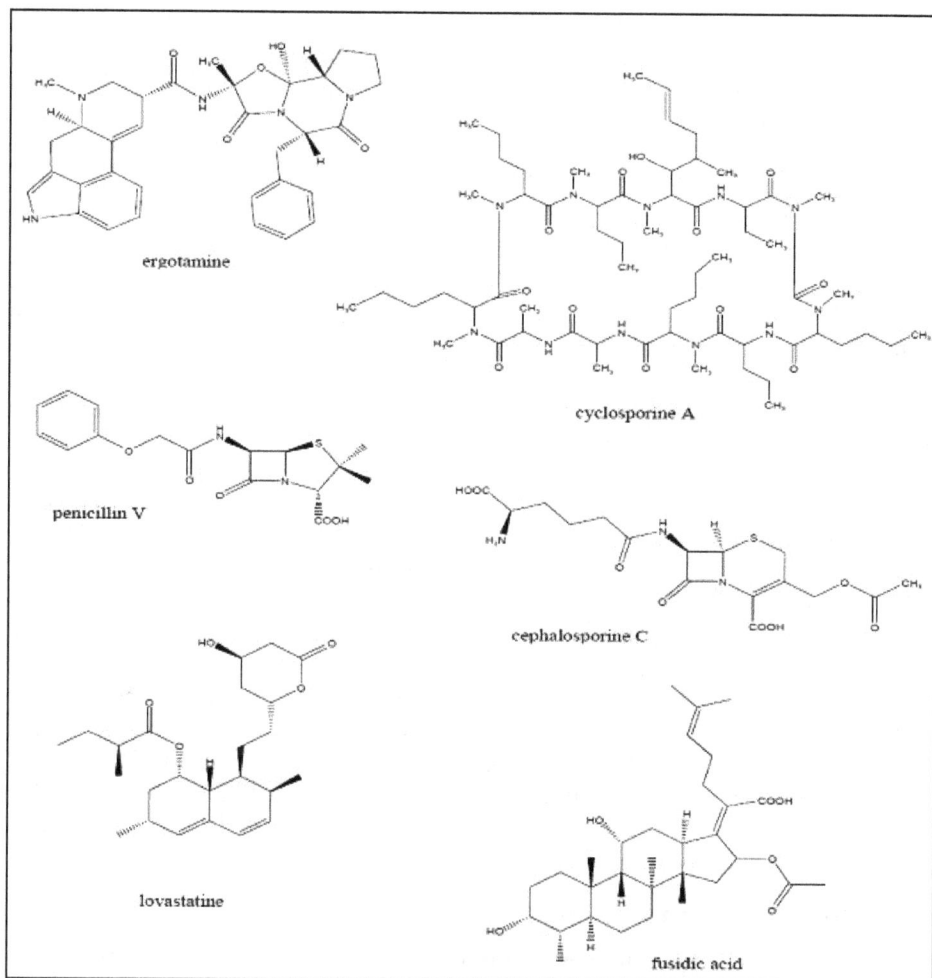

Figure 13.2: Some Important Biological Active Metabolites Derived from Fungi.

the activity of the vasomotor centre and stimulating the sympathetic structures in the midbrain, thereby causing a hallucinogenic effect, and also inhibiting the secretion of prolactin, hyperthermia and vomiting (Fluckiger, 1980; Rehachek and Sajdl, 1990). Another group of biologically active compounds synthesized by penicillia are cyclic peptides diketopiperazines consisting of residues of two amino acids and mevalonic acid. The precursors of roquefortine and related alkaloids such as 3, 12-dihydroroquefortine, glandicoline A and B, meleagrine, oxalin are tryptophan, histidine and mevalonic acid (Kozlovskii, *et al.,* 1996). Tryptophan and mevalonic acid are also the precursors of diketopiperazine alkaloids fellutanines and isofellutanines (Kozlovskii, *et al.,* 1997). Brevianamides A and B and the new alkaloids piscarinines A and B are formed from tryptophan, proline and from one or more mevalonic acid molecules via the same pathway (Kozlovsky, 2000). Diketopiperazine alkaloids whose precursors are tryptophan and leucine include leucyl tryptophanyl diketopiperazine and verrucosine (Kozlovskii, *et al.,* 1996). Compounds formed from the residues of tryptophan and phenylalanine are represented by such alkaloids as rugulosuvine, isorugulosuvine, puberulin (= rugulosuvine A), puberulin A (= rugulosuvine B) (Kozlovsky, 2001).

Fungi as Producers of Organic Acids

Although many organic acids are made by living cells, few are produced commercially. Citric, gluconic, itaconic, and lactic acids are manufactured via large-scale bioprocesses. Oxalic, fumaric, and malic acids can be made through fungal bioprocesses, but the market demand is small, since competing chemical conversion routes are currently more economical. A few other organic acids have been explored for the development of novel processes. To date, the largest commercial quantities of fungal organic acids are citric acid and gluconic acid, both of which are prepared by fermentation of glucose or sucrose by *Aspergillus niger*. Another *Aspergillus* species, *A. terreus*, is used to make itaconic acid. A significant commercial source of lactic acid at the time of this writing is a bioprocess employing the Zygomycete fungus *Rhizopus oryzae*. These three species of fungi were initially chosen for process development because they exhibited the ability to produce large amounts of a particular organic acid. The four commercial organic acids produced by fungi are employed in high-volume, low-value applications. For example, they are used in industrial metal cleaning or other metal treatments and in the food and feed industry as flavor enhancers, acidifiers, stabilizers, or preservatives. The commercial success of fungal bioprocesses is ultimately based on rapid and economical conversion of sugars to acid, but that alone does not explain the commercial situation for each of these acids. An understanding of the economic and business parameters that have contributed to the success of these four products may be useful in development and commercialization of new organic acid products from filamentous fungi.

Fungi as Producers of Antibiotics

The term 'antibiotic' literally means 'against life'. An antibiotic was originally defined as a substance, produced by one micro-organism (Denyer *et al.,* 2004), or of biological origin (Schlegel, 2003) which at low concentrations can inhibit the growth of other micro-organisms or infectious organisms (Hugo and Russell, 1998). Antibiotics include a chemically heterogeneous group of small organic molecules of microbial origin that, at low concentrations, are deleterious to the growth or metabolic activities of other micro-organisms (Thomashow and Weller, 1995). According to Talaro and Talaro (2002), antibiotics substances produced by natural metabolic processes of some micro-organisms that can inhibit or destroy other micro-organism. It was not until 1940 with the discovery of penicillin, the first, best-known and most widely used antibiotic (Schlegel, 2003; Hugo and Russell, 1998; Berg *et al.,* 2002; Taylor *et al.,* 2003; Sommer, 2006) in 1928 by an English Bacteriologist, late Sir Alexander Fleming that the first clinical trials of penicillin were tried on humans. This antibiotic was obtained from a blue green mould of the soil called *Penicillium notatum* (Dutta, 2005). Penicillin was discovered accidentally in 1928 by Fleming, who showed its efficacy in laboratory cultures against many disease producing bacteria. This discovery marked the beginning of the development of antibacterial compounds produced by living organisms (Taylor *et al.,* 2003). Another antibiotic, streptomycin was isolated in 1944 by Waksman, a Microbiologist, from a species of soil bacteria, called *Streptomyces griseus*, particularly tubercle bacilli, and has proved to be very valuable against tuberculosis (Dutta, 2005). A vigorous search for more antibiotics was on at this time and in 1947, another antibiotic, chloromycetin was discovered by Burkholder (Dutta, 2005; Sommer, 2006). It was isolated from S. Venezuelae. It has a powerful action on a wide range of infectious bacteria both Gram positive and Gram negative (Dutta, 2005).

Although there are a large number of species of fungi, only a relative few have been found to produce antibiotics, and only seven antibiotics are produced commercially. The 1970 Information Bulletin, No. 8 (Delcambe, 1970) of the International Centre of Information on Antibiotics lists 338 species of fungi that produce antibiotics. According to 1967 (Perlman, 1967) and 1970 (Perlman, 1970) lists assembled by Perlman, the following antibiotics are produced commercially: Fusidic acid formed by *Fusidium coccineum* Hickel, griseofulvin formed by *Penicillium griseofulvim* Dierk, penicillins formed by *Penicillium chrysogenum* Thom, variotin formed by *Paecilomyces varioti* Bainier, derivatives of cephalosporin formed by *Emericellopsis* (*Cephalosporium*), and fumagillin produced by *Aspergillus fumigates* Fres. Since then, siccanin produced by *Helmintlwsporium siccans* Drechsler has been produced in Japan. According to Perlman (Perlman, 1968), only three groups of antibiotics are produced by fungi in the United States. These are the cephalosporins, penicillins, and fumagillin.

Fungi as Producers of Phytohormones

Phytohormones and their crucial role in the elicitation of plant physiological process have been known since 1937, when Went and Thimann published their book Phytohormones. At that time the term phytohormones was synonymous to auxin. Later on gibberellins, ethylene, abscisic acid and cytokinins together with the auxins were regarded as five classical phytohormones. Interestingly, phytohormones were not only produced by the plants only, but also by micro-organisms including bacteria (Costacurta and Vanderleyden, 1995) and fungi (Tudzynski, 1997). Reports on the production of substances with auxin activity by fungi date back to the early 1930's (Dolk and Thimann, 1932; Gruen, 1959; Tapani *et al.,* 1993). The major phytohormone auxin Indole 3-acetic acid (IAA) was first detected by Dolk and Thimann (1932) in culture filtrates of the fungus *Rhizopus suinus*, even before IAA was identified in plants. Since then, much research has been dedicated to the study of IAA in fungi. In most of the studies IAA was found in culture filtrates but there are also reports on the accumulation of IAA inside the mycelium. Tapani *et al.* (1993) reported that, the mycelium of *Botrytis cinerea* contained 128 mg/g IAA while less than 1 mg/g was detected in the medium. Ek *et al.* (1983) measured the amount of IAA produced by the mycorrhizal fungi. Seventeen out of 19 examined species produced detectable levels of IAA ranging from 0.016µg/mg to 20 µg/mg dry weight of mycelium. Ethylene is a simple gaseous phytohormone and first chemical proof of ethylene production was given by Gane (1934) by showing that ripped apples released the gas to atmosphere. Many fungi produce ethylene. *Pencillium digitatum* was first reported by Baile in 1940.

The presence of Abscisic acid (ABA) was first described in fungi was *Cercospora rosicola* (Assante *et al.,* 1977). Since, then ABA as described as fungal metabolite also from *Cercospora cruenta* (Oritani and Yamashita, 1985), *Botrytis cinerea* (MArumo *et al.,* 1982), and from other species of several genera of phytopathogenic and saprophytic fungi (Table 13.1.). Interestingly, ABA producing most of the fungi are phtopathogens. This suggests that, fungal ABA might have a possible function in phytopathogenesis. However, no reports are available on production of ABA during infection process. Another important and key phytohormone is gibberellins. These are group of terpenoid phytohormones capable of influencing many developmental processes in higher plants. The discovery of gibberellins (GAs) as natural phytohormone is a classical example of interaction between soil fungi and plants. GAs were first obtained from culture filtrates of the fungus *Gibberella fujikuroi.* Hori (1898) detected the causative agent of Bakane disease and identified this soil fungus as *Fusarium moniliformae.* The perfect stage of fungus is *Gibberella fujikuroi.*

Table 13.1: List of some Important Fungi Producing different Phytohormones

Fungus	Detection Methods	References
Alternaria alternata	HPLC, GC	Crocoll *et al.,* 1991
Alternaria brassicae	HPLC	Dahiya and Tewari 1991
Botrytis cinerea	TLC	Marumo *et al.,* 1982
Ceratocystis coerulescens	GC-MS	Dorfling and Peterson, 1984
Cercopsora pinidensiflorae	GC-MS	Okamoto *et al.,* 1984
Cercospora cruenta	TLC, GC-MS	Oritani *et al.,* 1984
Cercospora rosicola	TLC, HPLC	Assante *et al.,* 1977
Cytospora ciacta	HPLC	Vizarova *et al.,* 1997
Fusarium culmorum	HPLC	Michniewicz *et al.,* 1987
Fusarium oxysporum	GC-MS, HPLC	Dorfling and Peterson, 1984
Monilia spp.	HPLC	Vizarova *et al.,* 1997
Pleurotus florida	Spectrofluorimetrically	Unyayar *et al.,* 1997
Rhizoctonia solani	GC-MS, HPLC	Dorfling and Peterson, 1984
Rhizopus nigricans	HPLC, GC	Crocoll *et al.,* 1991

In 1972 Lozano at the Centro International de Agricultura Tropica (CIAT) in Colombia described a super elongation disease of cassava caused by the phytopathogenic fungus *Sphaceloma manihoticola.* Rademacher and Graebe (1979) and Zeigler *et al.* (9180) found that, the reason for the disease symptoms is the production of considerable amounts of GAs and some minor amounts of GA_9, GA_{13}, GA_{14}, GA_{15}, GA_{24}, GA_{36} and GA_{37} and some other species of the genus *Sphaceloma* also produce small amounts of GAs. (Rademacher, 1992). Kawanabe *et al.* (1983, 1985) have analysed the mycelium of *Neurospora crass* by GC-MS and identified small amounts of GA_3. Searching for new plant growth regulators, Sassa *et al.* (1989, 1994) detected GA_1 and GA_4 and smaller amounts of GA_9, GA_{12}, GA_{15}, GA_{20}, GA_{24}, GA_{25} and GA_{82} in the cultures of *Phaeosphaeria* sp., by GC-MS.

Fungi as Producers of Plant Pathogen Suppressors

Secondary metabolites play important role in the control of plant pathogens. These are used to control the phytopathogenic fungi and bacteria (Burge, 1988). *Trichoderma harzianum* produces an antifungal agent, alkyl-pyrone which is active against wide range of fungi and bacteria (Claydon *et al.,* 1987). *Pisolithus arhizus* produce compounds, hydroxy benzyl formic acid and R-(-)-p-hydroxymendelic acid active against *Truncatella hartigii* (Kope *et al.,* 1991). Most of the *Trichoderma* sps., produce gliotoxin active against root pathogenic fungus *Rhizoctonia solani* (Dennis and Webster, 1971; Harman *et al.,* 1980). Gliotoxin is also isolated from *Gliocladium* species active against wood-rotting fungus, *Armillarriab mellea* (Lumsden *et al.,* 1991, 1992). *Trichoderma viridae* and *Peniophora gigantean* are the two fungi successfully exploited and commercialized against plant pathogenic fungi. *Trichoderma viridae* has been used to control *Armillaria mellea* on trees, *Cerotocystis ulmion* elm, *Chondrosterium purpureumon* fruit trees and Eucalyptus and *Heterobasidium annosum* on pine (Ricard and Highley, 1988), which parasitizes hyphae of the pathogenic fungi and produces secondary metabolites (Dennis and Webster, 1971). *Peniophora gigantean* is used to control infestation caused by *Heterobasidium annosum*, root rot of pines (Lynch, 1990).

Coniothyrium mimitans parasitizes sclerotia of *Sclerotium trifoliarum, Sporidesmium sclerotivorum* and *Sclerotinia minor* (Ayers and Adams, 1979). *Rhizoctonia solina* was reported to be parasitized by *Gliocladium virens* (Howell, 1987) and it also protects damping off caused by *Pythium ultimum* and *Rhizoctonia solina* (Howell, 1982). *Ampelomyces quisqualis* is used to prevent powdery mildews (Chet, 1990). *Verticellium chlymadosporium* showed strong inhibitory effect on *M. phaseolina, R. solani* and *F. solani* both in vitro and in vivo (Ehteshamul-Haque., *et al.,* 1994). *Talaromyces flavus* which parasitizes hyphae of *Rhizoctonia solani* (Boosalis, 1956) also inhibits the growth of other fungi (Dwivedi and Garrette, 1968; Husain and McKeen, 1963). Similarly *Verticellium chlymadosporium* possess antibacterial activity (Marchisio, 1977) and was found to parasitize oospore of *Phytophthora cactorum* (Sneh *et al.,* 1977). Now a day's *Fusarium oxysporum* received attention as a biocontrol agent. Been rot caused by *Fusarium solani* and infection in red clover decreased by non pathogenic *Fusarium oxysporum* (Lechappe *et al.,*1988).

Soil contains various nematophagous fungi which are natural enemies to nematodes. Large number of nematophagous fungi with significant nematicidal activity have been discovered so far (Dayal, 2000). *Paeciomyces lilacinus* infects eggs of *Meloidogyne incognita* (Jatal, 1985 and 1986). *Vericillium chamydosporium* also showed significant activity against the cysts of *Meloidogyne* spp., and *Heterodera* spp. (De Leji, 1992; Kerry, 1990). The hatching of nematode cysta, *Globodra pallaada* was adversely affected by the fungi like *Ulocladium Botrytes, Drechslera* spp., *Gliocladium* spp., and *Trichurus* spp (Gonazales *et al.,* 1984). The fungus *Paecilomices lilacinus* was also widely tested for nematode control (Wainwright, 1992). *Penicillium anatolicum* produces compounds that acts on the eggs of nematodes (Jatala, 1986).

Table 13.2: List of Metabolites Isolated from Fungi and their Biological Activity

Metabolite	Produced by	Activity
Trichodenone	*Trichoderma harzianum*	Antitumor
Harzialactone	*Trichoderma harzianum*	Antitumor
Glisoprenin	*Gliocladium roseum*	Antifungal
Curvularin	*Aspergillus* sps	Biological
Longibrachin	*Trichoderma longibrachiatum*	Antibacterial
Trichorzin	*Trichoderma harzianum*	Antibacterial
Pyrenocine	*Penicillium uxiksmanii*	Antitumor
Aranochlor	*Pseudoarachniotus roseus*	Antifungal
Terprenin	*Aspergillus candidus*	Biological
Haematocin	*Nectria haematococca*	Antifungal
Arisugacin	*Penicillium* sps.	Biological
Curtisian	*Paxillus curtisii*	Biological
Melleolide	*ArmillarieUa mellea*	Antibacterial
Roseoferin	*Mycogone rosea*	Antifungal
Chaetoatrosin-A	*Chaetomium atrobrunneum*	Antifungal
Strobilurin	*Mycena galericulata*	Antiparasitic
Topopyrone	*Phoma* sps.	Antitumor
Phellinsin-A	*Phellinus* sps.	Antifungal
Xanthoepocin	*Penicillium simplicissimum*	Antifungal
Neobulgarone	*Neobulgaria pura*	Agricultural
Ampullosporin	*Sepedonium ampullosporum*	Biological
Phoenistatin	*Acremonium fusigerum*	Antitumor
Cephaibol	*Acremonium Tubakii*	Antiparasitic
Epicorazine-C	*Stereum Hirsutum*	Antibacterial
Spirobenzofuran	*Acremonium* sps.	Antibacterial
Cladospolide-D	*Cladosporium* sps.	Antifungal
Lucilactaene	*Fusarium* sps.	Antitumor
Bisabosqual-D	*Stachybotrys ruwenzoriensis*	Antifungal
Altersetin	*Altemaria* sps.	Antibacterial
Chrysoqueen	*Chrysosporium queenslandicum*	Antibacterial
Chrysolandol	*Chrysosporium queenslandicum*	Antibacterial
Thielavin	*Chaetomium carinthiacum*	Biological
Miyakamide	*Aspergillus flauus*	Others
Phenylpyropene	*Penicillium griseofuluum*	Biological
Ustilipid	*Ustilago maydis*	Biological

Contd...

Table 13.2–Contd...

Metabolite	Produced by	Activity
Coniosetin	*Coniochaeta ellipsoidea*	Antibacterial
Asperaldin	*Aspergillus niger*	Biological
Terreulactone	*Aspergillus terreus*	Biological
Quinocitrinine-B	*Penicillium citrinum*	Antibacterial
Monorden	*Humicola* sps.	Biological
Malbranicin	*Malbranchea cinnamomea*	Antifungal
Acremonidin	*Acremonium* sps.	Antibacterial
Exophillic acid	*Exophiala pisciphila*	Antiviral

Fungi as Producers of Antimicrobial Agents

Penicillium vericulum produce a vermiculin with antiprotozoal potential (Fuska *et al.*, 1972). The compound As-186 from *Penicillium asperosporium* exihibit acyl-CoA:cholesterol acyltransferase (ACAT) activity. Lovastatin reported from *Aspergilus terrus* is used in the inhibition of cholesterol biosynthesis (Kuroda *et al.*, 1994). Lovastatin related compounds were isolated from P. brevicompactum and *P. citrinum* (Endo *et al.*, 1976). Arohynapenes A and B, anticoccidial agents (Masuma *et al.*, 1994) were reported from *Penicillium* sps. Alfavarin and β-aflatrem, a new antisectan metabolite was isolated from the sclerotia of *Aspergillus flavus* (TePaske *et al.*, 1992). *Fusarium solani* is known to produce toxin like Neosalaniol, T-2 toxin, HT-2 toxin diacetoxyseripenol (Ueno and Nishimura, 1973) and fusarubin reported to have antimicrobial (Arnstein, *et al.*, 1946), antitumor (Issaq *et al.*, 1977), and phytotoxin activitivities (Kern, 1970). Apart from clinical uses the metabolites produced from micro-organisms are also used as growth stimulant and in the control of plant diseases (Evans, 1989).

Fungi produce various types of metabolites used in current chemotherapy, for example penicillin, cephalosporin and fusidic acid, which have antibacterial and antifungal activity (Lowe and Elander, 1983). After the discovery of penicillin in 1928 from *P. notatum*, the modern era in the research of antibiotics started. A number of antibiotics have been isolated from micro-organisms with a ratio of 70:20:10 from actinomycetes, fungi and eubacterials respectively (Bredy, 1974). *Tolypocladium inflatum* and *T. geodes* as well as strains of *Acremonium, Beauvaria, Fusarium, Paecilomyces* and *Verticillum* Sps (Dreyfus *et al.*, 1976; Good *et al.*, 1985) were know to produce Cyclosporine-A. Multipliolides-A and B, 10-membered lactose compounds were produced from *Xylaria mutiplex* (Boonphong *et al.*, 2001). Most of the β-lactam antibiotics like pencillin, cephalosporin and their relatives are produced from *Penicillium* and *Cephalosporium* group where as polyene antibiotics are produced from *Aspergillus* sps (Egoron, 1985). Griserofulvin is a metabolic product of *Penicillium nigricans, P. urticae* and *P. raistrickii* used as therapeutic agent in the treatment of dermatomycosis (Brain, 1960).

Fusarium solani and *Fusarium oxysporum* produces several napthaquinones with antibacterial properties (Baker, 1990). From *Talaromyces flavus*, azole fungicide FKI-0076 was isolated. Fusidic acid and fucidin produced from *Acromonium fusidioides* have antibacterial activity against gram-negative bacteria (Godtfredson and Lorck, 1963). An antibiotic, botrydiplodin producedby *Botryosphaeria rhodina* was found active against both Gram-positive and Gram negative bacteria (Sen Gupta *et al.*, 1966). Cochliodinol produced by *Chaetomium cochlioides* exhibit both anti bacterial and antifungal activities whereas chaetomin was active against gram positive bacteria (Brewer *et al.*, 1970). *Cryptosporiopsis guercina* produce crytocandin, a lipopeptide antibiotic showed strong inhibitory activity against human pathogenic fungi like *Candida albicans, Trichophyton mentagrophytes, Trichophyton rubrum* and plant pathogenic fungi like *Sclerotinia sclerotium* and *Botrytis cinerea* (Strobel *et al.*, 1999). Epidithiadiketo-piperazine and chaetomin produced by *Chaetomium globosum* showed strong antifungal activity (DiPietro *et al.*, 1992). *Glioclabium virens* produced compounds, gliovirin having antibiotic activity towards *Pythium ultimum* (Howell and Stipanovic, 1983) and viridian, fungistatic in nature (Jones and Hancock, 1987).

Mycotoxins and their Importance

Historically, mycotoxins were "discovered" following a sudden and fatal outbreak which occurred in 1960 on turkey farms in Great Britain (Asao *et al.*, 1963). This acute case led to the identification of aflatoxins and consecutively the relationship between moulds, their toxins and mycotoxicosis. Similarly, many cases of nephropathy in pigs were reported a few years later in Denmark due to barley being naturally contaminated by OT A and have led to the identification and characterisation of chronic adverse effects in animals related to the contamination of their feed (Krogh *et al.*, 1973). Other diseases associated with the ingestion of mycotoxin-contaminated feed by farm animals were observed and described a long time ago such as the common feed refusal when feed are contaminated by mycotoxins or the oestrogenic syndrome in pork; other syndromes such as the Equine LeucoEncephaloMalacie (ELEM) in horses due to fumonisins-contaminated-oats were only discovered in the late nineteen eighties (FAO, 2007; Gelderblom *et al.*, 1988). Nowadays acute mycotoxicosis episodes in livestock are very rare in Europe.

Mycotoxins are secondary metabolic products from moulds belonging in particular to the *Aspergillus, Penicillium* and *Fusarium* genera. More than 300 secondary metabolites have been identified although only around 30 have true toxic properties which are of some concern. Toxinogenic moulds may develop under all climatic conditions on any solid or liquid supports as soon as nutritional substances and moisture (water activity Aw over 0.6) are present, hence the wide variety of contaminated foodstuff substrates. These toxins are found as natural contaminants in many feedstuffs of plant origin, especially cereals but also fruits, hazelnuts, almonds, seeds, fodder and foods consisting of, or manufactured from, these products and intended for human or animal consumption. Two groups of toxinogenic (mycotoxin producing) fungi can be distinguished. The first one consists of fungi (such as *Fusarium*) which invade their substrate and produce mycotoxins on the growing plants before harvesting: this is the category of field (pre-harvest) toxins. Aflatoxins and *Fusarium* toxins are included in this group. The other group contains fungi which produce toxins after harvesting and during crop storage and transportation. These toxins are named storage (or post-harvest) toxins and ochratoxin A belongs to this group.

Table 13.3: Mycotoxins and their Physical Properties Producing by different Fungi

Mycotoxin	Main Producing Organism	Molecular Weight	Melting Point, °C
Aflatoxin B$_1$	*Aspergillus flavus, Aspergillus parasiticus, Aspergillus nomius, Aspergillus niger*	312	268-269
Aflatoxin B$_2$	*Aspergillus flavus, Aspergillus parasiticus, Aspergillus nomius, Aspergillus niger*	314	286-289
Aflatoxin G$_1$	*Aspergillus flavus, Aspergillus parasiticus, Aspergillus nomius. Aspergillus niger*	328	244-246
Aflatoxin G$_2$	*Aspergillus flavus, Aspergillus parasiticus, Aspergillus nomius, Aspergillus niger*	330	237-240
Sterigmatocystin	*Aspergillus vesicolor*	324	246
Ochratoxin A	*Aspergilliis ochraceus*	403	169
Aspergillic acid	*Aspergillus flavus*	224	93
Kojic acid	*Aspergillus fumigalus*	142	152.5
Terreic acid	*Aspergillus lerreus*	154	127
Fumagillin	*Aspergillus fumigatus*	459	190-192
Citreoviridin	*Penicillium citreoviride*	434	107-110
Citrinin	*Pen icilliuin cifrinum*	250	172
Penicillic acid	*Pemcillfum puherciilum*	170	83-87
Mycophenolic acid	*Penicillium brevicompactum*	320	138-142
Griseofulvin	*Penicillium janczewski*	353	217-224
Rubratoxin A	*Penicilliuin ruhruin*	442	210-214
Rubratoxin B	*Penicillium rubrum*	518	168-170
Cyclopiazonic acid	*Penicilliuin cyclopium*	336	246
Patulin	*Penicillium patuluin*	154	111
Brevianamide A	*Penicillium viridicatum*		
Deoxynivalenol	*Fusarium graminearum, Fusariiim culmorum, Fusarium crookwellense*	296	131-135
T-2 toxin	*Fusarium poae, Fusarium sporotrichioides*	466	150-151
Fumonisin B$_1$	*Fusarium moniliforme, Fusarium proliferatum*	721	Powder
Zearalenone	*Fusarium graminearum, Fusarium culmorum, Fusarium crookwellense*	318	164

Aflatoxins

Aflatoxins are metabolites produced by many strains: *Aspergillus flavus, Aspergillus parasiticus, Aspergillus nomius* and *Aspergillus niger* and less frequently by *Aspergillus bombycis, Aspergillus ochraceoroseus, Aspergillus nomius, Aspergillus pseudotamari* (Goto *et al.,* 1996) (Klich *et al.,* 2000) (Peterson *et al.,* 2001). The biosynthetic pathway of these secondary metabolites is the polyketide route. Aflatoxins commonly contaminate cereals, peanuts, soybeans, figs, dried fruits, chili peppers, green coffee beans and sometimes milk, eggs and meat (Diener *et al.,* 1987). Normally, aflatoxins are produced at 12°C-40°C and at pH from 3.5 to 8.0 (Bakutis *et al.,* 2006). Storage in conditions of high humidity and temperature can increase formation of aflatoxins on commodities.

Chemically, aflatoxins are polyciclic compounds containing a coumarin nucleus fused to a bifuran and a pentanone. These compounds fluoresce strongly in UV light. Originally they were chromatographically separate in four compounds B1, B2, G1 and G2. Later the metabolic derivatives M1 and M2 were detected in milk. The structure of aflatoxins B1 and M1 are shown in Figure 13.3.

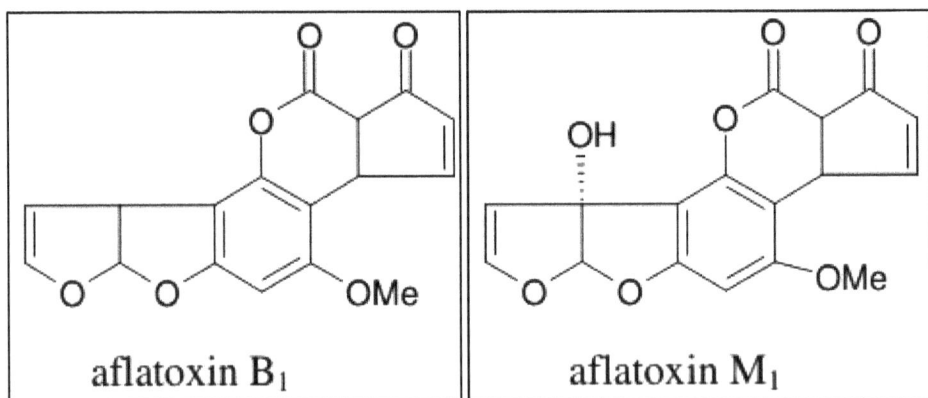

aflatoxin B_1

aflatoxin M_1

Figure 13.3: Chemical Structure of Aflatoxin B1 and Aflatoxin M1.

Aflatoxins are extremely toxic for all vertebrates from fishes to humans, aflatoxin B1 being the most naturally occuring carcinogen known (Squire, 1981), but toxicity varies strongly with animal species. Aflatoxins are well-known carcinogen of class 2B (possibly carcinogenic to humans), mutagen and teratogen agents. The toxicity widely varies with dose, age, sex, species, route of administration, exposure to microbial agents, nutritional state and composition of the diet. It was demonstrated that aflatoxins caused cancer in experimental animals and in humans from epidemiological studies. The carcinogenicity of aflatoxins B1 is due to metabolic activation by cellular enzymes to the corresponding 2, 3-epoxide, compound that binds covalently to DNA generating ireversible mutations (Eaton *et al.,* 1994). Aflatoxin B1-2, 3-epoxide can bind also to proteins determining toxic effects. Aflatoxins affect primarily the liver but produces damages also to other organs (kidney, heart, spleen, and pancreas).

Ochratoxins

Ochratoxins are fungal metabolites of many species of *Aspergillus* in particular *Aspergillus ochraceus*, but also *Aspergillus alliaceus*, *Aspergillus auricomus*, *Aspergillus carbonarius*, *Aspergillus glaucus*, *Aspergillus melleus*, *Aspergillus niger* and *Penicillium* sp. (Abarca *et al.*, 1994) (Bayman *et al.*, 2002) (Ciegler *et al.*, 1972). Normally, ochratoxins are produced at 12°C-37°C but some ochratoxins produced by Penicillium viridicatum occur at lower temperatures 4°C-31°C. These toxins consist of a group of seven compounds closely structural related. The most common metabolite and the most toxic is ochratoxin A with a chemical structure consisting of chlorodihydroisocoumarine linked to L-phenylalanine (Kurata, 1990), as shown in Figure 13.4. Ochratoxin A contaminates a wide range of commodities: cereals particularly barley but also nuts, moldy bread, porcine kidney, coffee beans, beer, wines and dried fruits.

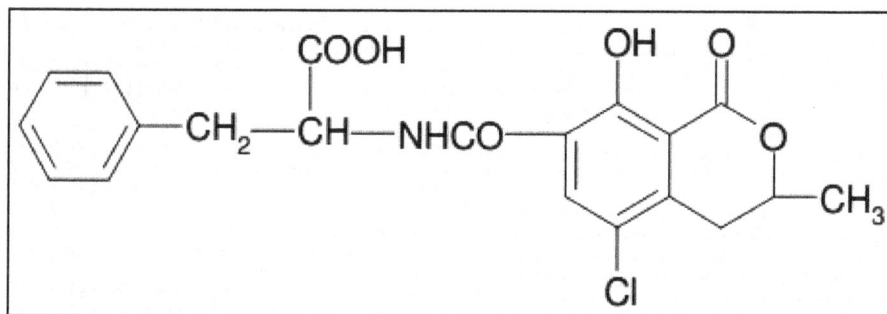

Figure 13.4: Chemical Structure of Ochratoxin A.

Citrinin

The organic anion citrinin is a benzopyran metabolite produced by toxic strains of *Penicillium* and *Aspergillus* species. As such, it can be coproduced with ochratoxin A and an isocoumarin ring is common to the structure of both (Figure 13.5). Both of these mycotoxins have been associated with the development of Balkan endemic nephropathy and urothelial tumors in humans. This is discussed in the section dealing with ochratoxin A and mycotoxin interactions (see below). Originally, citrinin was suggested for use as an antibiotic due to its marked antibacterial activity (Hetherington and Raistrick, 1931). However, animal tests demonstrated it to be severely nephrotoxic, with the detrimental effects far outweighing any potential benefits (Ambrose and DeEds, 1946; Blanpin, 1959). Citrinin has been demonstrated to be acutely toxic

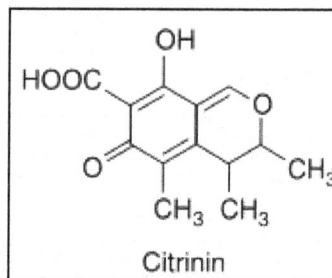

Figure 13.5: Chemical Structure of Citrinin.

in several species, including rabbit, rat, mouse, and hamster, with LD 50 values for intraperitoneal administration of 50, 64, 80, and 75mg/kg body weight, respectively (Hanika and Carlton, 1983; Jordan and Carlton, 1977, 1978; Jordan *et al.*, 1978b).

Toxicity varies considerably with route of administration, with 134mg/kg epresenting the oral LD 50 in rabbits.

Fumonisins

Fumonisins are a group of water-soluble bifuranocumarin mycotoxins (Figure 13.6) which, under suitable environmental or storage conditions, may be produced by several *Fusarium* species, in particular *Fusarium moniliforme* and *F. proliferatum* (Marasas *et al.*, 1988). Several fumonisins have been identified to date, however of these fumonisin B1and B2(FB1,FB2) are the most abundant, making up 70 per cent of the total concentration of fumonisins detected (Prozzi *et al.*, 2000), and also the most toxic and hence the most investigated of the group (Sydenham *et al.*, 1991). Commonly found as a contaminant of corn and in particular overwintered grain, it is thought that FB1is produced as a result of an endophytic relationship with grain (Bacon and Hinton, 1996), imparting an increased resistance to diseases or insects.

FB1has been determined to be neurotoxic, hepatotoxic, and toxic to the lung, and has also been associated with the development of esophageal cancer, particularly in areas where corn forms part of the staple diet (Gelderblom *et al.*, 1988; Haschek *et al.*, 1992; Norred, 1993). As in the case with many other mycotoxins, the concentrations required to cause these toxic syndromes vary both in effect induced and with the species tested.

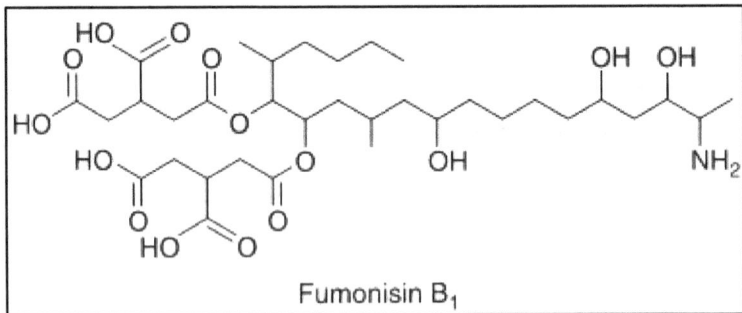

Figure 13.6: Chemical Structure of Fumonisin.

Conclusion

It has been noted since the earliest days of fungal manipulation that many species of filamentous fungi readily synthesize complex compounds that are putatively helpful but not necessary for survival and whose production is presumably costly to maintain. Furthermore, production is often linked to fungal development. Some compounds might function as virulence factors, or their presence could give a competitive edge to the producing organism or enhance the survivability of spores. Some secondary metabolites stimulate sporulation and therefore influence the development of the producing organism and neighboring members of the same species, perhaps enhancing the fitness of a community of related species. Natural products or secondary metabolites are often produced late in fungal development, and their biosynthesis is complex. This complexity is due to a number of factors that affect secondary metabolite production. These include (i) the influence of a number of external and internal factors on natural product biosynthesis, (ii) the involvement of many sequential enzymatic reactions required for converting primary building blocks into natural products, (iii) tight regulation of natural product enzymatic gene expression by one or more transcriptional activators, (iv) close association of natural product biosynthesis with primary metabolism, and (v) close association of natural products with later stages of fungal development, particularly sporulation. Furthermore, the genes required for biosynthesis of some natural products are clustered, perhaps as a consequence of these factors. Gene clusters contain all or most of the genes required for natural product biosynthesis, and logic suggests that their maintenance could only be selected for if the final natural product conferred some advantage to the producing organism.

References

Ainsworth, G.C. 1973. Introduction and keys to higher taxa. In: Ainsworth, G.C., Sparrow, F.K. and Sussman, A.S. (Eds) *The Fungi: An Advanced Treatise.* Vol. IV. Academic Press, New York.

Anke T, Thines E. 2007. Fungal metabolites as lead structures for agriculture. In: *Exploitation of Fungi* (Robson GD, VanWest P, Gadd GM, eds.): Cambridge University Press, Cambridge, UK. pp. 45–58

Barr, D.J.S. 1992. Evolution and kingdoms of organisms from the perspectives of a mycologist. *Mycologia.* 84: 1–11.

Berbee, M. and Taylor, J.W. 1999. Fungal phylogeny. In: *Molecular Fungal Biology* (Oliver, R.P. and Schweizer, M. eds.), Cambridge University Press. pp. 21–77.

Bérdy J. 2005. Bioactive microbial metabolites: a personal view. *Journal of Antibiotics* 58, 1–26.

Berdy, J., 1995. Are actinomycetes exhausted as a source of secondary metabolites? *Biotechnologia,* 7–8: 13–34.

Berg, J.M., Tymoczko, J.L. and Stryer, L., 2002. *Biochemistry,* 5th dn. W.H. Freeman and Company, New York,

Blain, J.A. 1975. Industrial enzyme production. In: *The Filamentous Fungi.* Smith, J.E. and De Berry (Eds) Arnold London. 1: 193–211.

Bode HB, Bethe B, Höfs R, Zeek A. 2002. Big effects from small changes: possible ways to explore nature's chemical diversity. *Chem Bio Chem.* 3: 619–627.

Bode, H.B., Bethe, B., Höfs, R., Zeek, A. 2002. Big effects from small changes: possible ways to explore natures chemical diversity. *Chem Bio Chem.* 3: 619–627.

Bruns, T.D., T.J. White, and J. W. Taylor. 1991. Fungal molecular systematics. Annu. *Rev. Ecol. Syst.* 22: 525–564.

Butler, M.S. 2004. The role of natural product in chemistry in drug discovery. *J. Nat. Prod.* 67: 2141–2153.

Chin, Y.W., Balunas, M.J., Chai, H.B. and Kinghorn, A.D. 2006. Drug discovery from natural sources. *The AAPS Journal* 8, 239–253.

Chun E, Han CK, Yoon JH, Sim TB, Kim YK, Lee KY. 2005. Novel inhibitors targeted to methionine aminopeptidase 2 (MetAP2) strongly inhibit the growth of cancers in xenografted nude model. *International Journal of Cancer* 114, 124–130.

Cole R. J. and Cox R. H. 1981. *Hàndbook of toxin fungal metabolites*. New York: Acad. Press,

Costacurta A, Vanderleyden J. 1995. Synthesis of phytohormones by plant–associated bacteria. *Crit Rev Microbiol.* 21: 1–18.

Cragg, G.M.; Newman, D.J. 2005. Biodiversity: A continuing source of novel drug leads. *Pure Appl. Chem.* 77: 7–24.

Curran, M.P. and Keating, G.M. 2005. Mycophenolate sodium delayed release: prevention of renal transplant rejection. *Drugs* 65: 799–805.

Darkes MJM and Plosker GL. 2002. Cefditoren pivoxil. *Drugs* 62, 319–336.

Delcambe, L. 1970. Infonnation BulletinNo.8.International Center of Information on Antibiotics, Etud'imprim, Liege, Belgium.

Denyer, S.P., Hodges, N.A. and German, S.P., Hugo 2004. *Russell's Pharmaceutical Microbiology*, 7 th edn. Blackwell Science, India,

Donadio S, Monicardini P, Alduina R, Mazzaa P, Chiocchini C, Cavaletti L, Sosio M, Puglia AM. 2002. Microbial technologies for the discovery of novel bioactive metabolites, *Journal of Biotechnology.* 99: 187–198.

Dutta, A.C. 2005. Botany for Degree Students 18th edn. Oxford University Press, New York,

Eveleigh, D.E. 1981. The microbial production of industrial chemicals. *Scientific American* 245: 155–178.

Fluckiger E., 1980. In *Ergot compounds and Brain function: neuroendocrine and neuropsychiatric aspects*,Goldstein M., Ed., New York: Raven Press,

Frattarelli DAC, Reed MD, Giacoia GP, Aranda JV. 2004. Antifungals in systemic neonatal candidiasis. *Drugs* 64, 949–968.

Gunatilaka, A.A.L. 2006. Natural products from plant–associated micro-organisms: Distribution, structural diversity, bioactivity, and implications of their occurrence. *Journal of Natural Products.* 69: 509–526.

Haefner, B. 2003. Drugs from the deep: Marine natural products as drug candidates. Drug Discov. *Today.* 8: 536–544.

Hawksworth, D.L. 2001. The magnitude of fungal diversity: the 1.5 million species estimate revisited. *Mycological Research* 105: 1422–1432.

Hugo, W.B. and Russell, A.D. 1998. *Pharmaceutical Microbiology*, 5th edn. Blackwell Science, U K.

Kirk PM, Cannon PF, David JC, Stalpers JA (eds) 2001. *Ainsworth and Bisby's Dictionary of the Fungi*. 9th edition. CABI Publishing, Wallingford.

Kock, JLF, Strauss T, Pohl CH, Smith DP, Botes PJ, Pretorius EE, Tepeny T, Sebolai O, Botha A, Nigam S. 2001. Bioprospecting for novel oxylipins in fungi: the presence of 3–hydroxy oxylipins in Pilobolus. *Antonie van Leeuwenhoek* 80: 93–99.

Kock, JLF, Strauss T, Pohl CH, Smith DP, Botes PJ, Pretorius EE, Tepeny T, Sebolai O, Botha A, Nigam S. 2001 Bioprospecting for novel oxylipins in fungi: the presence of 3–hydroxy oxylipins in *Pilobolus. Antonie van Leeuwenhoek* 80: 93–99.

Kozlovskii A. G., Vinokurova N. G., Adanin V. M., and Sedmera P. 1997. Prikl. Biokhim. Mikrobiol. 33, 408–414 [*Appl. Biochem. Microbiol.*(Engl.Transl.).33: 364–369].

Kozlovskii A. G., Vinokurova N. G., Solov'eva T. F., and Buzilova I. G. 1996. Prikl Biokhim Mikrobiol. 32, 43–52 [*Appl. Biochem. Microbiol.*(Engl. Transl.), 1996, 32, 39–48].

Kozlovsky A. G., Adanin V. M., Daze Kh. M., and Grefe U. 2001. Prikl. Biokhim. Mikrobiol. 37, 292–296. [*Appl. Biochem. Microbiol.*(Engl. Transl.), 37, 253–256]

Kozlovsky A. G., Vinokurova N. G., Adanin V. M., and Grafe U. 2000. *Nat. Prod. Lett.* 14, 333–340.

Kurtzman, C.P. 1983. Fungi as source of food, fuel and biochemicals. *Mycologia.* 75 (2): 374–382.

Liu CH, Zou XW, Lu H, Tan RX. 2001. Antifungal activity of Artemisia annua endophyte cultures against phytopathogenic fungi. *Journal of Biotechnology* 88: 277–282.

Liu CH, Zou XW, Lu H, Tan RX. 2001. Antifungal activity of *Artemisia annua* endophyte cultures against phytopathogenic fungi. *Journal of Biotechnology.* 88: 277–282.

McMorris TC, Kelner MJ, Wang W, Yu J, Estes LA, Taetle R. 1996. Hydroxymethyl acylfulvene: an illudin derivative with superior antitumour properties. *Journal of Natural Products* 59, 896–899.

Mishra, B.B.; Tiwari, V.K. 2011. Natural products: An evolving role in future drug discovery. *Eur. J. Med. Chem.* 46, 4769–4807.

Mitchell AM, Strobel GA, Hess WM, Vargas PN, Ezra D. 2008. Muscodor crispans, a novel endophyte from *Anans ananassoides* in the Bolivian Amazon. *Fungal Diversity* 31: 37–43.

Peláez F. 2005. Biological activities of fungal metabolites. In: *Handbook of Industrial Mycology* (An Z., ed.): 49–92. Marcel Dekker, New York, USA.

Perlman, D.1967. Clzel71.Week101,82–85,88,9.3–95,98, and 100–101.

Perlman,D.1968. *Process Bioche.* 171.3: 54–58.

Perlman, D.1970. Fallerstein Lab.COl71l71un.33: 165–175.

Rehachek Z., and Sajdl P. 1990. *Ergotalkaloids. Chemistry, Biological Effect, Biotechnology.* Praha: Academia,.

Rey–Ladino, J.; Ross, A.G.; Cripps, A.W.; McManus, D.P.; Quinn, R. 2011. Natural products and the search for novel vaccine adjuvants. *Vaccine*, 29: 6464–6471.

Schlegel, H.G., 2003. *General Microbiology*, 7th ed. Cambridge University Press, Cambridge.

Scott LJ, Curran MP, Figgitt DP. 2004.Rosuvastatin: a review of its use in the management of dyslipidemia. *American Journal of Cardiovascular Drugs* 4, 117–138.

Sommer, C.V., 2006. Antibiotics. In Shapp MG, Gerald FC, Feder B, and Martin LA (eds), *The New Book of Knowledge*, pp. 306–312.

Stadler M, Keller NP. 2008. Paradigm shifts in fungal secondary metabolite research. *Mycological Research*. 112: 127–130.

Suryanarayanan TS, Thirunavukkarasu N, Govindarajulu MB, Sasse F, Jansen R, Murali TS. 2009. Fungal endophytes and bioprospecting. *Fungal Biology Reviews* 23: 9–19.

Suryanarayanan, TS, Hawksworth DL. 2005. Fungi from little explored and extreme habitats. In: *Biodiversity of Fungi; Their Role in Human Life*. (Deshmukh SK, Rai MK, eds.): Oxford and IBH Publishing Co. Pvt. Ltd., New Delhi, India. 33–48.

Talaro, K. and Talaro, A. 2002. *Foundations in Microbiology*, 4th edn. McGraw Hill, New York.

Tapan, T., Livesoksa, J., Laasko, S. And Rosenquist, H. 1993. Interaction of Abscisic acid and Indole–3–acetic acid producing fungi with Salix leaves. *Journal of Plant Growth Regulators*. 12: 149–156.

Taylor, D.J., Green, D.P.O. and Stout, G.W., 2003. *Biological Science*, 3rd edn. Cambridge University Press, Cambridge.

Thomashow, L.S. and Weller, D.M. 1995. Current Concepts in the Use of Introduced Bacteria for Biological Disease Control: Mechanisms and Antifungal Metabolites. In Stacey G and Keen N (eds), *Plant Microbe Interactions*. Chapman and Hall, New York, pp. 187–235.

Tudzynsk, B. 1977. Biosynthesis of gibberellins in Gibberelle fujikuroi: Biochemical aspects. *Applied Microbiology and Biotechnology*. 52: 298–310.

2015, Recent Trends in Microbiology, Mycology and Plant Pathology   Pages 197–211
Editor: Dr. H.C. Lakshman
Published by: DAYA PUBLISHING HOUSE, NEW DELHI

Chapter 14

Fungal Pectinases and their Biotechnological Applications

D.K. Dushyantha and K.S. Jagadeesh*

Department of Agricultural Microbiology,
University of Agricultural Sciences, Dharwad – 580 005, Karnataka, India
*E-mail: jagsbio@gmail.com

ABSTRACT

Pectinases are one of the upcoming enzymes of fruit and textile industries. These enzymes break down complex polysaccharides of plant tissues into simpler molecules like galacturonic acids. The role of acidic pectinases in bringing down the cloudiness and bitterness of fruit juices is well established. Recently, there has been a good number of reports on the application of alkaline pectinases in the textile industry for the retting and degumming of fiber crops, production of good quality paper, fermentation of coffee and tea, oil extractions and treatment of pectic waste water. This review discusses various types of pectinases, their production under solid substrate fermentation conditions and applications in various industries.

Introduction

Advent of biotechnology helped to unlock novel food ingredients through the use of biotechnologically-derived industrial enzymes. Microbes have played pivotal role in this progress and currently the fermented products contribute adequate share. Their role in bioconversion of waste commodities into value added products has been highlighted in the recent decades (Ahuja *et al.,* 2004). One of the significant applications of agro-industrial wastes is biotechnological production of enzymes such as pectinases for their food applications. There is an increasing tendency among the people to use chemical free foods. The use of enzymes like pectinases in the food processing can meet such public demands (Pandey *et al.,* 2000). Tremendous amounts of agro-based wastes/by- products are generated every year over the globe and their improper management causes environmental pollution. Being a co-partner, food industry is contributing a significant share in producing agro-based biological wastes. The elimination of wide range of pollutants and wastes is an absolute requirement to promote a sustainable and friendly environment. Biotechnology plays a major role in the bioconversion of agro-industrial wastes and is taking advantage of the greater adaptability of micro-organisms to degrade/convert such compounds into value added products.

Pectin: The Miracle Molecule

All terrestrial plants contain pectin, which binds with cellulose, and creates protopectin, a substance that gives plants their structure. Pectin is found primarily in the plant cell walls, and in the region between cell walls, called the lamella, where it assists in the binding of one cell wall to another. In addition to giving plants their structure, pectin has other important roles, such as determining how porous the cell is, and its pH (Voragen, *et al.,* 2009).

Structure of Pectin

Chemically, pectic substances are complex colloidal acid polysaccharides, with a backbone of galacturonic acid residues linked by α (1±4) linkage. Based on the type of modifications of the backbone chain, pectic substances are classified into protopectin, pectic acid, pectinic acid and pectin (Be Miller, 1986).

As a plant matures, the protopectin, which is insoluble, is broken down by the enzymes pectinase and Polygalacturonase by hydrolysis. This process, started by the enzyme pectin methylesterase, creates low methyl pectin which is the substrate for polygalacturonase.

Classification of Pectic Enzymes

Pectinases are group of enzymes that attack pectin and depolymerise it by hydrolysis and transelimination as well as by deesterification reactions, which hydrolyse the ester bond between carboxyl and methyl groups of pectin (Ceci and Loranzo, 1998). Figure 14.1 shows the action mode of the most studied pectinases.

Pectinases are classified under three headings according to the following criteria: whether pectin, pectic acid or oligo-D-galacturonate is the preferred substrate, whether pectinases act by trans-elimination or hydrolysis and whether the cleavage is at random (endo-, liquefying of depolymerizing enzymes) or endwise (exo- or saccharifying enzymes). Table 14.1 gives the classification of pectinolytic enzymes.

Protopectinases

They solubilize protopectin forming highly polymerized soluble pectin. They are classified into two types: one reacts with the polygalacturonic acid region of protopectin, A type; the other with the polysaccharide chains that may connect the polygalacturonic acid chain and cell wall constituents, B type.

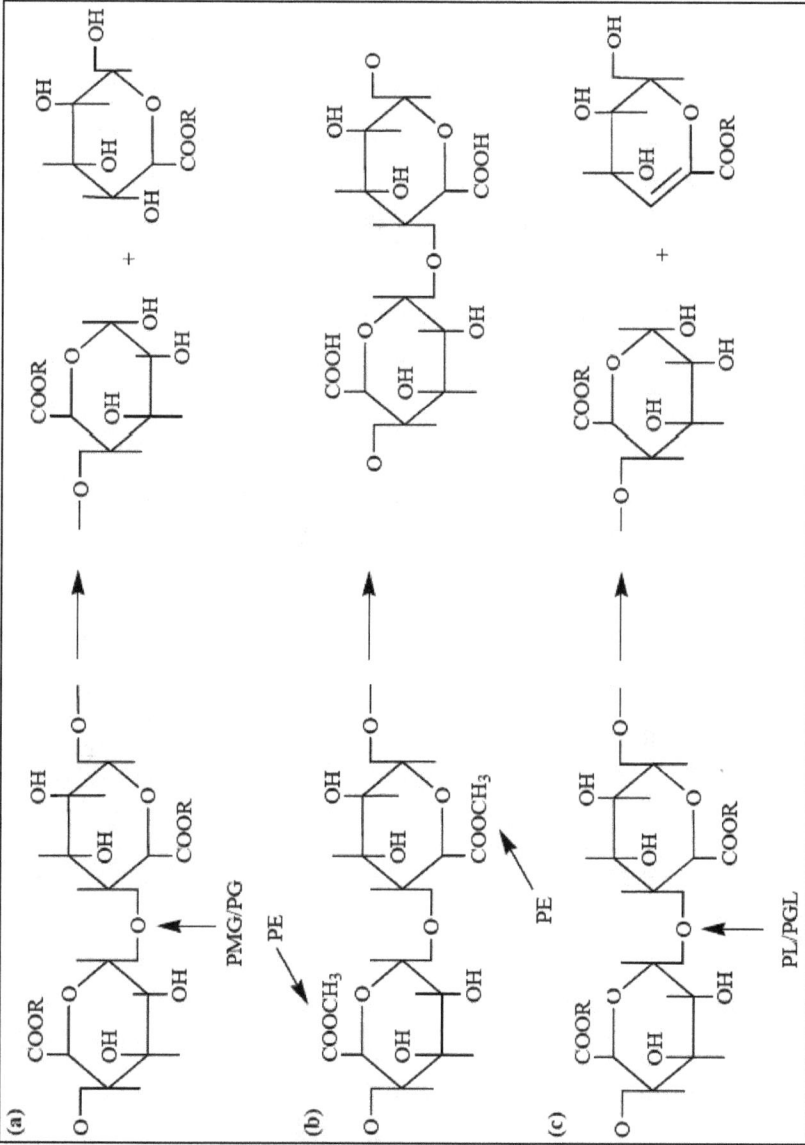

Figure 14.1: Mode of Action of Pectinases: (a) R = H for PG and CH₃ for PMG; (b) PE; and (c) R = H for PGL and CH₃ for PL. The arrow indicates the place where the pectinase reacts with the pectic substances. PMG: Polymethylgalacturonases; PG: Polygalacturonases (EC 3.2.1.15); PE: Pectinesterase (EC 3.1.1.11); PL: Pectinlyase (EC-4.2.2.10), Lang and Dornenburg (2000).

Table 14.1: An Extensive Classification of Pectinolytic Enzymes (Jayani *et al.,* 2005)

Enzyme	E.C. no.	Modified EC Systematic Name	Action Mechanism	Action Pattern	Primary Substrate	Product
Esterase						
Pectin methyl esterase	3.1.1.11	–	Hydrolysis	Random	Pectin	Pectic acid + methanol
Depolymerizing enzymes: Hydrolases						
1. Protopectinases	–	–	Hydrolysis	Random	Protopectin	Pectin
2. Endopolygalacturonase	3.2.1.15	Poly-(1-4)-α-D-galactosidu-ronate Glycanohydrolase	Hydrolysis	Random	Pectic acid	Oligogalacturonates
3. Exopolygalacturonase	3.2.1.67	Poly-(1-4)- α -D-galactosidu-ronate Glycanohydrolase	Hydrolysis	Terminal	Pectic acid	Monogalacturonates
4. Exopolygalacturonan-digalacturono hydrolase	3.2.1.82	Poly-(1-4)- α -D-galactosidu-ronate Digalacturonohydrolase	Hydrolysis	Penultimate-bonds	Pectic acid	Digalacturonates
5. Oligogalacturonate hydrolase	–	–	Hydrolysis	Terminal	Trigalacturonate	Monogalacturonates
6. D4:5Unsaturated oligo-galacturonate hydrolases	–	–	Hydrolysis	Terminal	D4:5 (Galacturo-nate)n	Unsaturated monogalac-turonates and saturated (n-1)
7. Endopolymethyl-galacturonases	–	–	Hydrolysis	Random	Highly esterified pectin	Oligomethylgalacturona-tes
8. Endopolymethyl-galacturonases	–	–	Hydrolysis	Terminal	Highly esterified pectin	Oligogalacturonates
Lyases						
1. Endopolygalacturonase lyase	4.2.2.2	Poly-(1-4)- α -D-galactosidu-ronate Lyase	Trans-elimination	Random	Pectic acid	Unsaturated oligogala-cturonates
2. Exopolygalacturonase lyase	4.2.2.9	Poly-(1-4)- α -D-galactosidu-ronate Exolyase	Trans-elimination	Penultimate-bond	Pectic acid	Unsaturated digalacturo-nates

Contd...

Table 14.1–Contd...

Enzyme	E.C. no.	Modified EC Systematic Name	Action Mechanism	Action Pattern	Primary Substrate	Product
3. Oligo-D-galactosiduronate lyase	4.2.2.6	Oligo-D-galactosiduronate lyase elimination	Trans-	Terminal	Unsaturated digalacturonates	Unsaturated monogalacturonates
4. Endopolymethyl-D-galactosiduronate Lyase	4.2.2.10	Poly (methyl galactosiduronate) Lyase	Trans-elimination	Random	Unsaturated poly-(methyl-D-digalacturonates)	Unsaturated methyloligogalacturonates
Exopolymethyl-D-galacto siduronate Lyas	–	–	Trans-elimination	Terminal	Unsaturated poly-(methyl-D-digalacturonates)	Unsaturated methyl-monogalacturonates

Pectin Methyl Esterases (PME)

They catalyze deesterification of the methoxyl group of pectin forming pectic acid and methanol. The enzyme acts preferentially on a methyl ester group of galacturonate unit next to a non-esterified galacturonate unit. It acts before polygalacturonases and pectate lyases which need non-esterified substrates.

Pectin Acetyl Esterases (PAE)

They hydrolyse the acetyl ester of pectin forming pectic acid and acetate. It is classified into carbohydrate esterase families 12 and 13.

Polymethylgalacturonases (PMG)

They catalyze the hydrolytic cleavage of alpha-1,4-glycosidic bonds in pectin backbone, preferentially highly esterified pectin, forming 6-methyl-D-galacturonate.

Polygalacturonases (PG)

They catalyze hydrolysis of alpha-1,4-glycosidic linkages in polygalacturonic acid producing D-galacturonate. It is classified into glycosyl-hydrolases family 28. Both groups of hydrolase enzymes (PMG and PG) can act in an endo- or exo- mode. Endo-PG and endo-PMG catalyze random cleavage of substrate, exo-PG and exo-PMG catalyze hydrolytic cleavage at substrate nonreducing end producing monogalacturonate or digalacturonate in some cases. Hydrolases are produced mainly by fungi, being more active on acid or neutral medium at temperatures between 40 °C and 60 °C.

Pectate Lyase

They cleave glycosidic linkages preferentially on polygalacturonic acid forming unsaturated product (delta-4,5-D-galacturonate) through transelimination reaction. PGL has an absolute requirement of Ca2+ ions. Hence it is strongly inhibited by chelating agents as EDTA. Pectate lyases are classified as endo-PGL that acts towards substrate in a random way, and exo-PGL that catalyze the substrate cleavage from nonreducing end.

Pectin Lyases (PL)

They catalyze the random cleavage of pectin, preferentially high esterified pectin, producing unsaturated methyloligogalacturonates through transelimination of glycosidic linkages. PLs do not have an absolute requirement of Ca^{2+} but they are stimulated by this and other cations. Up until now, all described pectin lyases are endo-PLs.

Table 14.2 furnishes the Biochemical and physico-chemical properties of pectinases.

Pectinolytic Fungi from different Sources

The ability of filamentous fungi to grow on rather simple and inexpensive substrates as well as their capacity to produce a wide range of commercially interesting metabolites have attracted considerable interest to exploit them as production organisms in biotechnology. Diverse compounds ranging from simple organic acids to complex secondary metabolites are produced for the use in various

Table 14.2: Biochemical and Physico-chemical Properties of Pectinases

Source of Pectinase	Nature	Molecular Weight (kDa)	pI	Specific Activity (U mg⁻¹)	Km	Optimum Temp. (°C)	Optimum pH	Temp. Stability (°C)	pH Stability	Reference
Polygalacturonases										
Aspergillus japonicus	Endo (PG I)	38	5.6	–	–	30	4.0–5.5	–	–	Hasunuma *et al.* (2003)
	Endo (PG II)	65	3.3	–	–	30	4.0–5.5	–	–	
Mucor flavus	–	40	8.3	–	–	45	3.5–5.5	40	2.5–6.0	Margo *et al.* (1994)
Aspergillus niger	Endo (PG I)	61	–	982	0.12	43	3.8–4.3	50	–	Singh and Rao (2002)
	Endo (PG II)	38	–	3750	0.72	45	3.0–4.6	51	–	
Penicillium frequentans	Exo	63	–	2571	1.60	50	5.0	–	–	Favey *et al.* (1992)
	Exo	79	–	185	0.06	50	5.8	–	–	
Aspergillus awamori	Endo	41	6.1	487	–	40	5.0	50	4.0–6.0	Nagai *et al.* (2000)
Fusarium oxysporum. sp. lycopersci	Exo	38	–	209	–	69	11.0	–	7.0–11.0	Pietro and Roncero (1996)
Pectin lyases										
Penicillium italicum	PMGL	22	8.6	–	3.20	40	6.0–7.0	50	8.0	Alana *et al.* (1990)
Aspergillus japonicus	PMGL	–	7.7	–	0.16	55	6.0	–	–	Dinnella *et al.* (1994)
Penicillium adametzii	PMGL	–	–	–	–	60	8.0	40	7.0	Sapunova *et al.* (1995)
P. citrinum	PMGL	–	–	–	–	45	7.0	40	7.0	
P. janthinellum	PMGL	–	–	–	–	40	6.5	40	7.0	
Fusarium monoliforme	PGL	–	–	–	–	–	8.5	75	–	Dixit *et al.* (2004)
Pectin esterases										
Aspergillus niger	–	–	–	–	1.01	45	5.0	–	–	Maldonaldo and Saad (1998)
A. japonicus	–	46(PE I)	3.8	–	–	–	4.0–5.5	50	–	Hasunuma *et al.* (2003)
	–	47(PE II)	3.8	–	–	–	4.0–5.5	50	–	

market segments. Due to their exceptional high capacity to express and secrete proteins, filamentous fungi have become indispensable for the production of enzymes of fungal and non-fungal origin. Filamentous fungi exhibit characteristics that make them good models for industrial applications. Noteworthy, among some of these are their capacity for fermentation, the production of large quantities of extracellular enzymes (*e.g.* several grams per liter in strains of *Aspergillus*), the feasibility of cultivation, and the low-cost of production in large bioreactors (Aro *et al.*, 2005). The most active producers were selected by screening among the cultures of thermophilic, alkali, acidophilic and halophilic microscopic fungi strains. The cultures have been selected from the following genera *Aspergillus, Penicillium, Mucor, Trichoderma, Rhizopus, Sporotrichum, Chaetomium* and *Stempillium.*

Different Agro-wastes for Pectinase Production under Solid State Fermentation (SSF) Conditions

Agro-industrial residues are generally considered the best substrates for the SSF processes, and use of SSF for the production of enzymes is no exception to that. A number of such substrates have been employed for the cultivation of micro-organisms to produce host of enzymes. Table 14.1 gives an account of various substrates that have been used for pectinase production under SSF.

Solid-state fermentation (SSF) has been defined as the fermentation process which involves solid matrix and is carried out in absence or near absence of free water; however, the substrate must possess enough moisture to support growth and metabolism of the micro-organism. The solid matrix could be either the source of nutrients or simply a support impregnated by the proper nutrients that allows the development of the micro-organisms. (Pandey, 2003). The potential of SSF lies in bringing the cultivated micro-organism in close vicinity of substrate and achieving the highest substrate concentration for the fermentation. SSF resembles the natural habitat of micro-organism and is, therefore, preferred choice for micro-organisms to grow and produce useful value added products.

Though historically known since centuries, SSF gained a fresh attention from researchers and industries all over the world since recent few years, mainly due to few advantages it offers over liquid (submerged) fermentation, particularly in areas of solid waste management, biomass energy conservation and its application to produce high value–low volume products such as biologically active secondary metabolites, etc., apart from the production of food, feed, fuel and traditional bulk chemicals (Pandey *et al.*, 2007). Capability of genetic manipulation of fungal strains has broadened horizon for SSF enabling the technology for the production of recombinant proteins and value added chemicals.

Pectin lyase and polygalacturonase production by newly isolated *Penicillium viridicatum* strain RFC3 was carried out by means of solid state fermentation using orange bagasse, corn tegument, wheat bran and mango and banana peels as carbon sources (Silva *et al.*, 2002). The maximal activity value of polygalacturonase (PG) (30 Ug^{-1}) was obtained using wheat bran as carbon source while maximal pectin lyase (PL) (2000 Ug^{-1}) activity value was obtained in medium composed of orange bagasse. Mixtures of banana or mango peels with sugar cane bagasse resulted in increased

PG and PL production compared to fermentations in which this residue was not used. The mixture of orange bagasse and wheat bran (50 per cent) increased the production of PG and PL to 55 Ug^{-1} and 3540 Ug^{-1} respectively. Maximal activity of PG and PL fractions was determined at 55°C and 50°C respectively. PG was stable in neutral pH range and at 40 °C whereas PL was stable in acidic pH and at 35°C, for 1 h.

Palaniyappan *et al.* (2009) conducted an experiment on different natural substrates such as wheat flour and corn flour in comparison with synthetic pectin for the production of pectinase using *Aspergillus niger* (MTCC 281). The work involves optimizing various parameters like substrate concentration, pH, temperature, rpm, time of fermentation for the production of pectinase and the effect of carbon sources on the synthesis of the pectinase enzyme to suggest a plausible, commercially suitable substrate than the standard. The experimental studies indicate that maximum synthesis of pectinase (6.1 U/mL) was obtained with *A. niger* (MTCC 281) by using wheat as substrate under the influence of the additional carbon source, starch. The optimal conditions are found to be substrate concentration–wheat (1 per cent), pH 5.5, temperature 30°C, time 72 h, rpm 170, and carbon source starch (0.025 per cent).

Possibility of producing pectinase utilizing fruit wastes of cashew, banana, pineapple, and grape under controlled fermentation with *Aspergillus foetidus* was studied (Venkatesh *et al.,* 2009). Among the different media composition tried, medium containing 5 g fruit waste + 0.05 g urea + 0.25 g ammonium sulphate supported better growth of the micro-organism. Enzyme production was maximum in the medium with grape waste. As an extractant, distilled water was better than CaCl$_2$. The ideal temperature and duration of fermentation were 40°C and 8 days respectively.

Pectinase production studies were carried out in submerged and solid-state conditions from deseeded sunflower head employing *Aspergillus niger* (Patil and Dayanand, 2006). The two potential strains of *A. niger*, DMF 27 for submerged and DMF 45 for solid-state were isolated by multi-step screening technique based on coefficient of pectolysis and capability of pectinase production. Process variables such as size of inoculum, pH, temperature, particle size and moisture content were optimized with an aim to achieve the maximum production of pectinases. The increased level of pectinase production was recorded at pH 5.0 and temperature 34 °C in submerged and solid-state conditions. The optimum inoculum size was 1X10^5 ml^{-1} for submerged and 1X10^7 g^{-1} for solid-state conditions. Five hundred micrometer particle size and 65 per cent moisture content of the substrate were optimum for the maximum production of pectinases in solid-state condition. Under optimum conditions, maximum production of exo-pectinase was 34.2U/g in SSF and endo-pectinase was 12.6U/ml in SmF.

SSF process was described for the production of pectinase by *Aspergillus niger* Aa-20 and lemon peel pomace (LPP) as support and carbon source in a solid-state bioreactor (Ruiz *et al.,* 2012). Invasion ability of selected fungal strain was analyzed on four particle sizes of LPP. The SSF process was operated in a column-tray bioreactor at 30 °C and 70 per cent moisture content, 194 mL/min of air flow rate and substrate particle size (2–0.7 mm) of LPP for 96 h. Results showed that high levels of pectinase activities were obtained. The maximum pectinase activity obtained was 2181 U/L. Maximum biomass and maximum specific growth rate of *A. niger* Aa-20 were V_{max} = 8 mg glucosamine/g of LPP and V_{max} = 0.127 1/h. The LPP and the use of *A. niger* Aa-20 in SSF was suggested as a very promising process for pectinase production.

Endo-polygalacturonase (endo-PG), exo-polygalacturonase (exo-PG) and pectin lyase (PL) were produced by solid-state fermentation of a mixture of orange bagasse and wheat bran (1:1) with the filamentous fungus *Penicillium viridicatum* RFC3 (Silva *et al.*, 2005). This substrate was prepared with two moisture contents, 70 per cent and 80 per cent, and each was fermented in two types of container, Erlenmeyer flask and polypropylene pack. When Erlenmeyer flasks were used, the medium containing 80 per cent of initial moisture did not afford higher PL production while exo- nor endo-PG production was influenced by substrate moisture. The highest enzyme activities obtained were 0.70 U mL^{-1} for endo-PG, 8.90 U mL^{-1} for exo-PG, and 41. 30 U mL^{-1} for PL. However, when the fermentation was done in polypropylene packs, higher production of all three enzymes was obtained at 70 per cent moisture (0.7 and 8.33 U mL^{-1} for endo- and exo-PG and 100 U mL^{-1} for PL). An increase in the pH and decrease in the reducing sugar content of the medium was observed.

The synthesis of pectolytic enzymes appears to be considerably influenced by the glucose content of the cultivation medium containing mixed carbon sources in a predetermined optimal ratio. Studies conducted to examine the effect of glucose showed that glucose severely reduced the synthesis of polygalacturonase and polymethylgalacturonase at concentrations above 5 g/l while pectinlyase was unaffected (Panda *et al.*, 2004). *In vitro* studies showed that the inhibition of the two pectin hydrolases was reversible and competitive in nature. The results of the study also indicated induction of pectolytic enzymes by corn in *Aspergillus niger* NCIM 548.

Fruit-processing wastes including apple pomace, cranberry pomace and strawberry pomace were used as substrates for polygalacturonase (PG) production by *Lentinus edodes* through solid-state fermentation (Zheng and Shetty, 2000). Strawberry pomace was the best substrate for highest PG yield, followed by apple pomace, while cranberry pomace was not a suitable substrate for PG production in this study. The highest PG activity was obtained after 40 days of culture and the yields from strawberry pomace, apple pomace and cranberry pomace were 29.4 U, 20.1 U and 14.0 U per gram of pomace, respectively. PG activity was increased by the addition of polygalacturonic acid in the apple pomace and cranberry pomace media, but was not affected in strawberry pomace medium. The PG produced by *L. edodes* from strawberry pomace exhibited a maximal activity at 50 °C and at pH 5.

Purification and Characterization of Pectinase Enzyme

During enzyme production process, in addition to the target enzyme, growth medium may have some undesirable metabolites of the micro-organisms. The purified enzymes exhibit higher activity, lesser risk of harmful substances and thus better affectivity for the specific product.

Removal of Microbial Cells and other Solid Matter

Microbial cells and other insoluble materials are separated from the harvested broth by filtration or centrifugation. Filtration separates particles simply on the basis of their size. Commonly used filter is diatomaceous earth. Ultrafiltration is one of the purification step whereby water and low molecular weight materials are removed by passage through a membrane under pressure, while the enzyme being retained. Enzymes can be concentrated by precipitation, and this is generally used as the initial step of purification. Salting out ammonium sulphate is the best known method for concentration and purification of the enzyme. Some enzymes do not survive with the ammonium sulphate precipitation. In such cases organic solvents such as ethanol, propanol and acetone can be used as the alternative. For further purification, the ammonium sulphate present in the protein precipitate is to be removed, this can be achieved by dialysis, ultrafiltration or sephadex G-100 coloumn.

An extracellular pectinase (PECI) was purified to apparent homogeneity from liquid state cultures of the thermophilic fungus *Acrophialophora nainiana* by ultrafiltration and a combination of gel filtration and ion-exchange chromatographic procedures (Celestino *et al.,* 2006). The molecular masses of PECI were 35,500 and 30,749 Da, as determined by SDS-PAGE and mass spectrometry, respectively. It was more active at 60 °C and pH 8.0 and showed high stability at 50 °C with half-life of 7 days.

Al-Najada *et al.* (2012) worked on partial purification and characterization of polygalacturonases from fruit spoilage *Fusarium oxysporum* and *Aspergillus tubingensis* isolated from banana and peach, respectively. By using diethylaminoethyl (DEAE)-Sepharose column, one and two forms of polygalacturonases were separated from *F. oxysporum* (PGase) and *A. tubingensis* (PGaseI and PGaseII), respectively. The polygalacturonases examined had higher affinity toward various polygalacturonic acids and pectins. The apparent K_m and V_{max} values were reported for the enzymes. Acidic pH optima (4.0 to 6.0) was also reported for the enzymes. Optimal temperature and thermal stability of the enzymes showed a range from 40 to 60°C.

Kant *et al.* (2013) conducted an experiment on the production of polygalacturonase from *Aspergillus niger* (MTCC 3323). The enzyme precipitated with 60 per cent ethanol resulted in 1.68-fold purification. The enzyme was purified to 6.52-fold by Sephacryl S-200 gel-filtration chromatography. On SDS–PAGE analysis, enzyme was found to be a heterodimer of 34 and 69 kDa subunit. Homogeneity of the enzyme was checked by native PAGE and its molecular weight was found to be 106 kDa. The purified enzyme showed maximum activity in the presence of polygalacturonic acid at temperature of 45°C, pH of 4.8, reaction time of 15 min. The enzyme was stable within the pH range of 4.0–5.5 for 1 h. At 4°C it retained 50 per cent activity after 108 h but at room temperature it lost its 50 per cent activity after 3 h.

Applications of Pectinases

Pectinases are widely used for a number of purposes in various industries.

Fruit Juice Extraction

The largest industrial application of pectinases is in fruit juice extraction and clarification. Pectins contribute to fruit juice viscosity and turbidity. A mixture of pectinases and amylases is used to clarify fruit juices. It is known to decrease filtration time up to 50 per cent. Treatment of fruit pulps with pectinases also showed an increase in fruit juice volume from banana, grapes and apples. Pectinases in combination with other enzymes, *viz.*, cellulases, arabinases and xylanases, have been used to increase the pressing efficiency of the fruits for juice extraction.

Textile Processing and Bioscouring of Cotton Fibers

Pectinases have been used in conjunction with amylases, lipases, cellulases and hemicellulases to remove sizing agents from cotton in a safe and ecofriendly manner, replacing toxic caustic soda used for the purpose earlier. Bioscouring is a novel process for removal of noncellulosic impurities from the fiber with specific enzymes. Pectinases have been used for this purpose without any negative side effect on cellulose degradation.

Degumming of Plant Bast Fibers

Bast fibers are the soft fibers formed in groups outside the xylem, phloem or pericycle, *e.g.* Ramie and sunn hemp. The fibers contain gum, which must be removed before its use for textile making. The chemical degumming treatment is polluting, toxic and non-biodegradable. Biotechnological degumming using pectinases in combination with xylanases presents an ecofriendly and economic alternative to the above problem.

Retting of Plant Fibers

Pectinases have been used in retting of flax to separate the fibers and eliminate pectins.

Wastewater Treatment

Vegetable food processing industries release pectin containing wastewaters as by product. Pretreatment of these wastewaters with pectinolytic enzymes facilitates removal of pectinaceous material and renders it suitable for decomposition by activated sludge treatment.

Coffee and Tea Fermentation

Pectinase treatment accelerates tea fermentation and also destroys the foam forming property of instant tea powders by destroying pectins. They are also used in coffee fermentation to remove mucilaginous coat from coffee beans.

Paper and Pulp Industry

During papermaking, pectinase can depolymerise pectins and subsequently lower the cationic demand of pectin solutions and the filtrate from peroxide bleaching.

Animal Feed

Pectinases are used in the enzyme cocktail, used for the production of animal feeds. This reduces the feed viscosity, which increases absorption of nutrients, liberates nutrients, either by hydrolysis of non biodegradable fibers or by liberating nutrients blocked by these fibers, and reduces the amount of faeces.

Improvement of Stability of Red Wines

Pectinolytic enzymes added to macerated fruits before the addition of wine yeast in the process of producing red wine resulted in improved visual characteristics (colour and turbidity) as compared to the untreated wines. Enzymatically treated red wines presented chromatic characteristics, which are considered better than the control wines. These wines also showed greater stability as compared to the control.

Table 14.3: Application of Pectinases in Clarification of Various Fruits

Raw Materials	Applications	Final Products
Apple, pear	Mash, juice, liquefaction	Clear juice concentrate
Grape juice and wine	Mash, must, juice, wine	White, rose, red wines
Strawberry, raspberry blackberry	Juice, pulp	Juice concentrates, flavor, aroma
Orange, lemon	Juice, pulp, peeled segments	Concentrates essential oils
Mango, guava, papaya pineapple, banana	Pulp, juice, puree	Pulp wash, puree, clear juice, concentrates

Acknowledgement

The first author acknowledges the financial assistance received from Indian Council of Agricultural Research (ICAR), in the form of a Senior Research Fellowship (SRF).

References

Ahuja SK Ferreira GM and Moreira AR. 2004. Utilization of enzymes for environmental applications. *Crit Rev Biotechnol* 24(2–3): 125–154.

Alana A, Alkorta I, Dominguez JB, Llama MJ and Serra JL. 1990. Pectin lyase activity in a *Penicillium italicum* strain. *Appl Environ Microbiol* 56: 3755–3759.

Al–Najada AR, Al–Hindi RR and Mohamed SA. 2012. Characterization of polygalacturonases from fruit spoilage *Fusarium oxysporum* and *Aspergillus tubingensis. African J Biotechnol* 11(34): 8527–8536.

Aro N Pakula T and Penttila M. 2005. Transcriptional regulation of plant cell wall degradation by filamentous fungi. *FEMS Microbiol Rev* 29: 719–739.

Be Miller JN 1986. An introduction to pectins: Structure and properties. In: Fishman ML Jem JJ. (Eds.) *Chemistry and Functions of Pectins*, ACS Symposium Series 310. American Chemical Society, Washington, DC.

Ceci L and Loranzo J 1998. Determination of enzymatic activities of commercial pectinases for the clarification of apple juice. *Food Chem* 61(1/2): 237–241.

Dinnella C Lanzarini G and Stagni A. 1994. Immobilization of an endopectin lyase on g–alumina: study of factors influencing the biocatalytic matrix stability. *J Chem Technol Biotechnol* 59: 237–41.

Dixit VS, Kumar AR Pant A and Khan MI, 2004. Low molecular mass pectate lyase from *Fusarium moniliforme*: similar modes of chemical and thermal denaturation. Biochem *Biophys Res Commun* 315: 477–84.

Favey S Bourson C Bertheou Y Kotoujansky A Boccora M. 1992. Purification of the acidic pectate lyase of *Erwinia chrysanthemi* 3937 and sequence analysis of the corresponding gene. *J Gen Microbiol* 138: 499–508.

Hasunuma T Fukusaki EI Kobayashi A. 2003. Methanol production is enhanced by expression of an *Aspergillus niger* pectin methylesterase in tobacco cells. *J Biotechnol* 6: 45–52.

Jayani RS Saxena S and Gupta R 2005. Microbial pectinolytic enzymes: A review, *Process Biochemistry* 40: 2931–2944.

Kant S Vohra A and Gupta R. 2013. Purification and physicochemical properties of polygalacturonase from *Aspergillus niger* MTCC 3323. *Protein Expr Purif* 87: 11–16.

Lang C and Dornenburg H 2000. Perspectives in the biological function and the technological application of polygalacturonases. *Appl Microbiol Biotechnol* 53: 366–75.

Maldonaldo MS and Saad AMS. 1998. Production of pectinesterase and polygalacturonase by *Aspergillus niger* in submerged and solid state systems. *J Ind Microbiol Biotechnol* 20: 34–8.

Margo P Varvaro L Chilosi G Avanzo C Balestra GM. 1994. Pectinolytic enzymes produced by *Pseudomonas syringae* pv. Glycinea. *FEMS Microbiol Lett* 117: 1–6.

Nagai M Katsuragi T Terashita T Yoshikawa K Sakai T. 2000. Purification and characterization of an endo–polygalacturonase from *Aspergillus awamori*. *Biosci Biotechnol Biochem* 64: 1729–32.

Palaniyappan M, Vijayagopal V, Viswanathan R and Viruthagiri T. 2009. Screening of natural substrates and optimization of operating variables on the production of pectinase by submerged fermentation using *Aspergillus niger* MTCC 281. A*fr J Biotechnol* 8(4): 682–686.

Panda T Nair SR and Prem Kumar M. 2004. Regulation of synthesis of the pectolytic enzymes of *Aspergillus niger*. *Enz Micro Technol* 34: 466–473.

Pandey A Soccol CR and Larroche C. 2007. Current Developments in Solid–state fermentation, In: Pandey A Soccol CR and Larroche C. (Ed), Current Developments in Solid–state Fermentation, Springer Science/Asiatech Publishers, Inc., New York, USA/New Delhi, India, 2007, pp. 3–25.

Pandey A, Soccol CR, Nigam P and Soccol VT. 2000, Biotechnological potential of agro–industrial residues I sugar cane bagasse. *Biores Technol* 74: 69–80.

Pandey A. 2003. Solid–state fermentation. *Biochem Eng J* 13: 81–84.

Patil SR and Dayanand A. 2006. Optimization of process for the production of fungal pectinases from deseeded sunflower head in submerged and solid–state conditions, *Bioresour Technol* 97: 2340–2344.

Pietro AD and Roncero MIG. 1996. Purification and characterization of an exo–polygalacturonase from the tomato vascular wilt pathogen *Fusarium oxysporum* f.sp. lycopersici. *FEMS Microbiol Lett* 145: 295–8.

Ruiz HA, Rodriguez–Jasso RM, Rodriguez R, Contreras–Esquivel RC and Aguilar CN. 2012. Pectinase production from lemon peel pomace as support and carbon source in solid–state fermentation column–tray bioreactor. *Biochem Eng J* 65: 90– 95.

Sapunova LI Mikhailova RV and Lobanok AG. 1995. Properties of pectin lyase preparations from the genus *Penicillium*. *Appl Microbiol Biochem* 31: 435–8.

Silva D, Martins ES, Silva R and Gomes E. 2002. Pectinase production by *Penicillium viridicatum* RFC3 by solid state fermentation using agricultural wastes and agro–industrial by–products. *Braz J Microbiol*. 33: 318–324.

Silva D, Tokuioshi K, Martins ES, Silva RD and Gomes E. 2005. Production of pectinase by solid–state fermentation with *Penicillium viridicatum* RFC3. *Process Biochem* 40: 2885–2889.

Singh SA Rao AGA. 2002. A simple fractionation protocol for, and a comprehensive study of the molecular properties of two major endopolygalacturonases from *Aspergillus niger*. *Biotechnol Appl Biochem* 35: 115–23.

Venkatesh M, Pushpalatha PB, Sheela KB and Girija D. 2009. Microbial pectinase from tropical fruit wastes. *J Tropical Agricul* 47(1–2): 67–69.

Voragen AG, Coenen GJ, Verhoef RP and Schols HA. 2009. Pectin, a versatile polysaccharide present in plant cell walls. *Structural Chemistry* 20 (2), 263–275.

Zheng Z and Shetty K. 2000. Solid state production of polygalacturonase by *Lentinus edodes* using fruit pressing waste. *Process Biochem* 35: 825–830.

2015, Recent Trends in Microbiology, Mycology and Plant Pathology   *Pages* 213–241
Editor: **Dr. H.C. Lakshman**
Published by: **DAYA PUBLISHING HOUSE, NEW DELHI**

Chapter 15

Role of Fungi in the Contribution of Pharmaceuticals and Enzymes

Jayshree M. Kurandawad* and H.C. Lakshman

Department of Botany Microbiology Laboratory,
Karnatak University Dharwad – 580 003, Karantaka, India
*E-mail: jayashreekurandawad@gmail.com

What would the world be like without fungi? What if fungi didn't degrade dead plant material or help to manufacture many of our most indispensable antibiotics, medicines, and vaccines? What if fungi did not live on and inside our bodies? And what if fungal diseases didn't stalk crops, wildlife, and humans alike?

Mycologists-scientists who study fungi-find it difficult to envision a world without fungi, because they know these organisms are integral to almost every facet of ecology, agriculture, and medicine. However, most people, including many scientists, are largely unaware of fungi and the roles of fungi play in the world around us. What are fungi? Fungi are eukaryotic, heterotrophic organisms (they consume organic forms of carbon for energy). They come in three basic shapes: unicellular yeasts, filamentous hyphae (molds) and, among the most basal groups, flagellated, swimming, unicellular organisms that encyst to form sporangia. The yeasts, hyphae, and sporangia have cell walls that contain at some of the rigid polysaccharide chitin, along with a variety of glucans. If you think you know what a fungus looks like, you may want to think again. Fungi are wildly diverse in appearance, ranging from the well-known mushrooms and molds to the less familiar smuts, rusts, truffles, yeasts, and others. In the environment, fungi are the primary degraders of organic matter, responsible for turning dead plants into the small nutrient building blocks other organisms can use. In tropical rain forests, ~50 per cent of the dead plant and animal matter (by weight) is degraded by fungi. Without fungi, dead plants wouldn't break down promptly, and dead material would gradually accumulate, eventually choking off living plants. Mycorrhizae represent another way for fungi to impact ecosystems. These associations, in which fungi live intimately with plant roots, allow plants better access to nutrients in the soil and provide these plants a competitive advantage over plants that lack mycorrhizal fungi. Almost all vascular plants interact with mycorrhizal fungi, and some, such as orchid species, are totally dependent on their fungal partner to germinate and grow. Reforestation requires special mycorrhizae to be successful. In the rumens of cattle and other livestock, fungi help bacteria symbiotically by breaking down cellulose that bacteria can then degrade into even smaller molecules so that the animal can harvest energy from cellulose-rich grasses. Certain fungi are also responsible for causing disease in humans, plants, animals, and insects. In humans, fungal diseases, which are also called mycoses, can range from merely aggravating (athlete's foot) to life-threatening (*Candida albicans*, *Aspergillus*, and *Cryptococcus*), and perhaps because the incidence of fungal disease is under reported, they are more much common than most people think. For example, each year about 200,000 Americans contract coccidioidomycosis-20,000 seek medical treatment, 2,000 are hospitalized and 200 die. There are approximately 5,000 cases of cryptococcosis per year, and in Africa and Asia, as many as 30 per cent of AIDS patients are afflicted with cryptococcosis. Fungal molds have also been implicated in sick building syndrome, in which building occupants experience a range of illnesses or discomforts upon exposure to a given dwelling. And fungi are suspected to potentially play a role in asthma as well. Most plant diseases are caused by fungi making these organisms very important in agriculture. The repercussions of managing fungal pathogens on crops-the money and effort spent, the numerous pesticide applications, the consequences of these applications for surface water and soil quality, and the impacts on crop yields- are extraordinary. Notwithstanding their negative impacts on health, fungi are invaluable resources. Their value is particularly clear when you consider the usefulness of the many secondary metabolites fungi produce. Secondary metabolites are compounds that are not necessary for growth and reproduction, and fungi make them for a variety of reasons, including self-protection from predators, such as mites and amoeba, killing competing fungi and bacteria, and signaling to nearby microbes. We use fungal secondary metabolites as life-saving medications (including *penicillin*, the first antibiotic ever discovered), as well as drugs that facilitate organ transplants (cyclosporine) and that reduce cholesterol (statins), and insecticides. As more fungal species are discovered and characterized, the opportunities to capitalize on new compounds to improve human health or contribute to industrial or biotechnological uses will increase exponentially. The food industry also relies on fungi to do some of the heavy lifting in manufacturing. All leavened bread, all alcoholic beverages, vinegars, citric-acid-based beverages, the Roquefort and camembert cheese families, and many Asian foods, such as tempeh, soy sauce and miso, are among the many foods produced with the assistance of fungi. Fungi, including mushrooms and truffles, are also used as food sources themselves. Although fungi have been used traditionally in food and alcoholic fermentations, in more recent times recombinant yeasts have been used to produce biologically-active proteins and other compounds in an efficient and cost-effective way. Due to the fact that yeasts are eukaryotic organisms, the production of proteins with the correct post-translational modifications making them biologically active has proven to be of tremendous benefit in the production of therapeutic proteins. The protein secretion pathway for the yeast *Pichia pastoris* has been successfully humanized through genetics and engineering and enables precise control over modifications of proteins. These organisms also make it easier to produce drugs, biologically active products, diagnostics, and vaccines. Yeasts and other fungi are presently being used as production machines for vaccines, vitamins, monoclonal antibodies for use in immune therapy, and other therapeutics. In the future, industry will undoubtedly expand on current strategies to produce ever more complex molecules with great efficiency and accuracy, reducing the costs of many therapies currently in use today. Fungi play an integral role in the function of ecosystems, especially in relation to plants and soil. Fungi are used in industry, and fungal products have a profound impact in human health. Fungi cause economically important reductions in plant and animal production. In addition, fungi have been used to elucidate the complex biochemistry and genetic principles of eukaryotes. The importance of fungi transcends the simple description of their form and function. Thus we have provided this introduction to their biology.

Mycology was a branch of botany because, although fungi are evolutionarily more closely related to animals than to plants, this was not recognized until a few decades ago. Pioneer *mycologists* included Elias Magnus Fries, Christian Hendrik Persoon, Anton de Bary and Lewis David von Schweinitz. Mycology (from the Greek μýêçò, mukçs, meaning "fungus") is the branch of biology concerned with the study of fungi, including their genetic and biochemical properties, their taxonomy and their use to humans as a source for tinder, medicine (*e.g.*, penicillin), food (*e.g.*, beer, wine, cheese, edible mushrooms) and entheogens, as well as their dangers, such as poisoning or infection. From mycology arose the field of phytopathology, the study of plant diseases, and the two disciplines remain closely related because the vast majority of "plant" pathogens are fungi. A biologist who studies mycology is called a mycologist. Many fungi produce toxins, antibiotics and other secondary metabolites. For example the cosmopolitan (world wide) genus *Fusarium* and their toxins associated with fatal outbreaks of alimentary toxic aleukia in humans were extensively studied by Abraham Joffe. Fungi are fundamental for life on earth in their roles as symbionts, *e.g.* in the form of mycorrhizae, insect symbionts and lichens. Many fungi are able to break down complex organic biomolecules such as lignin, the more durable component of wood, and pollutants such as xenobiotics, petroleum, and polycyclic aromatic hydrocarbons. By decomposing these molecules, fungi play a critical role in the global carbon cycle. Some fungi can cause disease in humans or other organisms. The study of pathogenic fungi is referred to as medical mycology.

History

Humans probably started collecting mushrooms as food in Prehistoric times. Mushrooms were first written about in the works of Euripides (480-406 B.C.). The Greek philosopher Theophrastos of Eressos (371-288 B.C.) was perhaps the first to try to systematically classify plants; mushrooms were considered to be plants that were missing certain organs. It was later Pliny the elder (23–79 A.D.), who wrote about truffles in his encyclopedia Naturalis historia. The Middle Ages saw little advancement in the body of knowledge about fungi. Rather, the invention of the printing press allowed some authors to disseminate superstitions and misconceptions about the fungi that had been perpetuated by the classical authors (Ainsworth *et al.,* 1976).

" Fungi and truffles are neather herbs, nor roots, nor flowers, nor seeds, but merely the superfluous moisture or earth, of trees, or rotten wood, and of other rotting things. This is plain from fact that all fungi and truffles, especially those that are used for eating, grow most commonly those that are used for thundery and wet weather." Jerome Bock (Hoeronymus Tragus), the start of the modern age of mycology begins with Pier Antonio Micheli's 1737 publication of *Nova plantarum genera*. Published in Florence, this seminal work laid the foundations for the systematic classification of grasses, mosses and fungi. The term *mycology* and the complementary *mycologist* were first used in 1836 by M.J. Berkeley.

The Importance of Fungi

☆ Fungi are one of the most important groups of organisms on the planet. This is easy to overlook, given their largely hidden, unseen actions and growth. They are important in an enormous variety of ways. Such as, Looking at the above list, it is clear that fungi play a role in just about every part of our daily lives!

☆ **Recycling:** Fungi, together with bacteria, are responsible for most of the recycling which returns dead material to the soil in a form in which it can be reused. Without fungi, these recycling activities would be seriously reduced. We would effectively be lost under piles many metres thick, of dead plant and animal remains.

☆ **Mycorrhizae and plant growth**: Fungi are vitally important for the good growth of most plants, including crops, through the development of mycorrhizal associations. As plants are at the base of most food chains, if their growth was limited, all animal life, including human, would be seriously reduced through starvation.

☆ **Food**: Fungi are also important directly as food for humans. Many mushrooms are edible and different species are cultivated for sale world wide. While this is a very small proportion of the actual food that we eat, fungi are also widely used in the production of many foods and drinks. These include cheeses, beer and wine, bread, some cakes, and some soya bean products. While a great many wild fungi are edible, it can be difficult to correctly identify them. Some mushrooms are deadly if they are eaten. Fungi with names such as 'Destroying Angel' and 'Death Cap' give us some indication that it would not be a terribly good idea to eat them! In some countries, collecting wild mushrooms to eat is a popular activity. It is always wise to be totally sure that what you have collected is edible and not a poisonous look-a-like.

☆ **Medicines**: Penicillin, perhaps the most famous of all antibiotic drugs, is derived from a common fungus called *Penicillium.* Many other fungi also produce antibiotic substances, which are now widely used to control diseases in human and animal populations. The discovery of antibiotics revolutionized health care world wide. Some fungi which parasitise caterpillars have also been traditionally used as medicines. The Chinese have used a particular caterpillar fungus as a tonic for hundreds of years. Certain chemical compounds isolated from the fungus may prove to be useful treatments for certain types of cancer. A fungus which parasitises Rye crops causes a disease known as Ergot. The fungus can occur on a variety of grasses. It produces small hard structures, known as sclerotia. These sclerotia can cause poisoning in humans and animals which have eaten infected material. However, these same sclerotia are also the source of a powerful and important drug which has uses in childbirth.

☆ **Biocontrol**: Fungi such as the Chinese caterpillar fungus, which parasitise insects, can be extremely useful for controlling insect pests of crops. The spores of the fungi are sprayed on the crop pests. Fungi have been used to control Colorado potato beetles, which can devastate potato crops. Spittlebugs, leaf hoppers and citrus rust mites are some of the other insect pests which have been controlled using fungi. This method is generally cheaper and less damaging to the environment than using chemical pesticides.

☆ **Crop disease**: Fungal parasites may be useful in biocontrol, but they can also have enormous negative consequences for crop production. Some fungi are parasites of plants. Most of our common crop plants are susceptible to fungal attack of one kind or another. Spore production and dispersal is enormously efficient in fungi and plants of the same species crowded together in fields are ripe for attack. Fungal diseases can on occasion result in the loss of entire crops if they are not treated with antifungal agents.

☆ **Animal disease**: Fungi can also parasitise domestic animals causing diseases, but this is not usually a major economic problem. A wide range of fungi also live on and in humans, but most coexist harmlessly. Athletes foot and Candida infections are examples of human fungal infections.

☆ **Food spoilage**: It has already been noted that fungi play a major role in recycling organic material. The fungi which make our bread and jam go mouldy are only recycling organic matter, even though in this case, we would prefer that it didn't happen! Fungal damage can be responsible for large losses of stored food, particularly food which contains any moisture. Dry grains can usually be stored successfully, but the minute they become damp, moulds are likely to render them inedible. This is obviously a problem where large quantities of food are being produced seasonally and then require storage until they are needed.

Role of Fungi in the Production of Pharmaceuticals

Fungi make an extraordinarily important contribution to managing disease in humans and other animals. At the beginning of the 21st century, Fungi were involved in the industrial processing of more than 10 of the 20 most profitable products used in human medicine. Two anti-cholesterol statins, the antibiotic penicillin and the immunosuppressant cyclosporin A are among the top 10. Each of these has a turn over in excess of $1 billion annually. Drug discovery continues. The following have recently been approved for human use: Micafungin is an antifungal agent; mycophenolate is used to prevent tissue rejection; rosuvastatin is used to reduce cholesterol; and cefditoren as an antibiotic.

Fungi are extremely useful organisms in biotechnology. Fungi construct unique complex molecules using established metabolic pathways. Different taxa produce sets of related molecules, each with slightly different final products. Metabolites formed along the metabolic pathway may also be biologically active. In addition, the final compounds are often released into the environment. Manipulation of the genome, and environmental conditions during formation of compounds, enable the optimisation of product formation.

On the negative side, single isolates of fungi in manufacture may lose their capacity to form or release the target molecules. Indeed, the target compound may only be expressed under specific conditions, or at a specific point in the life cycle of the fungus. It is amazing that so many biologically active compounds have been discovered and taken to the point where they are medically important. Indeed, attempts to 'discover' new and exciting molecules remains the core activity of many research groups.

The role of fungi was established early in history. Yeasts have been used in the making of bread and alcohol since the beginning of civilisation. In modern times, the discovery of penicillin marked the beginning of a new approach to microbial diseases in human health. More recent approaches include the application of hydrophobins to surfaces leading to biocompatibility of implants, and to emulsion formation improving drug delivery. The established importance of fungi is being expanding way beyond their capacity to transform and protect. a green mildew, be longing to deuteromycetes.

From a pharmaceutical point of view, mushrooms are extremely interesting. Fungi have recently helped to produce other innovative and important drugs, such as cyclosporin, an anti-rejection substance that has aided the development of organ-transplant surgery over the last few years. Penicillin, the first antibiotic ever, is made from a fungus of the genus *Penicillium*, a green mildew, be longing to deuteromycetes. Fungi produce a wide variety of substances that are not only important for the human organism. An ascomycete, Giberella Fujikuroi, can secrete a plant-growth hormone known as gibberellin.For centuries, certain mushrooms have been documented as a folk medicine in China, Japan, and Russia. Although the use of mushrooms in folk medicine is largely centered on the Asian continent, people in other parts of the world like the Middle East, Poland and Belarus have been documented using mushrooms for medicinal purposes. Certain mushrooms, especially polypores like Reishi were thought to be able to benefit a wide variety of health ailments. Medicinal mushroom research in the United States is currently active, with studies taking place at City of Hope National Medical Center,(Di Rado, 2008) as well as the Memorial

Sloan–Kettering Cancer Center. Current research focuses on mushrooms that may have hypoglycemic activity, anti-cancer activity, anti-pathogenic activity, and immune system enhancing activity. Recent research has found that the oyster mushroom naturally contains the cholesterol-lowering drug lovastatin, mushrooms produce large amounts of vitamin D when exposed to UV light, and that certain fungi may be a future source of taxol. To date, penicillin, lovastatin, ciclosporin, griseofulvin, cephalosporin, ergometrine, and statins are the most famous pharmaceuticals which have been isolated from the fungi kingdom.

Fungi have been used medicinally since ancient times. Ergotamine, produced by *Claviceps purpurea* is used to facilitate delivery of babies and can also be used to relieve migraine headaches. The steroids in "the pill" are produced industrially by a fungus, *Rhizopus nigricans,* as are the steroids cortisone and prednisolone. Penicillin is one antibiotic that kills bacteria. It is produced by *Penicillium chrysogenum* and related species. Cephalosporins are another class of antibiotics produced by *Cephalosporium acremonium* and related species. For a fungal infection you might be taking griseofulvin, produced by *Penicillium griseofulvum.* The use of antibiotics in medical practice dates from recognition of the antibiotic properties of penicillin. For the mass production of penicillin, there began a search for antibiotics from other fungi, bacteria and algae. Many are found to be just as important and even more commonly used than penicillin. However, few of these antibiotics discovered since penicillin have been fungal in origin. The fungus being tested is inoculated with various pathogenic bacteria to determine if an antagonistic response occurs between the two organisms. Should the fungus demonstrates anti-bacterial activities, it would be further determine to see if its metabolites are harmful to animals. Most fungal metabolites tested had proven to be as lethal as the disease that they are trying to stop. Only a few very useful antibiotics were discovered in fungi. Other Antibiotics recovered from Fungi, Many antibiotics today are produced by non fungal micro-organisms. Griseofulvin, however, is an antifungal antibiotic formed by several species of a genus of fungi. The immunosuppressant drug Cyclosporin is used in organ transplantation, is also derived from fungi.

Fungi in the Antimicrobial Activities

The first antimicrobial agent (antibiotic) to be produced was Penicillin, and it was discovered through the sheer serendipity of Alexander Fleming in 1928. This was derived from the ascomycetous fungus *Penicillium notatum.* The antibiotic was put into mass production and large scale therapeutic use because of the scale up work subsequently carried out by Howard Florey and Ernst Chain in the 1940s, and this work was spurred by the necessity to cure wounded soldiers of infections during the II world war (Abraham, 1981; Bottcher, 1964; Jacobs, 1985). With the discovery of Streptomycin from an actinomycete, by Selman Waksman in 1944, the era of antibiotics truly began, and during this period extending over two decades, more than 1000 antibiotics were discovered, many of them from actinomycetes, and some from fungi.

As on today, the important antibiotics derived from fungi, other than *Penicillin,* are: Cephalosporin from *Cephalosporium* spp., Griseofulvin from *Penicillium griseofulvum,* Lentinan from *Lentinus* sp., and Schizophyllan from *Schizophyllum commune*. Penicillin and Cephalosporin are antibacterial antibiotics acting against Gram-positive bacteria, whereas, Griseofulvin is an antifungal antibiotic useful in treating dermatophyte infections. Lovastatin is a cholesterol biosynthesis inhibitor derived from *Aspergillus terreus*. It is one of the many drugs used as a cholesterol reducing agent. A similar cholesterol reducing drug is produced from *Penicillium citrinum*, and it is called Pravastatin. Lentinan is active against Mycobacterium tuberculosis, *Listeria* sp., and Herpes Simplex Virus-1 (HSV-1). *Schizophyllan* is both antibacterial and antifungal in activity. It is useful in controlling *Candida albicans* and *Staphylococcus aureus*.

Mushrooms and polypores are rich source of natural antibiotics. The cell wall glucans are well known for their immune modulatory properties, and the secondary metabolites are active against bacteria (Benedict and Brady, 1972; Kupra *et al.,* 1979) and viruses (Suzuki *et al.,* 1990; Brandt and Piraino, 2000). Exudates from mushroom mycelia are active against protozoa such as the malarial parasite *Plasmodium falciparum* (Lovy *et al.,* 1999; Isaka *et al.,* 2001). Since humans and fungi share common microbial antagonists such as *Escherichia coli*, *Staphylococcus aureus* and *Pseudomonas aeruginosa*, humans can benefit from the natural defense strategies of fungi to produce antimicrobials (Hardman *et al.,* 2001). The general hypothesis increasingly substantiated is that polypores provide a protective immunological shield against a variety of infectious diseases (Chihara, 1992; Hobbs, 1986; Mizuno *et al.,* 1995).

Two other polypores are notable, *Fomes fometarius* and *Piptoporus betulinus*, both of which were found in the high Alpines near the border of Italy, buried along with the legendary 'ice man' 5300 years ago. Scientists believe that the use of these fungi was for their antimicrobial properties.

In a recent in vitro study, more than 75 per cent of polypore species surveyed showed antimicrobial property (Suay *et al.,* 2000). In particular, this study showed species of the genus Ganoderma such as *G. applanatum*, *G. lucidum*, *G. pfeifferi* and *G. resinaceum* to be effective against Gram-positive bacteria. In contrast, gilled mushrooms such as *Psylocibe semilanceata*, *Pleurotus eryngii*, and *Lactarius delicious* all strongly inhibited the growth of *Staphylococcus aureus* (Smania *et al.,* 2001). One could understand that the mushroom and polypore genome stands out as a virtually untapped resource for novel antimicrobials.

Non-antibiotic Therapeutics from Fungi

There are non-antibiotic therapeutic agents obtained from fungi that have revolutionized medical practice. Cyclosporin is an important immunosuppressant drug that is used in organ transplantation surgery. Cyclosporin-A is derived from *Tolypocladium inflatum*, and *Aspergillus* sp. Isopenicillin-N is a common precursor of Penicillin and Ensuphulosporan antibiotics and this is produced by *T. inflatum*. About 20 per cent of the drugs produced by pharmaceutical industry today are derived from fungi (Churchill, 2001).

Lovastatin is a cholesterol biosynthesis inhibitor derived from *Aspergillus terreus*. It is one of the many drugs used as a cholesterol reducing agent. A similar cholesterol reducing drug is produced from *Penicillium citrinum*, and it is called Pravastatin.

Fungi are the source of vitamin B12 (*Saccharomyces cerevisiae*) and other vitamins (*S. cerevisiae, Ashbya* sp., *Blakeslea* sp.), hallucinogens (*Psylocybe* sp.), and steroids useful in fertility regulation (*Rhizopus* spp.).

Antibiotics from Fungi

In 1941, penicillin from the fungus *Penicillium chrysogenum* was first used successfully to treat an infection caused by a bacterium. Use of penicilin revolutionised the treatment of pathogenic disease. Many formally fatal diseases caused by bacteria became treatable, and new forms of medical intervention were possible. When penicillin was first produced, the concentration of active ingredient was approximately 1 microgram per ml of broth solution. Today, improved strains and highly developed fermentation technologies produce more than 700 micrograms per ml of active ingredient. In the early broths, several closely related molecules were present. These molecules are beta lactam rings fused to five-membered thiazolidine rings, with a side chain. The side chain can be chemically modified to provide slightly different properties to the compound. The natural penicillins have a number of disadvantages. They are destroyed in the acid stomach, and so cannot be used orally. They are sensitive to beta lactamases, which are produced by resistant bacteria, thus reducing their effectiveness. Also, they only act on gram positive bacteria. Modifications to manufacturing conditions have resulted in the development of oral forms. However, antibiotic resistance among bacteria is becoming an extremely important aspect determining the long-term use of all antibiotics.

Cephalosporins

First discovered from *Acremonium chrysogenum* in 1953. This species actually produced several antibiotics: cephalosporin C, cephalosporins P1–P5 and penicillin N. Cephalosporin has also been identified from other fungi such as *Emericellopsis* and *Paecilomyces*, two genera that are morphologically similar to *Penicillium*. also contain the beta lactam ring. The original fungus found to produce the compounds was a Cephalosporium, hence the name. As with penicillin, the cephalosporin antibiotics have a number of disadvantages. Industrial modification of the active ingredients has reduced these problems. The only broadly useful antifungal agent from fungi is griseofulvin. The original source was Penicillium griseofulvin. Griseofulvin is fungistatic, rather than fungicidal. It is used for the treatment of dermatophytes, as it accumulates in the hair and skin following topical application.

More recently, several new groups have been developed. Strobilurins target the ubihydroquinone oxidation centre, and in mammals, the compound from fungi is immediately excreted. Basidiomycetes, especially from tropical regions, produce an enormous diversity of these compounds.

Sordarins are structurally complex molecules that show a remarkably narrow range of action against yeasts and yeast-like fungi. The compounds inhibit protein biosynthesis and so may become important agents against a number of fungal pathogens of humans.

Echinocandins are cyclic peptides with a long fatty acid side chain. They target cell wall formation. Semi-synthetic members of the group of compounds include pneumocandins which are in use in humans.

Economic Importance of Cephalosporins

Cephalosporin is one of the most widely used antibiotics, and economically speaking, has about 29 per cent of the antibiotic market. The structure and mode of action of the cephalosporins are similar to that of penicillin. Like penicillin, cephalosporins are valuable because of their low toxicity and they broad spectrum action against various diseases. Cephalosporins affect bacterial growth by inhibiting cell wall synthesis, in Gram-positive and -negative bacteria.

Griseofulvin

Griseofulvin was first isolated from *Penicillium griseofulvum*. The commercial production of griseofulvin is derived from a much mutated strain of *P. patulum*. The antibiotic is the only fungal antibiotic that is effective against fungal infections of hair, nail, skin and athlete's foot and ringworm. Griseofulvin is a fungistatic antibiotic. It only inhibits fungal growth but does not kill fungi. It affects a wide range of fungi but is limited to those with chitinous cell walls.

The original source was *Penicillium griseofulvin*. *Griseofulvin* is fungistatic, rather than fungicidal. It is used for the treatment of dermatophytes, as it accumulates in the hair and skin following topical application.

Tolypocladium inflatum

Not all antibiotic that we derive from fungal metabolites are extracted in the desired form from the fungus. Often chemical modifications will have to be carried out before it can be utilized for the desired purpose. Cilofungin is one such example. It is not the natural product extracted from fungi. The antibiotic is developed from a metabolite extracted from several species of *Aspergillus* such as *A. nidulans* and *A. rugulosus*. It is thus a semi-synthetic antibiotic.

Economic Importance of Cilofungin

Cilofungin is specifically developed to treat candiasis (which commonly manifests as thrush) caused by the fungus *Candida albicans*.

Immune Suppressants

Cyclosporin A

is a primary metabolite of several fungi, including *Trichoderma polysporum* and *Cylindrocarpon lucidum*. Cyclosporin A has proven to be a powerful immunosuppressant in mammals, being widely used during and after bone marrow and organ transplants in humans. Cyclosporin A is a cyclic peptide consisting of 11 mainly hydrophobic amino acids. Its inhibition of lymphocytes was first discovered during the 1970s. Subsequently, the mode of action was elucidated.

Cyclosporin A binds to a cytosolic protein called cyclophilin. Cyclophilin is found amongst many different organisms and its form appears highly conserved. Cyclophilin is involved with folding the protein ribonuclease. However, the Cyclosporin A/cyclophilin complex also binds to calcineurin. Calcineurin dephosphorylates a transcription factor, thereby triggering transcription of numerous genes associated with T cell proliferation. When the complex binds to calcineurin, T cell proliferation is suppressed. The inhibition of T cells proliferation results in the suppression of the activation process associated with invasion by foreign bodies. As a consequence, transplant tissues, which are foreign bodies, are not rejected.

Calcineurin is also highly conserved amongst phylogenetically diverse organisms. In fungi such as the human pathogen *Cryptococcus neoformans*, calcineurin is necessary for recovery from cell cycle arrest, growth in hypertonic solutions and regulation of the calcium pump. Thus the interaction of the Cyclosporin complex with calcineurin in Cryptococcus will result in death of the pathogen. However, in humans, cyclosporin also suppresses the immune system. The side effect is an unacceptable risk, and Cyclosporin A is not used as a fungicide in humans at present.

If you've had an organ transplant you might be taking Cyclosporin, produced by the fungus *Tolypocladium inflatum*. This is one of the most commonly used immunosuppressant drugs in organ transplantations.

Gliotoxins

Also have immunological and antibiotic activity. Produced by many fungi including *Aspergillus fumigatus*, gliotoxins belong to a class of compounds called epipolythiodioxopiperazines. The antibiotic activity is widely recognised and considered uninteresting. However, its effect on the immune system, especially macrophages, is being re-examined.

A wide range of other compounds with antibiotic activity are also known. They have been rejected for use in medicine because of unwanted side effects, or instability of the active compound.

Ergot Alkaloids

Claviceps purpurea is the causal agent of St Anthonies fire, a scourge of the middle ages when ergots contaminated flour. The ergots contain many alkaloids. Their effects are quite variable. They act on the sympathetic nervous system resulting in the inhibition of noradrenaline and sclerotin, causing dilation of blood vessels. They also act directly on the smooth muscles of the uterus causing contractions, thus their early use to induce abortion. Their strongest effect is intoxication, caused by lysergic acid amides, one of which is the recreational (and illegal) drug, LSD.

Seed of Paspalum Replaced by *C. paspalli*

Ergot alkaloids have a number of medicinal uses. Perhaps the most widespread use is in the treatment of migraines. The vasodilator activity reduces tension during an attack. The drugs also reduce blood pressure, though with untoward side effects. Alkaloids are now produced in culture by strains of *C. fusiformis* and *C. paspalii*.

Statins

Aspergillus terreus, a soil-borne fungus, produces a secondary metabolite called lovastatin and *Phoma* sp produces squalestatin. Statins have been used to reduce or remove low density lipoproteins from blood vessels in humans. In fact, the compounds all act via an enzyme in the liver that makes cholesterol, lovastatin inhibits HMG CoA reductase and squalestatin inhibits squalene synthase. By blocking the enzyme, the body removes cholesterol complexes from the inside of blood vessels. This has the effect of reducing or removing blockages in arteries, and thereby reducing the chance of a heart attack, strokes and diabetes. In addition, statins have been implicated in attracting stem cells to damaged tissues. The stem cells then appear to regenerate the tissue. Some statins induce problems. One form of the drug has been associated with muscle wastage. Others appear to lack side effects and have been recommended for wide spread use to control heart disease. (Wainwright *et al.,* 1995, Suryanarayanan *et al.,* 2009, Anke *et al.,* 2007).

Steroids

Microbial biotransformation of steroids is very important in the pharmaceutical industry. Steroids are used in the treatment of various disorders and also involve in regulation of sexuality. Chemical synthesis of steroids is very complex because of the requirement to achieve the necessary precision of substituent location. For example cortisone can be synthesised chemically from deoxycholic acid (Figure 15.1) but the process requires 37 steps, many of which must be carried out under extreme conditions of temperature and pressure with the resulting product costing over $200 per gram. The most difficult is introduction of oxygen atom at number 11 position of the steroid ring, but this can be accomplished by some micro-organisms. The fungus, *Rhizopus nigricans* for example hydroxylates progesterone, forming another steroid with the introduction of oxygen at the number 11 position (Figure 15.2). The fungus *Cunninghamella blakesleeana* similarily can hydroxylate the steroid cortexolone to from hydrcortisone with the introduction of oxygen at number 11 position. Other transformations of the steroid nucleus carried out by microbes include hydrogenatons, dehydrogenations, epoxidations and removal and addition of the side chains. The use of such microbial biotransformations in the formation of cortisone has lowered the original cost over 400- fold, so that in 1980 the price of cortisone in the U. S. was less than 50 cents per gram, compared to the original $200.

In a typical steroid transformation process, the microbe, such as *Rhizopus nigricans* is grown in a fermentation tank using an appropriate growth medium and incubation conditions to achive a high biomass. In most cases agitation and aeration are done to have rapid growth. After the growth of the microbe, the steroid to be transformed is

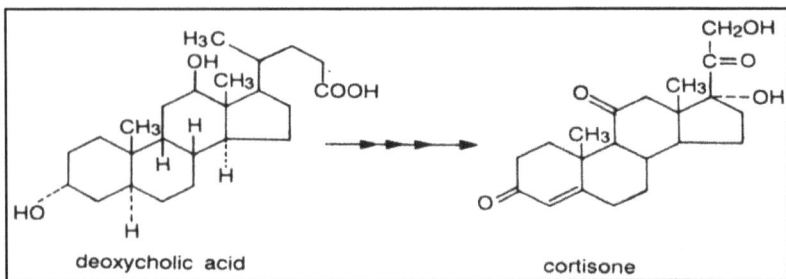

Figure 15.1: By Chemical Synthesis, Conversion of Deoxycholic Acid to Cortisone Requires 37 Steps, and Extreme Reaction Conditions and Oxygen is Needed to be Introduced at 11 Position, a difficult Step. Microbes can introduce oxygen at this position in a single reaction.

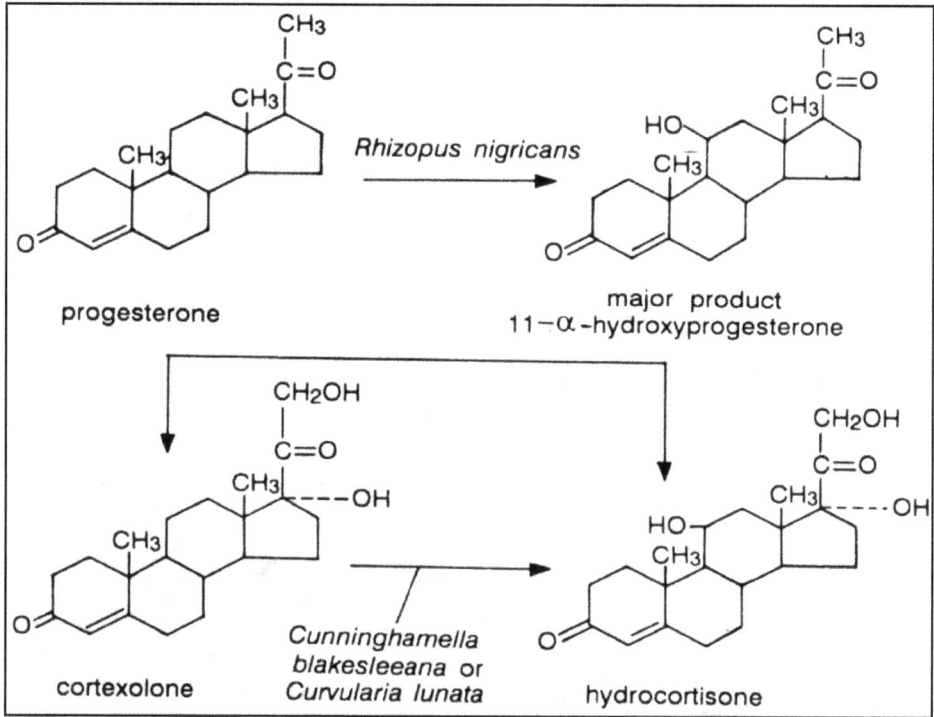

Figure 15.2: Microbial Transformation of Cortisone Derivatives.
Hydroxylation of (A) Progesterone by *Rhizopus arrhizus* used to
hydroxylate the cortisone ring to form new cortisone products.

added (as progesterone here) to the fermentor containing *R. nigricans* that has been growing for one day or so and the steroid is hydroxylated at number 11 to form 11-α-hydroxyprogesterone. Product is recovered by extraction with methylene chloride or other solvents, purifdied chromatographically and recovered by crystallization. A large number of similar transformations are carried out to produce a great variety of steroid derivatives for different medicinal uses (Figure 15.3) (P. D Sharma, 1999).

Selection of Industrially useful Fungal Organisms

The right micro-organism for an industrial process is first selected by screening the commercially useful one from the rest. This involves assays of the useful product from a number of micro-organisms isolated from soil or other sources. The classical example is the screening of thousands of soil samples for micro-organisms that can produce antibiotics. Several strains of actinomycetes isolated from soil have been laboriously tested for production of antibiotics using sensitive test organisms. The inhibition zone produced by on a plate culture of the test organism by the metabolite of the organism being screened indicates that the organism produces an antibiotic against the test organism. Different test organisms can be used to determine the antimicrobial spectrum of the micro-organism isolated. Analytical methods are used

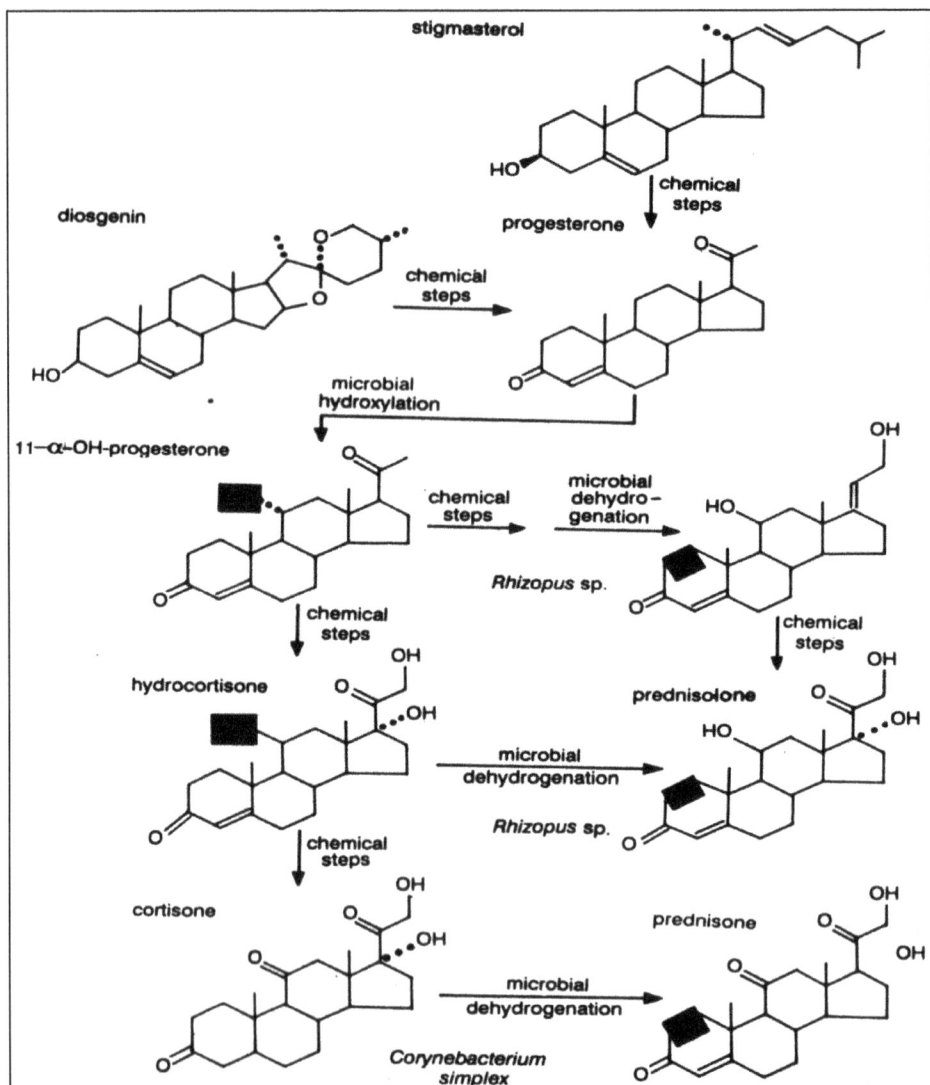

Figure 15.3: Examples of Steroid Transformations.
Some steps in these are chemically induced and other are microbial.

to determine the chemical nature of the inhibitory substance. The screening programme should identify the optimal incubation condition for maximal economic yield of the product. Toxicity test should be performed to determine whether the product can selectively inhibit pathogens without causing side effects that could preclude its therapeutic use. Once a species having industrial applications have been found, a research programme is undertaken to increase the capacity of the micro-organism to produce the desired product. For example, the *pencillium* species observed by Alexander Flaming to inhibit *Staphylococcus* had potential for commercial exploitation but did not produce sufficient quantities of pencilliun to permit industrial production. Extensive screening of soil samples from around the world led to the isolation of a potentially useful strain from soil in Peoria, IIIinois USA. From this strain through a series of induced mutations, a commercially useful strain was developed which cold yield 100 times more penicillin than the original strain. Mutation and screening approach has been most important. Industrial important strains of micro-organisms should be preserved and protected from further mutation and loss of the genetic trait. Lyophilization is the best way to preserve them as the process keeps the organisms metabolically inert. Other methods involve maintaining stock cultures in cold storage and checking periodically, which is a laborious process (Sharma, 1999).

Industrial Production of Pharmaceuticals

Fermatation technology has contributed to the production of various chemicals such as antibiotics, vitamins, steroids and vaccines, thus being the mainstay of pharmaceutical industry. Antibiotics production itself is a majar industry. Some of the important industrial processes of the production of pharmaceuticals are discussed here. Antibiotics are antimicrobial agents of microbial origin. Most antibiotics are industrially produced by microbial fermentation though some are now synthetically produced (*e.g.,* chloramphenicol).

Methods to Produce Mass Quantities of Penicillin

There are many species of *Penicillium*. A search was conducted to find other species that could be tested for penicillin production. *Penicillium chrysogenum* was found to produce approximately 200 times more penicillin than *P. notatum.* Scientist also began to try to increase the amount of penicillin produced by *P. chrysogenum,* by irradiating it with X-rays and UV rays in order to induce mutations of this species. A mutant form was formed and it could produce 1000 times more penicillin than Fleming's original culture. In addition to the development of this mutant, a new means of growing the mould was also perfected. Previously, penicillin was grown in flask. Hundreds of bottles of *Penicillium notatum* were needed to produce enough penicillin for one person. The new method involved growing the mould in large metal tanks, holding 25,000 gallons of nutrients, were aerated so that the mould could grow throughout the entire tank rather than on top. Aeration was the key to growing it in such large tanks. Previously, this had not been tried because it was known that the mold would only grow on the surface of the liquid medium. Thus, utilization of a large tank, under such circumstances would be highly inefficient in terms of cost, space and penicillin production. With this new method, production quantity began to rise.

Role of Fungi in the Contribution of Enzymes

Enzyme is a biocatalyst which accelerates biologyical reactions. However, the concept of biocatalysts is very wide. It includues the pure enzyme, crude cell extract, viable plant cells, viable animal cells, vaible microbial cells and intact non-viable microbil cells. Source of enzymes used in commerce is plant and animal cells. The sources of enzymes are micro-organisms, higher plants and animals. (Sasson A.,

Table 15.1: Important Antibiotics and their Producers

Antibiotic	Producing Micro-organism
Penicillin	*Pencillium chrysogenum*
Cephalosporin	*Cephalosporium acremonium*
Griscofulvin	*Pencillium griseofulvum*
Choramphenicol	*Streptomyces venezuelae*
Tetracyclines	
Tetracycline	*Streptomyces aureofaciens*
Chlortetracycline	*S. aureofaciens*
Oxytetracycline	*S. rimosus*
Polypeptides	
Polymyxin-B	*Bacillus polymyxa*
Bacitracin	*B. licheniformis*
Glutarimides	
Cycloheximide	*Streptomyces griseus*
Aminoglycosides	
streptomycin	*Streptomyces griseus*
Kanamycin	*S. kanamyceticus*
Neomycin	*S. fraadiae*
Polyenes	
Nystatin	*Streptomyces noursei*
Hamycin	*Streptomyces* sp.
Aureofungin	*Streptomyces* sp.
Amphoterecin-B	*Streptomyces nodosus*
Macrolides	
Erythromycin	*Streptomyces erythreus*
Oleandomycin	*Streptomyces antibioticus*
Carbomycin	*Streptomyces halstedii*
Novobiocin	*Streptomyces niveus*
Blasticidin-s	*Streptomyces* sp.
Vira-A	*Streptomyces antibioticus*
(Adenine arabinoside)	

1984). In addition, microbial enzymes have gained much popularity. Production of primary and secondary metabolites by micro-organisms is possible only due to involvement of various enzymes. They are of two types, the extracellular and the intracellular enzymes. The former is secreted out the cell and the later remain within the cell. There is a wide range of extracellular enzymes produced by pathogenic and saprophytic micro-organisms such as cellulose, polymethylgalacturonase, polyglacturonase, pectinmethylesterase, etc. these enzymes help in establishment in host tissues or decomposition of organic substrates. The intracellular enzymes such as invertase, uric oxidase, asparginase are of high economic value and difficult to extract as they are produced inside the cell (Riviere, 1977). They can be obtained by breaking the cells by means of a homogenizer or a bead mill and extracting them through the biochemical processes. The process of enzyme purification is difficult as the cell debris and nucleic acid are not easily removed.

Microbial enzymes have two advantages over the animal and plant enzymes. Firstly, they are economical and can be produced on large scale within the limited space and time. The amount produced depends on size of fermenter, type of microbial strain and growth conditions. It can be easily extracted and purified. Secondly, there is a technical advantage in producing enzymes via using micro-organism as

1. They are capable of producing a wide variety of enzymes
2. They can grow in a wide range of environmental conditions
3. They show genetic flexibility that is why they can be genetically manipulated to increase the yield of enzymes, and
4. They have short generation times (Trevan M. D, 1987).

Bulk Enzymes from Fungi

There are several multinational companies having stake in manufacturing industrial enzymes from fungi. Biocon India Ltd. is a major bulk enzyme producer in India, but not a major at global level. The top 14 companies at the global level are listed in Table 15.2.

Table 15.1: Major Bulk Enzyme Producing Companies
(from Ratlege and Kristiansen, 2001)

Sl.No.	Company Name	Country
1.	Solavy Enzymes gmbh	Germany
2.	Amano Pharmaceutical Co.	Japan
3.	Biocatalysis Ltd.	Wales
4.	Enzyme Development Corp.	USA
5.	Danisco Cultar	Finland
6.	DSM-GIST	Netherlands
7.	Meito Sankyo Co.	Japan
8.	Nagase Biochemicals Ltd.	Japan
9.	Novo Nordisk	Denmark
10.	Rhone-Poulenc	England
11.	Rohm gmbh	Germany
12.	Sankyo Co.	Japan
13.	Shin-Nihon Chemical Co.	Japan
14.	Yakult Biochemical Co.	Japan

The bulk enzymes produced have found use mainly in the areas mentioned in Table 15.3 (from Ratlege and Kristiansen, 2001).

Table 15.3: Use of Enzymes for different Purposes

Food	45 per cent
Detergent	34 per cent
Textile	11 per cent
Leather	3 per cent
Pulp/paper	1 per cent
Others	6 per cent

The major enzymes sourced from fungi are listed in Table 15.4.

Table 15.4: The Major Enzymes Sourced from Fungi

Enzyme	Source
Acid, alkaline and neutral proteases	*Aspergillus oryzae; A. niger A. flavus; A. sojae*
Cellulase	*Trichoderma koningi*
Diastase	*Aspergillus oryzae*
Glucoamylase	*Aspergillus niger; A. oryzae*
Invertase	*Saccharomyces cerevisiae*
Lactase	*S. lactis; Rhizopus oryzae*
Ligninase	Phanerochaete chrysosporium
Lipase	*Rhizopus* spp.
Pectinase	*A. niger; Sclerotinia libertine*

Fungal Resources for Industrial Enzyme Technology

Industrial enzymes occupy an important niche in the area of microbial fermentation technology and the search for novel enzymes suited to diverse applications from different microbial sources has been on the increase. Bacteria, Actinomycetes and fungi have been successfully employed in optimising fermentation processes for the manufacture of large quantities of enzymes for applications in the food, pharmaceutical, textile, leather, detergent and many other industries. Enzyme application for mankind's benefit predate the technical knowledge on the nature of the enzymes. For example, starch saccharification in the traditional brewing was carried out by the use of indigenous enzymes in malted barley, while in the oriental fermented foods, microbial enzymes native to the substrates were responsible for the fine flavor changes effected by the fermentation. Use of calf rennet for cheese making or the use of papain for the tenderising of meat during cooking are other well-known examples of enzyme applications in traditional food processing.

Fungi have received considerable attention as sources of large-scale production of industrially useful enzymes. From the classical process for the manufacture of

'Taka Diastase' on wheat bran koji developed by Jokichi Takamine using *Aspergillus oryzae* (US patent 525823, 1894), fungal- based enzyme technology has come a long way and today the emphasis is exploring naturally occurring fungal strains for a variety of enzymes, notably hydrolases of carbohydrates, Proteins and fats. Amylase, Protease, Pectinase, Cellulase and Lipase are among the diverse enzymes manufactured by fungal fermentation. Attempts to find substitutes for useful enzymes have also been successful. For example the production of milk clotting enzymes suitable for cheese making from mucoralean fungi *Rhizomucor miehei* and *R. pusillus* and more recently the cloning and expression of Chymosin as a heterologous protein in *Aspergillus niger* are significant developments from an industrial point of view. Compared to bacterial fermentations in which the process technology involves the use of sophisticated equipments for getting clear filtrates from the colloidal broths, fungal broths can be easily filtered by filter press or similar simple equipment saving considerable investment cost for equipment.

Hydrolytic enzymes are produced by micro-organisms to break down complex naturally occurring nutrient molecules and assimilate them for metabolism and growth. For example when inoculated on a starch-containing medium, amylases are secreted to degrade the starch molecules to glucose and utilize it. Likewise protease secretion is induced to degrade complex proteinaceous nutrients into amino acids before taking them up for metabolism and cellular biomass build up. Overproduction of enzymes and secreting them in quantities far beyond the amounts necessary for the hydrolysis of complex nutrient substrates is a trait associated with select number of species and in a sense these secreted high levels of enzymes can also be regarded as secondary metabolites. Often the levels of enzyme activity reach a peak level after much of the component substrate(s) has been substantially degraded. Fungi have been found to be extremely efficient secrete of soluble protein and under optimized conditions of fermentation, mutant strains secrete up to 30 grams per litre of extracellular protein. In the strains selected for enzyme fermentations, the desired enzyme constitutes the only component or at least forms the major ingredient of the secreted proteins with high specific activities. It is this trait of high-level protein secretion besides their eukaryotic nature that has made the fungi as favourite hosts for heterologous expression of high value mammalian proteins for manufacturing by fermentation.

Factors Regulating Extracellular Enzyme Production

All important industrial enzymes are hydrolases and in the natural environment micro-organisms catalyse the breakdown of complex substances through hydrolytic enzymes to derive essential nutrients for metabolism and growth. Because their function is remote and once released from the cell they cannot be controlled by the cell, the extracellular enzymes tend to be very stable and also tend to be produced in relatively large amounts. This means that even wild strains isolated from the significant level of enzyme for commercial manufacture. Ryu in 1979, analysed experimental data on cellulose production in a two-stage continuous culture. According to their estimates with specific activity of cellulose of 0.6 unit per mg protein and for a yield of enzyme of 1610 units per gram nitrogen consumed, it can be calculated that approximately 6 per cent of the total nitrogen is contributed to the enzyme molecule. The higher stability of extracellular enzyme and their broad tolerance to pH and temperature variations make them very useful for practical commercial applications. Complex regulatory mechanisms control biosynthesis and secretion of these enzymes. They are induced by low levels of the hydrolysis products of the polymers they hydrolyse and sometimes the polymers themselves seem to function best as inducers. Sometimes compounds, which are not substrates for the enzyme, can also act as alternate inducers. For example, cellulase is induced by cellobiose or lactose or even by a disaccharide sophorose. Enzyme production is subject to catabolite repression. While assimilable low molecular weight nutrients are available, the cells will grow without producing extracellular hydrolases and the production machinery will get activated only when the assimilable nutrients are exhausted. For example in a medium containing glucose and cellulose, cellulase induction can be observed only after exhaustion of supply of the readily metabolisable glucose. Extracellular enzymes account for about two thirds of total world market for industrial enzymes. Certain properties of enzymes necessitate special consideration in the design of suitable media for their production. These related to the phenomena of induction, catabolite repression, product inhibition and in the case of extracellular enzymes, protein release mechanisms. In some cases the main carbon sources can also serve as inducer, as in the case of cellulase induction by cellulose. Use of analogues or other compounds which can serve as inducers are however too expensive for practical use in the commercial manufacture of these enzymes. Mutation at a regulatory site to eliminate the requirement for an inducer, elimination of the ability of the operator gene to bind a repressor molecule or prevention of the formation of a repressor molecule are some of the molecular mechanisms that can lead to the development of the so- called 'constitutive' mutants producing significant level of the enzymes in the absence of an inducer or even in media containing compounds exerting repression of enzyme synthesis. Catabolite repression is a process in which synthesis of key enzymes in a metabolic sequence are suppressed when an easily metabolised carbon source is present. Glucose commonly represses the formation of catabolic enzymes and this is termed the 'glucose effect'. This type of repression may be prevented in the design of the fermentation medium by avoiding the use of repressive sources of carbon. Providing the metabolic carbon in the form of slowly metabolisable compounds like starch or lactose rather than glucose would largely overcome catabolite repression.

An alternative approach would be to feed the potential repressive substrate slowly into the fermentation broth so that its concentration during fermentation is kept at a low and non repressive level. Repression of enzyme synthesis by end product is another factor to be considered as seen in the case of protease synthesis repression by amino acids. The design of the fermentation medium must take into account the potential of such repression mechanisms to exert deleterious effects. For example, fermentation medium and operated the fermentation conditions under sulphate limitation since protease synthesis in *Aspergillus niger* was strongly influenced by the level of sulphate in the medium. An alternative approach would be to prevent the formation of the end products which are inhibitory by incorporating inhibitors of the end-product metabolic pathway into the medium.

Increase secretion of extracellular enzymes has been observed under the influence of surface active agents. In presence of 0.1 per cent Tween-80, fifteen to twenty- fold enhanced secretion of cellulase, invertase, β – glucosidase and xylanase. The effect may be due to the modification of cell membrane by surfactants to increase its permeability and "leakiness" to the enzyme. For optimising process technology, the composition of the medium is important and must take into account the cost and availability of the raw materials and also the ease of downstream processing including filtration and post- harvest processing of the effluents. The medium optimisation should also be related to the control of growth morphogenesis to favour filamentous or pellety growth and in relation to the levels of enzyme productivity. Also the medium optimisation for development of an economically viable process proceeds along with continuous strain improvement programmes.

Enzyme fermentations have been carried out by submerged culture in batch, fed-batch or continuous mode of operation. Batch fermentations in which the nutrient addition is carefully monitored and dosed have yielded the most promising results. The principal factors which influence the dissolved oxygen tension which is itself a function of the aeration rate, agitation rate and gas phase pressure applied to the fermenter system. For example temperature profiling has been found to be beneficial in some cases. In *Trichoderma reesei* mutant RUT-C-30, optimum cellulase production was obtained when the temperature was maintained at 31°C during the first 48 hours while the remaining period of the fermentation was carried out at 28°C. With the availability of instrumentation control facilities as well as online monitoring of fermentations, accurate assessment of different variables in the fermentation parameters has become possible and enhanced enzyme productivities could be standardised through careful studies to arrive at the optimal parameters with reference to individual production strains.

Hydrolytic Enzymes

A. Lipases

Lipase as fermentation products have received considerable attention and several fungi are identified as source material for such fermentations. These include species of *Aspergillus* such as *A. niger* and *A. luchuensis*, *Rhizopus* delemar and *R. arrhizus*, *lanuginosa*. Lipases are a complex of enzymes with varied substrate specificities and in fermentations more than one lipase may be produced by a fungus, which may have different genetic origins or may be produced as a result of modifications to a single enzyme protein. *Pencillium crustosum*, for example, produced two lipases with different relative activities towards tributyrin and olive oil. Lipase production has been studied in both submerged and SSF fermentation conditions and the choice of the method varied with the strains employed. The presence of an inducer like triglyceride, long chain fatty acid ester or free fatty acid was favourable for lipase synthesis. In *Candida cylindracea*, extensively studied for its lipase activity, highest titres were obtained when a sterol and a fatty acid or its derivative were both present. In other cases, requirement for an oil fatty acid derivative did not appear to be absolute and in a few instances marked decrease in lipase production was observed when oil was incorporated. It has been reported in many fungi that the production of lipase has shown little sensitivity to the nature of the carbohydrate used and on media containing simple sugars, lipase yields were independent of the type of sugar used. In the light of recent advances and newer applications of lipases for specific objectives, search for lipases with specific properties to meet the requirements and otimisation of the fermentation conditions to obtain the desired lipase in high yields need to be investigated and standardised.

B. Starch Degrading Enzymes

Fungal alpha amylases differ from the bacterial amylases in the spectrum of sugars produced from starch yielding a product containing mainly maltose and some glucose. While *Bacillus* amylases are highly thermostable and act in the endo mode to release dextrins, the fungal alpha amylases are also less thermostable. Alpha amylase from *Aspergillus oryzae* was the first microbial enzyme to be manufactured for sale and was made by solid state cultivation for many years. *A. oryzae* is almost universally used as the source of commercial fungal alpha amylase. Rapid induction of alpha amylase by non-growing mycelia of *A. oryzae* and observed maltose was giving the best activity followed by alpha-methyl β-D- glucoside and starch. However natural media containing starch or brans as sources of growth yielded higher enzyme titres compared to maltose, sugars or dextrins. These are alpha-linked glucose substrates favoured amylase production compared with monosaccharides and the production is enhanced by excess ammonium ions. They also claimed that solid state cultivation gave higher productivity of the enzyme compared with submerged culture. In addition to *A. oryzae*, strains of *A. oryzae* have also been widely investigated for alpha amylase production. The production of an acid stable alpha amylase from *A. niger*. By mutation and selection of suitable cultural conditions, suppression of the accompanying amloglycosidase and transglucosidase activities was achieved facilitating the use of the enzyme in the production of high maltose syrups. For complete hydrolysis of starch to glucose the combined action of alpha amylase and amyloglucosidase is necessary. Amyloglucosidase (glucoamylase) attacks the non-reducing chain ends of starch and release β-glucose. It also slowly attacks the alpha-1, 6 branch points of amylopectin so that almost quantitative yields of glucose can be obtained from starch subjected to thinning by alpha amylase treatment. Strains of aspergilla notably *A.* niger and species of *Rhizopus* such as *R. delemar* are rich sources of amyloglucosidase. Large-scale manufacture of alpha amylase and amyloglucosidase has enabled the manufacture of glucose from starch through enzymatic hydrolysis word wide completely replacing the earlier technologies based on the use of acid for effecting the hydrolysis. The enzymes from *R. delemar* and *A. niger* have been characterized as exo-alpha 1,4 glucohydrolases and a strain of *A. niger* designated NRRL 337 has been widely used in the commercial manufacture of amyloglucosidase. Use of transglucosidase negative mutants of *A. niger* have made the process technology even more suited for large scale applications. Amyloglucosidase production is mostly carried out by submerged culture on media containing maize starch with or without pretreatment with bacterial alpha amylase supplemented with cornsteep liquor and inorganic salts. Multiple forms of amyloglucosidases differing in their isoelectric points have been identified and separated from commercial samples of *A. niger* amyloglucosidase.

C. Cellulolytic and Related Enzyme

Many micro-organisms are capable of digesting cellulose and these include bacteria (*cellulomonas, cytophaga*), Actinomycetes (*Thermomonospora*) and moulds (*Chaetomium, Trichoderma, Aspergillus* and *Pencillium*). However only select fungi have been shown to be capable of secreting significant levels of extracellular cellulase enzymes suitable for enzymatic hydrolysis of cellulose to produce glucose. For complete hydrolysis of crystalline cellulose to glucose, it is usual for three types of enzymes to act synergistically, *viz.* endo β- 1, 4 glucanases which attack the bonds at random position in the cellulose chain, exo-β-1,4 glucanases which split cellobiose residues from the chain ends and β- glucosidases which split cellobiose to yield glucose. Multiple isoenzymes of endo and exoglucanases have been studied and characterised from cellulolytic fungi like *Trichoderma* and *Aspergillus*. Because cellulose is the most abundant and renewable natural carbohydrate of plant origin, there is a lot of interest in its breakdown to yield sugars for food, chemical feedstocks and fermentation. In particular, fermentation of cellulose-based feedstocks to produce liquid fuels has received considerable research. Development and optimistic projections of unlimited supply of renewable liquid fuels like ethanol to meet the demands in an era of depleting petroleum reserves continue to draw the attention of biotechnologists worldwide. A huge volume of research into cellulases and their potential applications has been undertaken during the last three decades. Considerable literature on various aspects of cellulose biotechnology is available. As yet however commercial- scale application of cellulase for conversion of cellulosic residues to ethanol have not met with commercial and economic success.

A strain of the fungus *Trichoderma*, isolated from the US army equipment deteriorating in storage received considerable attention. This fungus was selected as producer of an active complex of cellulases capable of digesting crystalline cellulose and was originally identified, as *Trichoderma viride*. In subsequent investigations, the strain has been placed under a new species, designated *Trichoderma reesi*, named in honour of the pioneering cellulase researcher Dr. Elwyn T. Reese. The work on this fungus including isolation of a large family of improved cellulolytic mutants and also the construction of a pilot plant where the enzyme was used to hydrolyse cellulose to glucose has been comprehensively reviewed by Reese and Mandels (Ann. Rep. fermentation process (edited by G. T. Tsao 7: 1-20, 1984). Several other fungi have been reported as source of extracellular cellulase complex of enzymes and these include *Penicillium funiculosum, Sclerotium rolfsii* and *Talaromyces* sp. Higher levels of β- glucosidase secretion compared to *T. reesei* has been reported in case of *Scytalidium lignicola* and *Penicillium funiculosum*. Economically viable process development for enzymatic conversion of the abundantly available natural cellulosic substrates to fermentable sugars has so far remained an unfulfilled dream. In enzymatic digestion, the complex structure of the cellulosic materials in which the cellulose is bound to lignin and may have low accessibility to the enzyme, dictates costly mechanical pretreatment and also alkali treatment to make the cellulose less crystalline and susceptible to the action of enzymes. Also the very low rate at which cellulase enzymes catalyse the digestion of natural cellulose is a major hurdle in the development of practical and viable technology. It is worthwhile to note that the activity of a pure cellulolytic enzyme on a natural substrate is less than one mole glycosidic bond cleaved per minute per mg protein, which is about 1 per cent of the corresponding rate of attack of amylolytic enzymes on starch.

Recent years have seen much interest in the development of cellulases stable to and active under alkaline pH conditions for potential applications in the detergent and textile industries. Alkalophilic *Bacillus* strains studied by the Kao corporation in Japan produce an alkaline cellulase and such cellulase has been incorporated in commercial detergent named ' Attack'. The cellulase introduced in laundry detergents exhibit fabric softening and colour brightening properties besides removing soil. Cellulases active at alkaline pH also find extensive application in the manufacture of denim jeans to produce the 'stone washed' effect and also 'biopolishing' of fabrics to achieve increased smoothness and softness of the finished product. Fungal sources of alkaline cellulase have not been extensively investigated. Alkalopihilic cephalosporium having arboxymethyl cellulase activity optimally between pH range of 7.5 and 9.5 and retaining 80 per cent of its activity at pH 11.0 for 24 hours. The enzyme was also compatible with the components of laundry detergents. According to a report on Industrial enzyme technology, cloning of the individual components of the *Trichoderma* cellulase complex (endoglucanase, cellobiohydrolase and β-glucosidase) has led to an improved understanding of cellulase activity related to the molecular structures of different cellulose polymers and the development of "customized cellulases for specific 'stone washing' process".

Natural cellulosic materials contain other carbohydrate polymers particularly the hemicelluloses which may also be hydrolysed by the cellulase. Beta glucans present in barley comprise glucose units linked by β-1, 3 as well as β- 1,4 linkages and in the brewing industry, the β-glucans impair the filtration of the wort. Enzymatic treatment with cellulases or β- 1, 4 glucanases is benificial in the speeding up of the filtration. Culture filtrates of cellulolytic fungi may have adequate levels of β- glucan hydrolysing activity to be of practical and commercial significance.

Xylans are xylose polymers, which are present in substantial amount along with cellulose, especially in the tropical plant residues. Many of the cellulase secreting fungi also elaboration media contain natural agro- residues such as wheat bran, which are also rich in hemicelluloses. Enzymatic hydrolysis of xylans to xylose and its fermentation to ethanol by xylose-fermenting yeasts like *Pachysolen tannophilus* or fungi like *Neurospora* (*Monilia*) *crassa* has also received considerable research attention but so far has eluded translation on to practical technologies for commercial ethanol production. In recent years there has been considerable interest evinced in discovering xylanase enzymes, which are not associated with high levels of cellulase activity. Such enzymes are recognised for their potential as 'environmentally friendly' alternatives for effective bleaching of paper pulp without the use of toxic chlorine compounds and without adversely affecting the quality of the paper pulp. Several alkalophilic *Bacillus* strains as well as actinomycetes secreting cellulase-free xylanases have been identified. Production and applications of cellulase- free xylanases in the paper industry. -Xylanases from fungal sources which are cellulase- free and active at high alkaline pH have not -been much explored. At the national chemical laboratory, Pune. Cephalosporium strain (NCL 87-11-9) which grew well at pH above 9.0 and secreted significant levels of cellulase-free xylanase active at high alkaline pH and this was the first report of such an enzyme. The process for production of the cellulase-free xylanase including strain isolation has been covered under a US Patent. A cellulase- free xylanase from an alkali tolerant *Aspergillus fischeri* stable at a pH range of 5.0-9.5 has been identified.

D. Pectolytic Enzyme

Pectolytic enzymes have been used for many years to reduce viscosity and improve clarity of fruit juices and vegetable juices and to increase their yields from their sources. Early attempts were made by growing moulds on moist bran heaps, which were dried and used directly as enzyme preparations. Presently solid state as well as submerged techniques under strictly aseptic and controlled fermentation conditions have ensured consistent and related species have been in commercial use for the manufacture of pectic enzymes. A complex of enzymes are involved in the degradation of pectin which include de-esterifying pectin methyl esterase followed by the chain cleaving endopolygalacturonase. Pectin transeliminase which cleaves the glycosidic bonds in the esterified pectn also plays an important role in pectin degradation. A strain of *A. niger* in submerged culture and found that while synthetic media with pectin gave poor yields, complex media with organic nitrogen such as peanut meal gave high activities. Besides aspergilli, other fungi such as *Coniothyrium diplodiella* produced high levels of pectic enzymes in solid statr fermentations using inexpensive beet pulp.

A complex mixture of pectinolytic and cellulolytic/hemicellulolytic enzymes referred to as Macerozyme has been marketed by Japanese manufacturers and find application in maceration of plant tissues in food industry applications as well as in basic studies in plant biotechnology in the preparation of plant protoplasts. The fungal sources for these enzymes have been strains of *Aspergillus niger* or mucoralean fungi such as *Rhizopus chinensis*. A cellulolytic enzyme of fungal origin termed Cellulase Onazuka, also of Japanese origin, is used in conjunction with the Macerozyme for protoplast isolation from diverse plant materials.

E. Proteolytic Enzymes

Proteinases are the largest category of commercial enzymes both in tonnage and in value terms. Numerous uses are made of them and several different types of proteolytic enzymes are manufactured to meet diverse demands. Four major types of proteinases, which are endopeptidases, are distinguished. These are serine proteinases which have a serine residue at the active site, and which exhibit activity at a pH range of 8.0- 11.0. The thiol proteinases use the – SH groups in catalysis and have a broad pH optima around neutrality. Carboxyl proteinases have acidic pH optima and use the carboxyl group, especially of aspartic acid in the catalytic mechanism. Metalloproteinases have bound metal atoms at the active sites which participate in catalysis and they are inhibited by chelating agents such as EDTA. They are normally most active in the neutral pH range. Microbial sources for proteinases have been identified for commercial manufacture, especially the serine proteinases which find application in food, pharmaceutical and detergent industries, while carboxyl proteinase include rennin from calf rennet used in cheese making and pepsin from pig stomach.

Bacillus cultures, especially strains which are alkalophilic, have been identified and widely used for the commercial manufacture of alkaline proteases in very large quantities for applications in the detergent industry. There are only a few fungal proteinases with high alkali stability. One of the significant findings on fungal alkaline proteinase has been from the National Chemical Laboratory, Pune where strains of saprophytic entomophthoralean fungi belonging to the genus *Conidiobolus* were studied and shown to secrete high levels of alkaline proteases active at and stable to high pH values. The *Conidiobolus* enzymes showed compatibility with several commercial detergents and also was potentially useful in animal cell cultures as a substitute for conventional trypsin. The enzyme appears also to hold promise and potential for applications in leather biotechnology and also in the degumming of silk for improving the quality and properties of silk fabrics. Species of *Aspergillus* are the principle sources of acid proteases. In *A. oryzae*, the predominant enzyme is a carboxyl proteinase with pH optimum at 4.0-5.0. The other important acid protease with commercial potential is the milk clotting enzyme identified from fungi like *Endothia parasitica* ("Sure Curd") and species of thermophilic Mucorales like *Rhizomucor miehei* and *R. pusillus* (referred to earlier as *Mucor miehei* and *M. pusillus*). The enzyme preparations have a high ratio of milk clotting to general proteolytic activity and have been tried out as substitutes for calf rennet in cheese making. The milk clotting activity is due to its selective attack on the k-casein fraction, which stabilises the casein micelle in milk, *R. pusillus* enzyme was stable from pH 3.0-6.0 and showed maximum activity at 55°C. Other sources of thermophilic fungi secreting acid proteases include *Pencillium duponti*, which secreted the enzyme in submerged fermentation in rice bran media at 50°C. Thermomycolase, an alkaline protease secreted by the thermophilic fungus, *Malbranchea pulchella* var. *Sulfurea* is a serine protease formed in presence of casein and was repressed by the presence of glucose, peptides, amino acids or yeast extract. Thermomycolase was active optimally at a pH of 8.5 and was stable over a pH range of 6.0 to 9.5 for over 20 hours at 30°C. Calcium ions stabilised the enzyme at high temperatures.

Molecular Approaches to the Study of Industrially Useful Enzymes

Molecular approaches to the study of extracellular enzyme secretion are aimed at getting a better insight into the various molecular aspects such as zymogen processing, folding of the protein in the heterologous host system, effect of glycosylation on the enzyme secretion etc. the *Rhizomucor miehei* protease has been subjected to molecular studies and the recombinant protein expressed in *Aspergillus nidulans* having similar properties. Using the alpha amylase promoter the enzyme protein has also been successfully expressed in *Aspergillus oryzae* with significant yields. A lipase gene from *Humicola lanuginose* (*Thermomyces lanuginosus*) has been cloned and expressed in *Aspergillus oryzae*. Similarly *Rhizomucor miehei* lipase is also processed and expressed in transformed *A. oryzae*. *H. lanuginosa* lipase is marketed by Novo industry under the trade name Lipolase as a detergent additive along with other hydrolases such as protease, amylase and cellulose. In a patent granted to Novo Nordisk, Denmark (Wo9630-502) (biotechnol. Abst. 96-15062, 1996), an alkaline lipase for detergent use from *Botryosphaeria* or *Guignardia* sp. and recombinant enzyme production in *Aspergillus* has been claimed. A method for producing recombinant alkaline lipase by isolating an alkaline lipase encoding DNA sequence from either fungus, combining the sequence with expression elements in a vector, subsequently transforming a host (*e.g. Aspergillus* sp.) and then recovering the alkaline lipase from the culture medium have been described.

The composition of the extracellular complex of cellulose enzymes secreted by *Trichoderma reesei* comprises 60-80 per cent of cellobiohydrolase, 20-30 per cent of endoglucanase and less than 1 per cent of the total secreted protein is β-glucosidase. Genetic engineering methods have been employed for producing strains with novel cellulose profiles. Use of general expression vector pAM- H11O containing the promoter and terminator sequences of the strongly expressed cellobiohydrolase I (cbh I) gene to over express a cDNA for the major endoglucanase (EG I). An *in vitro* modified cDNA of cbh I incapable of coding for active enzyme was used to inactivate the majar cellobiohydrolase gene. Thus new strains producing elevated amounts of all the specific endoglucanases and/or lacking the major cellobiohydrolase cbh I was produced. The finnish group have also successfully used the strong and highly inducible promoter of the gene encoding the major cellulose (cellobiohydrolase I) for the production of eukaryotic heterologous protein, chymosin in *Trichoderma*.

Hodgson 1994, reviewing the changing bulk biocatalyst market, pointed out that recombinant DNA techniques have changed bulk enzyme production dramatically and that 50 per cent or more of the industrial enzymes measured by value or mass are from organisms enzymes measured by value or mass are from organisms that have been genetically engineered. High level of confidentiality maintained by industrial firms about the manufacturing processes has resulted in lack of authentic information on the actual levels at which recombinant strains are currently in use for commercial manufacture. It can however be envisaged that this powerful technique along with those of protein engineering are and will continue to be judiciously employed in developing novel biocatalysts with augmented levels of desirable properties in industrial enzyme technologies (M. C. Srinivasan., 2004).

Conclusion

Fungi have traditionally been the source of several useful chemical substances starting with the well known ethyl alcohol from yeast, which continues to influence human civilization all over the world. In 1928 Alexander Fleming Discovered Penicillin opening up the era of antibiotics (Jacobs, 1985). Even though actinomycetes became the more important source of antibiotics in the years to come, some new antibiotics are still being sourced from fungi. Following the antibiotics era, in the latter part of 20th century, scientists isolated from fungi many more products important in agriculture, industry and medicine. However, emphasis started slowly shifting towards other groups of microbes, such as bacteria including actinomycetes. In the 21st century, there are reasons to believe that fungi can again occupy the center stage in bioprospecting novel chemicals and enzymes useful in industry. Protein based therapeutic agents are emerging as the largest class of new chemicals in drug industry. Most proteins being large molecules necessitate their production in living systems, mostly by recombinant DNA technology.

References

Abraham, E.P., 1981. The beta–lactam antibiotics. *Scientific American*, 244: 76–86.

Ainsworth, G. C., 1976. Introduction to the History of Mycology. Cambridge, UK: Cambridge University Press. ISBN 0–521–21013–5.

Anke, T and Thines, E., 2007. Fungal metabolites as lead structures for agriculture. In: Exploitation of Fungi. Eds: Robson G. D, van West P, Gadd GM. CUP.

Benedict, R.G. and Brady, L.R., 1972. Antimicrobial activity of mushrooms. *Journal of Pharmaceutical Science*, 61: 1820–1822.

Bottcher, H.M., 1964. Wonder Drugs: A History of Antibiotics. J.B. Lippincott, Philadelphia.

Bowerma and Susan., 2008. The Los Angeles Times, "If mushrooms see the light",

Brandt, C. R. and Piriano, F., 2000. Mushroom antivirals. Recent Research Developments for Antimicrobial Agents and Chemotherapy, 4: 11–26.

Chihara, G., 1992. Immunopharmacology of lentinan, a polysaccharide isolated from Lentinus edodes: Its application as a host defense potentiator. *Int. Journal Oriental Medicine*, 17: 55–77.

Churchill, A., Oct 4, 2001. Fungi: Uses as a Resource for Therapeutic Agents. *American Medical Association Briefing on Food Biotechnology*, 1–6.

Di Rado, 2008, Alicia. "Can a mushroom help fight lung cancer?".

Di Rado, 2008, Alicia, "A salad fixin' with medical benefits?".

Gerngross, T. U., 2004. Advances in the production of human therapeutic proteins in yeasts and filamentous fungi. *Nature Biotechnology*, 22: 1409–1414.

Hardman, A., Limbird, L., Gilman, A. (Eds.) 2001. *The Pharmacological Basis of Therapeutics*. Tenth edition, McGraw Hill, New York.

Hobbs, C., 1986. Medicinal Mushrooms. Interweave Press, Loveland CO.

Isaka, M., Tantichareon, M., Kongsaeree, P. and Thebtaranonth, Y., 2001. Structures of cordypyridones A–D, antimalarial N–hydroxy– and N–methoxy–2–pyridones from the insect pathogenic fungus *Cordyceps nipponica. Journal of Organic Chemistry*, 66: 4803–4808.

Jacobs, F., 1985. The True Story of Penicillin. Dodd Mead and Company, New York.

Kupra, J., Anke, T., Oberwinkler, G., Schramn, G. and Steglich, W., 1979. Antibiotics From basidiomycetes VII. *Crinipellis stripitaria* (Fr.) *Pat., Journal of Antibiotics*, 32: 130–135.

Lovy, A., Knowles, B., Labbe, R. and Nolan, L., 1999. Activity of edible mushrooms against the growth of human T4 leukemia cancer cells, and Plasmodium falciparum. *Journal of Herbs, Spices and Medicinal Plants*, 6: 49–57.

Srinivasan, M. C., 2004, *Practical Mycology for Industrial Biotechnologist*.

Mizuno, T., Saito, H, Nishitoba, T., and KAWAGISHI, H. 1995. Antitumor active substances from mushrooms. *Food Reviews International*, 111: 23–61.

Sharma, P. D., 1999. Microbiology and Plant Pathology

Ratlege, C. and Kristiansen, B., 2001. Basic Biotechnology. Cambridge University Press, Cambridge.

Reese and Mandels., 1984. Ann. Rep. fermentation process, edited by G. T. Tsao 7: 1–20.

Riviere, J., 1977. Industrial Apllication of microbiology, (Translated and edited by M. O. Moss and J. E. Smith), Surrey University, Press, London.

Sasson, A., 1984, Biotechnology: Challenges and promises, UNESCO, paris.

Smania, A., Monache, F.D., Loguericio–Lette, C., Smania, E.F.A., and Gerber, A.L. 2001. Antimicrobial activity of basidiomycetes. *Int. Journal Medicinal Mushrooms*, 3: 87–88.

Suay, I, Arenal, F., Asenio, F., Basilio, A., Cabello, M., and Diez, M.T., 2000. Screening of basidiomycetes for antimicrobial activities. *Antonie von Leeuwenhoek*, 78: 129–139.

Sullia S. B. and Shantharam S., 2000, *General Microbiology*.

Suryanarayanan, T. S., 2009. Fungal endophytes and bioprospecting. *Fungal Biology Reviews* 23, 9–19.

Suzuki, H., Iiyama, K., Yoshida, O., Yamazaki, S. Yamamoto, N. and Toda, S. 1990. Structural characterization of the immunoactive and antiviral water-solubilized lignin in an extract of the culture medium of Lentinus edodes mycelia (LEM). *Agric. Biol. Chem.*, 54: 479–487.

Trevan M. D, 1987. In Biotechnology: the biological principles. (Eds. Trevan, M. D.; Boffey, S.; S.; Goulding, K. H. and Stanbury, P.), pp. 155–228, (Indian Ed.), Tata McGraw Hill Publ. Co., New Delhi. Tulecke, W. (1953). *Science* 117: 599–600.

Wainwright, M., 1995. *An Introduction to Fungal Biotechnology*. Wiley, Chichester.

web.singnet.com.sg/~linlj/pharmace.htmý

www.countrysideinfo.co.uk/fungi/importce.

SECTION III
PLANT PATHOLOGY

2015, Recent Trends in Microbiology, Mycology and Plant Pathology   *Pages* **245–264**
Editor: **Dr. H.C. Lakshman**
Published by: **DAYA PUBLISHING HOUSE, NEW DELHI**

Chapter 16

Bioagents for the Management of Plant Diseases

Shripad Kulkarni* and S.P. Singh

Institute of Organic Farming, UAS, Dharwad, Karnataka
E-mail: shripadkulkarni@rocketmail.com

Plant diseases have been with mankind since agriculture began. The population explosion has caused a need for more food with increasing amount of land devoted to crop crop production. Intensification and monocropping have resulted in increasing disease pressure. Plant diseases cotributes 13-20 per cent of losses in crop production worldwide.Control of plant diseases by chemicals can be spectacular but this is relatively a short term measure and more over, the accumulation of harmful chemical residues some times causes serious ecological problems. Plant diseases are caused by various biotic and abiotic factors. But biotic factors viz; fungi bacteria, viruses, phytoplasmas, spiroplasmas, viriods, phanerogamic parasites, protozoan and nematodes are taking heavy toll of the crops. These pathogens are causing substantial yield losses in different crops and therefore need to be managed. In recent years, the increasing use of potencial hazardous pesticides and fungicides in agricultural has been the result of growing concern of both envoronmentlist available and public health authorities Moreover, use of such chemicals entails a substancial cost to the nation and developing country like India cannot afford it.

Several chemicals have been used in the past to manage diseases caused by various pathogens. No doubt, some degree of control was achieved but it posed a new problem of residual toxicity and development of resistant strains of the pathogens. The diseases were also managed through host plant resistance which took care of the residual toxicity problem and the cost of additional input. There is an urgent need to maintain the durability and stability of host plant resistance and check the variability in plant pathogens. Management of soil borne pathogens is difficult because of non-availability of desired level of resistance against major soil borne diseases caused by species of *Fusarium, Sclerotium, Macrophomina, Pythium, Phytophthora* and few others forced to search or resorting to new approaches to manage the diseases. Therefore, biocontrol agents or antagonists are means of plant disease control has gained importance in the recent years. The biocontrol agents multiply in soil and remain near root zone of the plants and offer protection even at later stages of crop growth.

Rosente 1968, stated that today we live in constant threat of man created irreversible phenomena. This statement is applicable to pesticides for in the case of DDT the time between its first use and the publics general awareness of the danger it presents to the environment has been in the order of a quarter of a century.Even with proper use, chemicals may be hazardous indirectly in various ways namly toxicity of residuals, injury to non target crops and endangered species, development of resistance and insurgence at more serious levels. The continued use of chlorinated pesticides has been responsible for the presence of residues in grain, milk, vegetables as well as in human tissues like the placenta, cord blood. As a result of outcries against the tragedies mentioned the discovery and development of alternatives to chemical control is an essential agenda for agricultural scientists. Biological control is the reduction of the amount of inoculum or disease producing activity of a pathogen accomplished by one or more organisms other than man.

So, Biological control have provided an effective and ecofriendly management of plant diseases. These are specific to the targeted pathogens and do not lead to residue problems and accumulation of toxic pollutants in the soil or underground water. One of the major challenges facing organic producers is disease management. The losses due to diseases can be significant and in some cases can devastate entire crops. Cultural methods of disease control are commonly used on organic farms. The application of organic chemicals is often a last resort and regulated while biological control is still not readily available.

Components of Biological control

The Host

The host plant is a main component in biological control of plant diseases, aimed at suppression of the disease producing activities of the pathogen and regulation of the amount of pathogen inoculum. The host plant also indirectly involved in biological control. The host also involved in suppression or termination of pathogenesis or reproduction of the pathogen by one or more mechanism of the host resistance. Plants also have general resistance to pathogens that cannot yet defined genetically which is controlled by the collective action of multiple physiological factors associated with active host metabolism and growth.

Antagonist

In biological control of plant pathogens antagonists are biological agents with the potential to interfere in the life processes of plant pathogens. Antagonists includes all classes of organisms *viz.,* fungi, bacteria, nematodes, viruses, viroids, protozoa and seed plants. Mild strain of plant virus may be classified as antagonists when introduced in to a plant, they cross protect that plant against severe strain and thereby biological control. The antagonist need not be a complete virus a small satellite self replicating RNA molecule when introduced into cucumber mosaic virus causes increased virulence of CMV on tomato but reduces symptoms caused by CMV in squash, tobacco, sweet corn referred to as biological control.

Pathogens

Pathogens include pathogenic fungi, bacteria, nematodes, seed plants, algae, viruses and viroids. The physical relationship of a pathogen to its host during pathogenesis may be either as an epiphyte, an endophyte or both. Most fungal pathogens have both epiphytic and endophytic phases. The vascular wilt fungi have a very brief epiphytic existence *i.e.,* the period of prepenetration growth between germination of the propagule and penetration of the host root, but there after they exist entirely as endophytes within the root cortex and vascular tissue.In general the more internal the pathogen during the host pathogen interaction the less vulnerable the pathogen to control by antagonist. The pathogen may be subjected to antagonism at any time during their life cycle whether dormant or active during their saprophytic or parasitic existence.In some cases the secondary colonists add to the plant damage and also have the potential to retard the spread of pathogens in the host. The longer the time between infection and sporulation the more vulnerable the pathogen displacement by nonpathogen(secondaries).

Biological Interactions

Severe disease and crop loss due to

☆ Susceptible crop, moderately adapted bioagent, Pathogen well adopted.

☆ Antagonists not well adopted and ineffective.

Slight disease loss due to

☆ Susceptible crop, antagonist well adopted to the environment, Pathogen poorly adopted,

☆ Antagonist moderately adopted and quite effective.

No disease loss

☆ Susceptible crop, Biological control agent operating well, antagonist and Pathogen well adopted to environment, Antagonist suppressing the Pathogen.

No disease loss

☆ Resistance genotype, antagonist and Pathogen well adopted to environment, resistant host preventing the disease.

Biocontrol must be Part of Agricultural Practices

☆ Method of tillage

☆ Planting date

☆ Irrigation timings

☆ Crop rotation

☆ Cultivar rotation

☆ Crop monoculture for suppressive soils

Biocontrol Methods

☆ Not a silver bullet.

☆ Gradually reduces the inoculum.

☆ Manages the disease over a period of time.

☆ Antagonist population has to be increased with organic matter as food base.

Identification and Development of Effective Strains

☆ Screening and selection of local strains.

☆ Improvement of strains through Bio technology,mutation

☆ Addition of genes which code for Biocontrol

☆ Deletion of genes which code for Pathogenecity.

Advantages of Biocontrol Agents

1. They avoid environmental pollution of soil, air and water as is being experienced in chemical control.

2. They avoid adverse effect on the beneficial microbes including antagonists in the soil, whereas, chemicals are lethal.

3. They are less expensive compared to chemical control.

4. Continuous use of bioagents avoids the development of resistance in pathogens.

5. Bioagents application is usually once and does not need repeated application.

6. Very high control potential by integrating fungicide resistant antagonist

7. Biopesticides help in induced system resistance among the crop species. *e.g., Trichoderma* sp. resistant to fungicide like, Benomyl and Metalaxyl etc.

8. Biopesticides are very effective for soil borne pathogen where fungicidal approach is not feasible

9. Biological control becomes part of modern large scale agriculture and helps in increasing crop production within existing resources maintaining biological balance.

Desirable Characters of Biological Control Agents

1. It should grow easily on available substrates.

2. It should grow and survive in the rhizosphere or spermosphere.

3. The antibiotics produced by one antagonist should not be inhibitory to other associated antagonists.

4. The antibiotic produced should not cause damage to host plants.
5. The antagonist should be adaptable to large scale commercial production and handling.
6. Spore germination should be quick and prolific.
7. It should be more adaptable to varied environmental extremes than pathogen.

Mechanisms of Disease Control by Bio-control Agents

1. Antibiosis
2. Competition
3. Hyper parasitism
4. Mycorrhizae
5. Plant Growth Promoting Rhizobacteria (PGPR)
6. Cross protection
7. Induced resistance

Table 16.1: Types of Interspecies Antagonisms Leading to Biological Control of Plant Pathogens

Type	Mechanism	Examples
Direct antagonism	Hyper parasitism/predation	Lytic/some nonlytic mycoviruses *Ampelomyces quisqualis* *Lysobacter enzymogenes* *Pasteuria penetrans* *Trichoderma virens*
Mixed – path antagonism	Antibiotics	2, 4-diacetylphloroglucimol Phenazines Cyclic lipopeptides
	Lytic enzymes	Chitinases Glucanases Hydrogen cyanide
	Unregulated waste products	Ammonia Carbon dioxide Hydrogen cyanide
	Physical/chemical interference	Blockage of soil pores Germination signals consumption Molecular cross-talk confused
Indirect antagonism	Competition	Exudates/leachates consumption Siderophore scavenging Physical niche occupation
	Induction of host resistance	Contact with fungal cell wals Detection of pathogen – associated, molecular patterns Phytohormone- mediated induction

1) Antibiosis

☆ Antibiosis – inhibition or destruction of one organism by a metabolite produced by another organism

- Antibiotics
- Other volatile compounds and enzymes

☆ Antibiosis can be an effective method for protection of germinating seeds.

☆ Examples – *Pseudomonas fluorescens* and *P. aureofaciens* – Take all of Wheat – *Gaeumannomyces graminis* var. *tritici*

Table 16.2: Antibiotics Produced by different Bioagents

Antagonist	Antibiotic Produced
Agrobacterium radiobacter	Agrocin 84
Glicladium virens	Gliotoxin, Viridin, Gliovirin
Trichoderma viride	Gliotoxins, Dermadine, Viridin, Trichodermin
Bacillus subtilis	Bulbiformin
Pseudomonas fluorescence	Phenazines, Pryoluteorin, Pyrrolnitrin
Streptomycin	Aureofungin, Kasugamycin, Streptomycin, Cyclohexamides
Penicillin spp.	Griseofulin, Penicillin

2) Competition

Two or more organisms trying to utilize the same carbon, nitrogen, or mineral source; occupy the same niche or infection site

Pseudomonas fluorescens

– Pseudobactin, a siderophore (extracellular, low molecular weight microbial compounds with a high affinity for iron), *Fusarium* spp.

Rhizosphere competence

– Rhizosphere competent organisms are those capable of colonizing the root surface or rhizosphere when applied as as seed or other point source at the time of planting (Ahmad and Baker, 1987).

Phyllosphere competence

– Phyllosphere competent organisms are those capable of colonizing micro niches and adaptive to the microclimate of aerial plant surfaces.

3) Hyper Parasitism

Micro-organisms that are themselves parasites may serve as a host for other parasites. A possible hyperparasite of sugarcane rust: *Cladosporium uredinicola.*

4) Mycorrhizae

Induced disease resistance of plant by Arbuscular Mycorrhizal Fungi (AMF) has become a hot spot in chemo-ecological study and in biocontrol of plant disease.

There were many reports indicating that AMF had antagonistic function to soil borne disease pathogen, or could suppress the growth of pathogen, and increase the resistance or tolerance of mycorrhizal plants to soilborne disease. In mycorrhizosphere, there are interactions among microbial community, in which, AMF could suppress the growth of pathogen and promote the growth of beneficial microbe. Thus, AMF may use as biocontrol fungi with other antagonism microbe. There were several hypotheses about the mechanisms of the increased resistance in mycorrhizal plants: (1) improvement of plant nutrient status; (2) competition; (3) changed roots morphology and structure; (4) changed microbial flora in rhizosphere; (5) induced resistance or systematic resistance in plant. After colonized by AMF, phenolic compounds accumulate in plant, and local defense response or systemic defense response occurs.

Soil application with *T. harzianum* or / and G. intraradices significantly reduced tomato seedlings damping-off incited by Rhizoctonia solani. Moreover, more pronounced disease suppression was obtained when both bioagents were applied together. Application of *T. harzianum* to healthy or inoculated seedlings significantly increased phosphorous supply, which resulted in higher yield, associated with the accumulation of high phosphorus levels in tissues of tomato plants (4.7- 6.5-fold), compared with low P supply. Inoculation with both bioagents in the presence or absence of the pathogen gave significant rise (2.1–2.2-fold), compared with low P levels. Root length of inoculated plants treated with *T. harzianum* or *G. intraradices* appeared longer than those of inoculated untreated plants at all P levels. Phosphorus uptake (mg P/plant) of tomato plant increased in all treatments with increasing of P levels with *R. solani, T. harzianum* or their combination and untreated plants have vigorous response to phosphorus fertilization.

5) Plant Growth Promoting Rhizobacteria (PGPR)

Plant growth-promoting bacteria (PGPB) are associated with many, if not all, plant species and are commonly present in many environments. The most widely studied group of PGPB are plant growth-promoting rhizobacteria (PGPR) colonizing the root surfaces and the closely adhering soil interface, the rhizosphere. Some of these PGPR can also enter root interior and establish endophytic populations. Many of them are able to transcend the endodermis barrier, crossing from the root cortex to the vascular system, and subsequently thrive as endophytes in stem, leaves, tubers, and other organs. The extent of endophytic colonization of host plant organs and tissues reflects the ability of bacteria to selectively adapt to these specific ecological niches. Consequently, intimate associations between bacteria and host plants can be formed without harming the plant. Although, it is generally assumed that many bacterial endophyte communities are the product of a colonizing process initiated in the root zone they may also originate from other source than the rhizosphere, such as the phyllosphere, the anthosphere, or the spermosphere.

Despite their different ecological niches, free-living rhizobacteria and endophytic bacteria use some of the same mechanisms to promote plant growth and control phytopathogens. The widely recognized mechanisms of biocontrol mediated by PGPB are competition for an ecological niche or a substrate, production of inhibitory allelochemicals, and induction of systemic resistance (ISR) in host plants to a broad spectrum of pathogens and/or abiotic. This review surveys the advances of plant-PGPB interaction research focusing on the principles and mechanisms of action of PGPB, both free-living and endophytic bacteria, and their use or potential use for the biological control of plant diseases.

6) Cross Protection

Cross Protection – a form of competition where an avirulent or weakly virulent strain of a pathogen is used to protect against infection by a more virulent strain of the same or closely related pathogen.

Example: Citrus tristiza virus

Table 16.3: Bacterial Determinants and Types of Host Resistance Induced by Bio-control Agents

Bacterial Strain	Plant Species	Bacterial Determinant	Type	Reference
Bacillus mycoides strain Bac J	Sugar beet	Peroxidase, chitinase and β-1, 3-glucanase	ISR	Bargabus *et al.* (2002)
Bacillus pumilus 203-6	Sugar beet	Peroxidase, chitinase and β-1, 3-glucanase	ISR	Bargabus *et al.* (2002)
Bacillus subtitles GB03 and *IN 937a*	Arabidopsis	2,3-butanediol	ISR	Ryu *et al.* (2004)
***Pseudomonas* putida strains**				
CHAO	Tobacco	Siderophore	SAR	Maurhofer *et al.* (1994)
	Arabidopsis	Antibiotics DAPG)	ISR	Iavicoli *et al.* (2003)
WCS374	Radish	Lipopolysaccharide	ISR	Leeman *et al.* (1995)
		Siderophore	ISR	Leeman *et al.* (1995)
		Iron regulated factor	ISR	Leeman *et al.* (1995)
WCS417	Cornation	Lipopolysaccharide	ISR	Van peer and schipper (1992)
	Radish	Lipopolysaccharide	ISR	Leeman *et al.* (1995)
		Iron regulated factor		Leeman *et al.* (1995)
	Arabidopsis	Lipopolysaccharide	ISR	Van Wees *et al.* (1997)
	Tomato	Lipopolysaccharide	ISR	Duijff *et al.* (1997)
***Pseudomonas* putida strains**				
WCS 358	Arabidopsis	Lipopolysaccharide	ISR	Meziane *et al.* (2005)
		Siderophore	ISR	Meziane *et al.* (2005)
BTPI	Bean	Z, 3 hexenal	ISR	ISR Orgena *et al.* (2004)
Serratia marcescens 90-166	Cucumber	Siderophore	ISR	Press *et al.* (2001)

7) Induced Resistance

Systemic acquired resistance is an important component of the disease resistance repertoire of plants. In this study, a novel synthetic chemical, benzo(1,2,3)thiadiazole-

7-carbothioic acid S-methyl ester (BTH), was shown to induce acquired resistance in wheat. BTH protected wheat systemically against powdery mildew infection by affecting multiple steps in the life cycle of the pathogen. The onset of resistance was accompanied by the induction of a number of newly described wheat chemically induced (WCI) genes, including genes encoding a lipoxygenase and a sulfur-rich protein. With respect to both timing and effectiveness, a tight correlation existed between the onset of resistance and the induction of the WCI genes. Induced resistance is due to a low level of persistent stress which can be induced chemically and not the result of a specific component of the inducing pathogen. Spraying aqueous solutions (20 or 50 mm) of oxalate, potassium phosphate dibasic (K_2HPO_4) or tribasic (K_3PO_4) on the upper surface of leaf 1 of cucumber plants, or inoculating leaf 1 with a spore suspension of *Colletotrichum lagenarium* induced systemic resistance to *C. lagenarium, Cladosporium cucumerinum, Dydimella bryoniae, Sphaerotheca fuliginea, Pseudomonas lachrymans, Erwinia tracheiphila*, tobacco necrosis virus (TNV) and cucumber mosaic virus (CMV) when plants were challenged on leaf 2 with the test pathogens 7 days after treatment of leaf 1. Plants were rated for disease 5–15 days after challenge-inoculation, depending on the pathogen. The level of protection induced by the test solutions varied and generally was lower than that induced by *C. lagenarium*. Each chemical solution tested caused chlorotic stippling on the induced leaf within 48 h after application. The stippling later developed into restricted necrotic lesions. The nature of damage in the inducer leaf was related to the level of protection. Plants with distinct but restricted necrotic lesions on inducer leaves were better protected than those with extensive damage or few lesions.

Approaches to Biological Control

Biological Control of inoculum, which includes

1. Destruction of propagules or biomass of pathogen by hyperparasites, hyper pathogens or predators
2. Prevention of inoculum formation
3. Weakening or displacement of pathogen in infested reside (food base) by antagonists.
4. Reduction of vigour of virulence of pathogen by agents such as mycoviruses, or hypovirulance determinants.

Bioagents Used in the Management of Plant Diseases

1. *Trichoderma viride*
2. *T. harzianum*
3. *T. haematum*
4. *T. koningii*
5. *Gliocladium virens*
6. *Pseudomonas fluorescens*
7. *Paecilomyces lilacinus*

8. *Bacillus subtilis*
9. *Glomus fasciculatum*
10. *Agrobacterium radiobacter* K-84 and K-1026
11. Mild strains of viruses
12. *Cladosporium herbarum*

Types of Formulations

1. Powder formulation
2. Encapsulations in organic polymer like sodium alginate.
3. Pelleting biomass and bran with sodium alginate.
4. Wheat bran saw dust water for soil application.
5. Molasses enriched (Kaolin) clay granules.
6. Liquid coating formulations bioprotectant as powder on which suspension of aqueous binder is sprayed on seeds for 0.1 mm. thick layer.
7. As spray from emulsifiable concentrate with 10 spores/1. cfu needed is 1-10 cfu/ml.

Methods of Application

1. Broadcast (125-150 kg/ha)
2. Furrow application (130-160 kg/ha)
3. Root zone application (Formulation mixed in soil @ 1 kg/plant)
4. Seed treatment (4 g/kg of seeds)
5. Wound application
6. Spraying ($10^6 - 10^8$ cfu/ml)

Key to Success of Biological Control

1. Selection of efficient strain of antagonist.
2. Adequate growth and sporulation on mass culture media.
3. Advance application to provide enough time for interaction.
4. Soil temperature, moisture and pH should be favourable for establishment and growth.
5. Rhizosphere competency and proper food base.
6. Monitoring of population of pathogen and antagonist on selective media.
7. Handling, production, storage should be easy, cheap and locally available product.
8. Integration of biocontrol with tolerant/resistant variety and/or chemical control may be more feasible in the integrated disease management system.

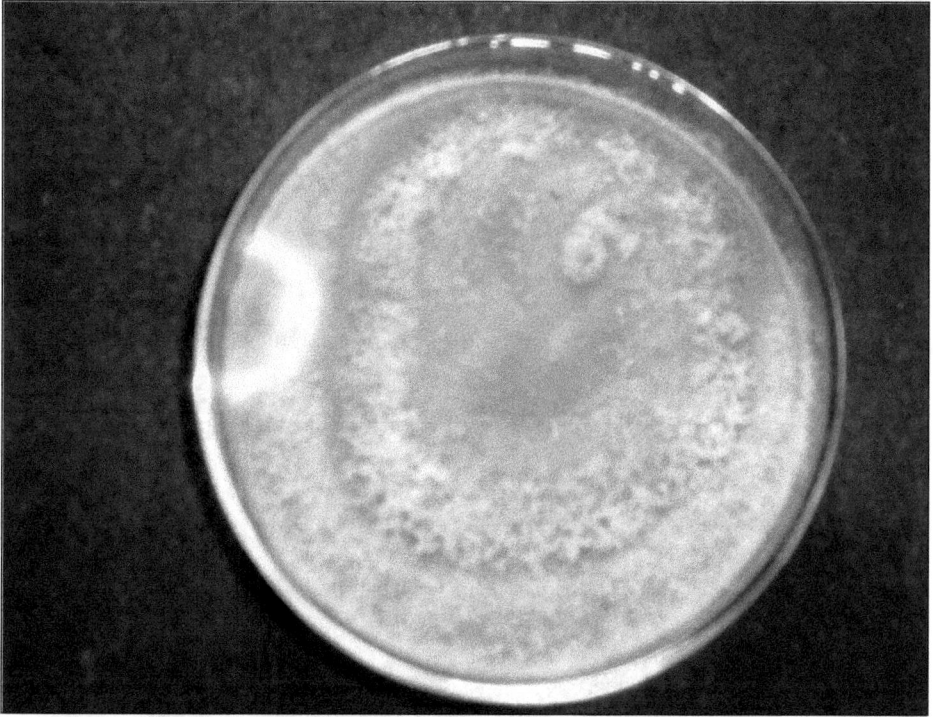

Figure 16.1: *Trichoderma* Suppressing *Sclerotium rolfsi.*

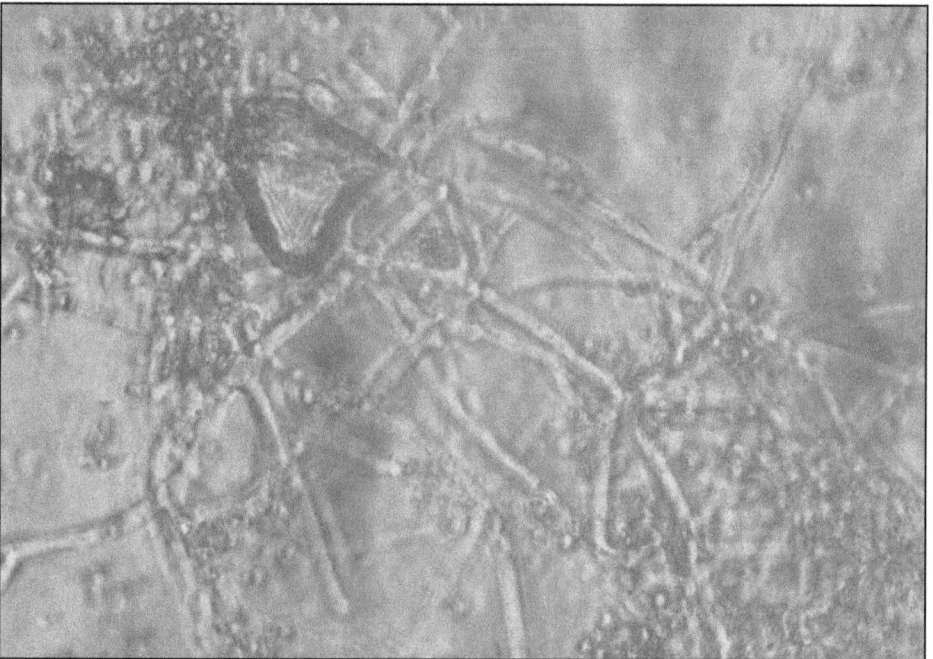

Figure 16.2: Microscopic Photograph of *Trichoderma.*

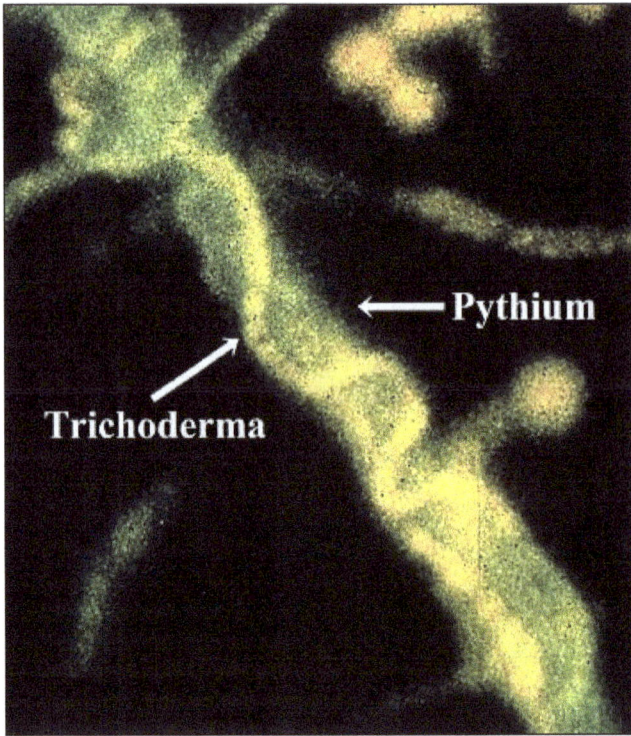

Figure 16.3: Microphotograph Depicting
Trichoderma mycilia **Inhabitating** *Pythium Mycilium.*

Figure 16.4: Bengalgram Field Infested with Fusarium Wilt

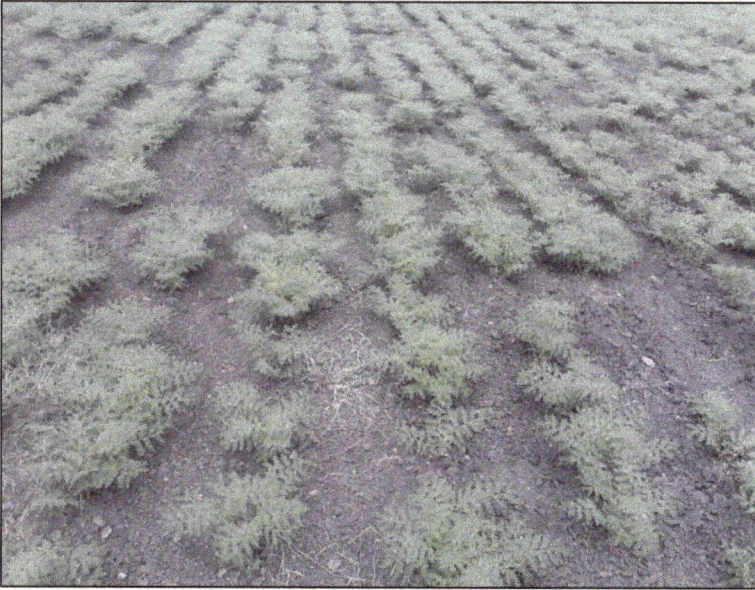

Figure 16.5: Disease Managed with *Trichoderma* Seed Treatment.

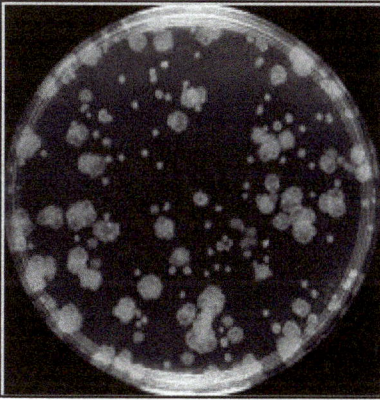

Figure 16.6: *Psuedomonas* Grown on Media.

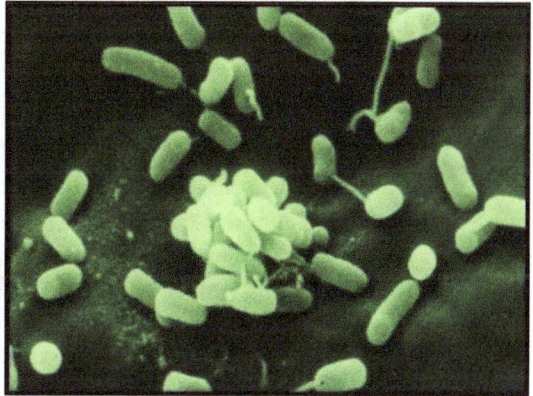

Figure 16.7: Microscopic Photograph of *Psuedomonas*.

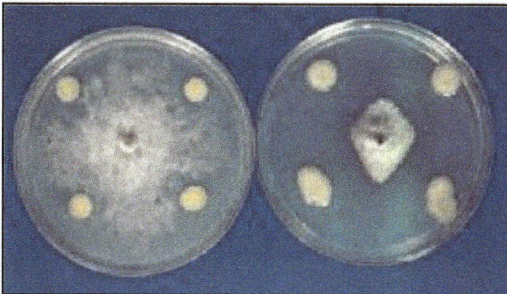

Figure 16.8: *Psuedomonas* Restricting Growth of Pathogen.

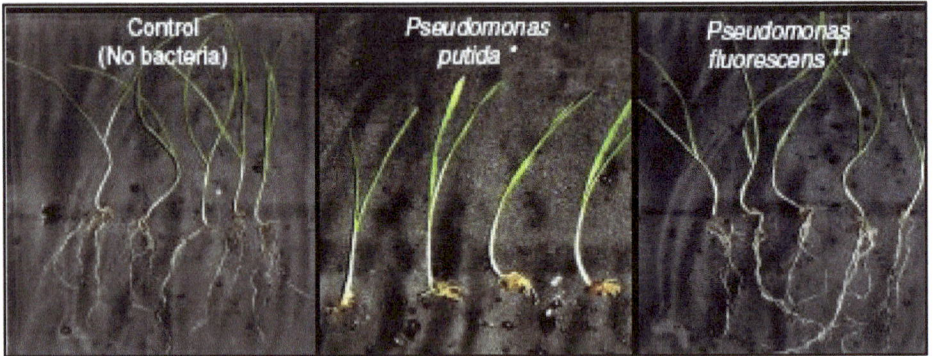

Figure 16.9: Enhanced the Root Growth due to *Psuedomonas.*

Figure 16.10: Blast Affected Paddy Field.

Figure 16.11: Managed by *Psuedomonas* Spray.

Table 16.4: Successful Control of Plant Diseases using *Trichoderma* Species

Crop/Diseases	
Chick pea wilt	Seed treatment (ST) with *T. harzianum* or neem cake + *T. harzianum* soil application
Redgram wilt	Res. var. ICP8863 + ST with *T. harzianum*
Foot rot of pepper	Oil solarization +Neem cake + *T. harzianum*
Cotton root rot	*T. harzianum* + Benomyl
Sclerotium wilt of sugarbeet	Captan + *T. harzianum*
Damping off of tobacco, tomato, egg plant	Metalaxyl MZ + *T. harzianum*
Sclerotium rolfsii in tomato, beans, groundnut, sugar beet and chick pea	Furrow application of 130-160 kg/ha with 4 g/kg of seed treatment of (*T. harzianum T. viride*)
Bean root rot *Macrophomina phaseolina*	Furrow application of *T. harzianum* (130-160 kg/ha)
Fusarium oxysporum in bean, chrysan-themum, cotton, melon and redgram	Soil application of *T. harzianum* at 130-160 kg/ha
Different wood rotting fungi in wood trees Grey mold, *Botrytis cinerea* in grapes	Dusting *T. viride* on wounds.Aerial spray with *T. viride* (106–108°C fu/ml)
Stalk rot of rabi sorghum	*T. harzianum* 50g/ha/8 kg seed)

Different Biocontrol Agents for the Management of Diseases

Pseudomonas fluorescens

It is a formulation of Pseudomonas flourescens–a very potent microbe that not only cures serious plant deseases like damping off, scab, root and stem rot and blights but also controls some species of nematodes.It also helps plants absorsb available phosphorous and works as catalyst for *Trichoderma viride*. Bacteria and fungal propagules on the leaves of crops often serve as nucleation sites for ice formation and ice crystals often form when they are present and the temperature falls below freezing, with resulting damage to the leaf. Other strains of *Pseudomonos fluorescens* are antagonistic to foliar or rhizosphere bacteria and fungi through the production of siderophores and antibiotics.

Crops

Paddy, Pomegranate, Potato, Eggplant, Tomatoes, Chilli, Cut flowers, Orchards, etc.

Target Diseases

Fusarium, Xanthomonas, Pythium spp., *Phytophtora* spp., *Rhizoctonia solani, Botrytis cinerea, Sclerotium* spp.

Dosage

Soil application: 2 kg/acre along with dried FYM

Seed treatment: @ 4-5 gm per kg of seeds as per standard wet treatment.

Seedling root dipping: @ 10 g/l prior to planting.

Over dosing does not cause any harmful side effects

Trichoderma harzianum

They are the front rank killer of harmful fungal diseases like *Fusarium, Rhizoctonia* and *Sclerotium* which cause great havoc to all important crops like soybean, cotton, sugarcane, cereals and many more. Formulation of potent antagonistic fungi, *Trichoderma viride* or *Trichoderma harzianum* successfully controls soil borne fungal pathogens that cause diseases in crops. Trichoderma parasitise pathogenic fungi limit their growth and activity. They also produce toxic metabolites and form a protective coating on seeds against soil borne pathogenic fungi. The hidden fungal diseases spring a surprise with their sudden attack before a farmer can do anything. It can be effectively used for seed treatment as well as soil application

Crops

Cereals, Oilseeds, Eggplant, Potato, Chilli, Tomatoes, Cucumbers, Cut and Pot flowers, Orchards, Vineyards Ornamentals in greenhouses; lawns nurseries etc.

Target Diseases

Pythium spp., *Ganoderma* spp., *Rhizoctonia solani, Fusarium* spp., *Botrytis cinerea, Sclerotium* spp., *Sclerotinia* sp.

Dosage

Soil application: 2 kg/Acre along with 200 kg FYM

Seed treatment: @ 4-6 gm per kg of seeds as per standard wet treatment.

Seedling treatment: @ 10 g/l prior to planting

Bacillus subtilis

It is an antagonistic bacterial Biocontrol agent, which controls many soil and air borne diseases of Paddy, Groundnut, Cotton, Vegetables, Soybean etc., Foliar application of *Bacillus subtilis* with *Pseudomonas fluorescence* control leaf diseases of many crops. The bacterium colonizes the developing leaf and root system of the plant and thus competes with and thereby suppresses plant diseases. The Plant Growth Promoting Rhizobacteria (PGPR) having an antagonistic interaction with various soil borne plant pathogens. It protects plants against seed and root diseases. It can be applied as a seed dressing, Seedling (Root dipping), Soil application and foliar spray.

Crops

Chilli, Grapes, Potato, Cucumbers, Eggplant, Tomatoes, Cut flowers, Orchards, Vineyards Ornamentals in greenhouses; lawns nurseries etc.

Target Diseases

Phytophtora spp., *Iodium* spp., *Alternaria, Aspergillus* spp., *Pythium* spp., *Rhizoctonia solani, Botrytis cinerea, Sclerotium rolfsii, Sclerotinia* spp., and many Powdery and Downey mildew causing organisms.

Dosage

Spray: 5g/litre

Seed treatment: @ 4-5 gm per kg of seeds as per standard wet treatment. Seedling root dipping: @ 10 g/l prior to planting.

Paecilomyces lilacinus

Paecilomyces are soil borne hyphomycetous fungi belonging to class dentromycetes and occur in the soil rhizhosphere. *Paecilomyces* is an egg parasite and referred as opportunistic fungus. Many enzymes produced by *P. lilacinus* have been studied and serine protease with biological activity against *Meloidogyne hapla* eggs. Strain of *P. lilacinus* shown to produce proteases and a chitinase,enzymes that could weaken a nematode egg shell.

Before infecting a nematode egg, *P. lilacinus* flattens against the egg surface and becomes closely appressed to it. *P. lilacinus* produces simple appressoria anywhere on the nematode egg shell either after a few hyphae grow along the egg surface, or after a network of hyphae form on the egg. The presence of appressoria appears to indicate that the egg is, or is about to be, infected. In either case, the appressorium appears the same, as a simple swelling at the end of a hypha, closely appressed to the eggshell. When the hypha has penetrated the egg, it rapidly destroys the juvenile within, before growing out of the now empty egg shell to produce conidiophores and to grow towards adjacent eggs.

Crops

Eggplant, Potato, Chilli, Tomatoes, Cucumbers, flowers, Orchards, Vineyards Ornamentals in greenhouses; lawns nurseries and landscape.

Target Nematodes

Plant parasitic nematodes in soil, Examples include *Meloidogyne* spp.(Root knot nematodes); *Radopholus similis* (Burrowing nematode); *Heterodera* spp. and *Globodera* spp. (Cyst nematodes); *Pratylenchus* spp. (Root lesion nematodes); *Rotylenchulus reniformis* (Reniform Nematode); *Nacobbus* spp. (False Root knot Nematodes).

Dosage of Application

Soil application: 5 kg/ha along with dry FYM.

Seed treatment: @ 4-5 g per kg of seeds as per standard wet treatment.

Seedling dip treatment: @ 10 g/l prior to planting

Different Microbial Pesticides for Pest Management

Beauvaria bassiana

Beauvaria bassiana is for protecting root and stem destroying insects. The entomopathogen invades the insect body. Fungal conidia become attached to the insect cuticle and the hyphae penetrate the cuticle and proliferate in the insect's body. The infected insect may live for three to five days after hyphal penetration and after death the conidiophores bearing conidia are produced on cadaver. Application rates depend upon the crop and the pests to be controlled. The normal application rate on commodity crops is 2g/litre with 500 litre solution per hectare. Useful against aphids, beetles, heliothis, etc. The products may be used alone or tank mixed with other products such as sticking agents, insecticidal soaps, emulsifiable oils, insecticides or used with beneficial insects. Do not use with fungicides and wait 48 hours after application before applying fungicides.

Crops

Cotton- *Helicoverpa* Cabbage-DBM Sugarcane- Rootgrub,Coffee- Berry borer Blackpepper- Pollu beetle Palms- Red weevil Paddy-BPH

Target Pests

Caterpillars, Weevils, Leafhoppers, Bugs, Grubs and Leaf-feeding insects

Application Dosage

Foliar spray: 1 kg/acre in 200 liters of water *i.e.*, 5 g per litre of water. The spray volume depends on the crop canopy.

Soil application: 5 kg/acre

Nomuraea rileyii

Nomuraea rileyii is a biological formulation based on friendly fungus *Nomuraea rileyii*. It acts on harmful caterpillars by gradually paralysing and thus killing them.It targets larvae and immature stages of aphids, root aphids (especially on grape vines), Thrips, mealy bugs, jassids, leaf hoppers, whiteflies, leaf rollers, leaf miners, cutworms, chaffer beetles (scarabids), etc.

Crops

Soyabean, Groundnut, Chilli, Tomato, Potato, Cauliflower, Paddy, Lucern and Maize

Target Pests

Leaf eating caterpillers, Borers

Application Dosage

Foliar spray: 200-400 g/Acre in 200 litres of water *i.e.*,1- 2 g per litre of water. The spray volume depends on the crop canopy.

Metarrhizium anisopliae

It is based on formulation of a friendly fungus *Metarrhizum anisopliae* which effectively controls White Grubs, brown Leaf Hoppers of paddy, Pyrilla of sugarcane, green semi looper, citrus mealybugs, tobacco cutworms and many more. Can be sprayed on standing crop or used as soil application.

Crops

White Grubs, brown Leaf Hoppers of paddy, Pyrilla of sugarcane, green semilooper, termites, Spotted pod borer of pigeonea, Coconut rhinicerous beetle,

Mango Hopper, Rhizome weevil of Banana, DBM of Cauliflower.

Target Pests

White Grubs, brown Leaf Hoppers, Pyrilla, semilooper, termites, spotted pod borer, Rhinocerous beetle, Rhizome weevil and Diamond back moth, Root weevils, Mango hoppers.

Application Dosage

Foliar spray: 400g/acre in 200 liters of water *i.e.*, 2 g per liter of water. The spray volume depends on the crop canopy.

Soil application: 5 kg/acre

Verticillium lecanii

Verticillium lecanii is another friendly fungus with talc based formulation. The fungus inserts its mycelium thru the body wall of all stages of insects and kills the insects by infecting it.

The death of the insect occurs because of release of certain toxic elements like Destraksin Desmethyl, Destraksin etc. which kill the hardy stage like pupae of the insects.

The fungus produces other insecticidal toxins such as dipicolinic acid. The activity of *V. lecanii* depends on the strain of the fungus. Formulated products from conidial production can last up to 1 year.

Crops

Ornamentals and vegetables in greenhouses; nurseries, lawns, landscape perimeters, Vegetables, field crops and other agricultural crops.

Target Pests

Whiteflies, Thrips, Aphids,Scales,Mites and Mealy bugs.

Dosage

Foliar spray: 400 g/Acre in 200 litres of water *i.e.*, 2 g per litre of water. The spray volume depends on the crop canopy.

References

Benhamou, N., and Chet, I 1997. Cellular and molecular mechanisms involved in the intersection between *Trichoderma horziamum* and *Phythium ultimum. Appl. Environ. Microbiol.* 63: 2095–2099.

De Meyer, G., and Hofte. M. 1997. Salicylic acid produced by the rhizobacterium pseudomonas aeruginosa 7NSK2 induces resistance to leaf infection by *Bitrytis cinerea* bean. *Phytopathology* 87: 588–593.

Fitter. A. H anf Garbaye, J. 1994. Interactions between mycoohizal fungi and other soil micro-organisms. *Plant soil* 159: 123–132.

Harman, G. E., Howell, C. R., Vitarbo, A., and Larito M., 2004. Trichoderma species – opportunistic avirulent plant symbionts. *Nature Rev. Microbial,* 2: 43–56.

Howell, C. r. and Stipanovic R. D. 1980. Suppression of Pythium ultimum induced damping off of cotton seedling by *Pseudomonas fluorescens* and its antibiotic, pyoluterin. *Phytopathology.* 70: 712–715.

Kilic – Ekici O and Yuen, G. Y., 2003. Induced resistance as a mechanisms of biological control by Lysobacter enzymogenes strain C3. *Phytopathology* 93: 1103–1110.

Leeman M. Van Pelt J. A., Den Ouden, F. M., heinbroek M and Bakker P. A. H. 1995. Induction of sysemic resistance by *Pseudomonas fluoresenes* in radish cultivars differing in susceptibility to *Fusarium* wilt using novel bioassay. *Eur J Plant Pathol.* 101: 655–664.

Paulitz T. C. and Belanger R. R. 2001. Biological control in greenhouse system. *Annu. Rev. Phytopatho*. 39: 103–133.

Weller, D. M. and Cook r. J. 1983. Suppression of take-all of wheat by seed treatments with fluorescent pseudomonades. *Phytopathology* 73: 463–469.

2015, Recent Trends in Microbiology, Mycology and Plant Pathology   *Pages* 265–282
Editor: **Dr. H.C. Lakshman**
Published by: **DAYA PUBLISHING HOUSE, NEW DELHI**

Chapter 17

Biological Control of Plant Pathogens

M. Nagalakshmi Devamma[1]*, Shaik Thahir Basha[2],
M. Kavya Deepthi[1] and H.C. Lakshman[3]

[1]Department of Botany,
[2]Microbiology Laboratory, Department of Virology,
Sri Venkateswara University, Tirupati – 517 502, Andhra Pradesh, India
[3]Microbiology Laboratory, P.G. Department of Studies in Botany,
Karnatak University, Pavate Nagar, Dharwad – 580 003, Karnataka, India
*E-mail: devi.bot@gmail.com

Biological Control–An Introduction

In plant pathology, the term biological control applies to the use of microbial antagonists to suppress diseases as well as the use of host specific pathogens to control weed populations. The organism that suppresses the pest or pathogen is referred to as the biological control agent (BCA). Biocontrol or Biological control can also be defined as the use of natural organisms or genetically modified, genes or gene products the effects of undesirable organisms to favor organisms' useful to human beings such as crop plants, trees, animals and beneficial micro-organisms. Biocontrol agents are widely regarded by the general public as natural and therefore non-threatening products. One of the most interesting aspects of the science of biological control is the study of the mechanisms employed by biocontrol agents to affect disease control. Biological control of soil borne pathogens by introduced micro-organisms has been studied over 80 years, but most of the time it has not been considered commercially feasible. Several companies now have programs to develop biocontrol agents as commercial products. Micro-organisms that can grow in the rhizosphere are ideal for use as biocontrol agents, since the rhizosphere provides the front-line defense for root against attack by pathogens. Pathogens encounter antagonism from rhizosphere micro-organisms before and during primary infection and also during secondary spread in the roots.

Disease suppression by biocontrol agents is the sustained manifestation of interactions among the plant, the pathogen, the biocontrol agent, the microbial community on and around the plant, and the physical environment. Even in model laboratory systems, the study of biocontrol involves interactions among a minimum of three organisms. Therefore, despite its potential in agricultural applications, biocontrol is one of the most poorly understood areas of plant-microbe interactions. Recently, substantial progress has been made in a number of biocontrol systems through the application of genetic and mathematical approaches that accommodate the complexity. Biocontrol of soil borne diseases is particularly complex because these diseases occur in the dynamic environment at the interface of root and soil known as the rhizosphere, which is defined as the region surrounding a root that is affected by it. The rhizosphere is typified by rapid change, intense microbial activity, and high populations of bacteria compared with non-rhizosphere soil. Plants release metabolically active cells from their roots and deposit as much as 20 per cent of the carbon allocated to roots in the rhizosphere, suggesting a highly evolved relationship between the plant and rhizosphere micro-organisms. It is the dynamic nature of the rhizosphere that makes it an interesting setting for the interactions that lead to disease and biocontrol of disease. The complexity of the root-soil interface must be accommodated in the study of biocontrol, which must involve whole organisms and ultimately entire communities, if we are to understand the essential interactions in soil in the field.

Lets us discuss a study focused on mutants of *Trichoderma* that are resistant to the fungicide benomyl. These mutants are dramatically increased in root colonisation and biocontrol ability, even in the absence of benomyl. Benomyl resistance correlates with several phenotypes, including increased cellulase production and altered morphology, making it difficult to determine the basis for increased colonisation. It is possible that increased cellulase production enhances root colonisation by enabling *Trichoderma* to utilize plant cell debris and that increased colonisation enhances biocontrol.

Why Biological Control?

Fungal diseases may be minimized by the reduction of the inoculums, inhibition of its virulence mechanisms and promotion of genetic diversity in the crop. The use of chemical fungicides in agriculture has been proven to bring about various benefits such as reducing the fungal infection that may rob water and nutrients from crop plants or may cause spoilage while the products are transported to the market. There are numerous classes of fungicides, with different modes of action as well as different potential for adverse effect on health and environment. There are 311 compounds registered and used as fungicides to control various plant fungal diseases. Of these, seven agents are antagonistic micro-organisms and only one agent is derived from plant extract, *i.e.*, extract of *Reynoutria sachalinensis* (Giant Knotweed). Most fungicides can cause acute toxicity, and some cause chronic toxicity as well. The use of chemical pesticides has been known to cause various environmental and health problems. The International Labor Organization (ILO) estimates that as much as 14 per cent of all occupational injuries are due to exposure to pesticides and other agrochemical constituents. The World Health Organization (WHO) and the United Nations Environment Programme estimates that each year, three million workers in agriculture in developing world experience severe poisoning from pesticides, about 18,000 of whom die. Appropriate technological improvement, which results in more effective use of natural resources, is required in agriculture. One of them is the use of microbial antagonists.

Many microbial antagonists have been reported to possess antagonistic activities against plant fungal pathogens, such as *Pseudomonas fluorescens, Agrobacterium radiobacter, Bacillus subtilis, B. cereus, B. amyloliquefaciens, Trichoderma virens, Burkholderia cepacia, Saccharomyces* sp., *Gliocadium* sp. Three species of rhizobacteria isolated from rhizospheres of rice grown in Bali, *i.e. Enterobacter agglomerans, Seratia liquefaciens* and *Xanthomonas luminescens* were found to effectively suppressed the growth of *Pyricularia oryzae* Cav. The cause of rice blast disease.

The antagonistic effect of *Pseudomonas* might be explained on the basis of its antifungal secondary metabolites that are capable of lysing chitin which is the most important component of fungal cell wall. *Trichoderma* spores or other biomass can be added to soil by a variety of methods. If the strain is rhizosphere competent, it colonizes root surfaces and the outer layers of the cortex. This establishes a zone of interaction into which the *Trichoderma* strain releases bioactive molecules. These include elicitors of resistance, such as homologues of avirulence (Avr) proteins and proteins with enzymatic or other functions. The fungi also produce enzymes that release cell-wall fragments, which also enhance plant resistance responses. The plants produce cell-wall deposits and biochemical factors that limit the growth of the *Trichoderma* strain and cause it to be avirulent. Pathogens can attack roots but, in the presence of *Trichoderma*, infection is reduced by the same or similar molecules and cell-wall alterations that result in the avirulence of the *Trichoderma* strains.

Mode of Action

Understanding the mechanism by which the biocontrol of plant diseases occurs is critical to the eventual improvement and wider use of biocontrol method. Over the past forty years, research has lead to the development of a small commercial sector that produces a number of biocontrol products. Throughout their lifecycle, plants and pathogens interact with a wide variety of organisms. These interactions can significantly affect plant health in various ways. In order to understand the mechanisms of biological control, it is helpful to appreciate the different ways that organisms interact. The types of interactions were referred to as mutualism, protocooperation, commensalism, neutralism, competition, amensalism, parasitism, and predation. From the plant's perspective, biological control can be considered a net positive result arising from a variety of specific and non-specific interactions.

Mutualism is an association between two or more species where both species derive benefit. Sometimes, it is an obligatory lifelong interaction involving close physical and biochemical contact, such as those between plants and mycorrhizal fungi. However, they are generally facultative and opportunistic. For example, bacteria in the genus Rhizobium can reproduce either in the soil or, to a much greater degree, through their mutualistic association with legume plants. These types of mutualism can contribute to biological control, by fortifying the plant with improved nutrition and/or by stimulating host defenses.

Commensalism is a symbiotic interaction between two living organisms, where one organism benefits and the other is neither harmed nor benefited. Most plant-associated microbes are assumed to be commensals with regards to the host plant, because their presence, individually or in total, rarely results in overtly positive or negative consequences to the plant. Neutralism describes the biological interactions when the population density of one species has absolutely no effect whatsoever on the other. Related to biological control, an inability to associate the population dynamics of pathogen with that of another organism would indicate neutralism.

Biocontrol can occur when non-pathogens compete with pathogens for nutrients in and around the host plant. Parasitism is a symbiosis in which two phylogenetically unrelated organisms coexist over a prolonged period of time. In this type of association, one organism, usually the physically smaller of the two benefits and the host is harmed to some measurable extent. The activities of various hyperparasites, *i.e.*, those agents that parasitize plant pathogens, can result in biocontrol. And, interestingly, host infection and parasitism by relatively avirulent pathogens may lead to biocontrol of more virulent pathogens through the stimulation of host defense systems. Lastly, predation refers to the hunting and killing of one organism by another for consumption and sustenance. While the term predator typically refer to animals that feed at higher trophic levels in the macroscopic world, it has also been applied to the actions of microbes, *e.g.* protists, and mesofauna, *e.g.* fungal feeding nematodes and microarthropods, that consume pathogen biomass for sustenance. Biological control can result in varying degrees from all of these types of interactions, depending on the environmental context within which they occur. Significant biological control most generally arises from manipulating mutualisms between microbes and their plant hosts or from manipulating antagonisms between microbes and pathogens.

Modes of Action of Post-harvest Biocontrol Agents

The modes of action of beneficial micro-organisms can be divided into direct antagonism and indirect antagonism. These mechanisms do not exclude each other since they are frequently described as co-occurring within the activity of the same BCA. Direct antagonism comprises those mechanisms that are a direct result of the action of an individual BCA: competition for space and nutrients, secretion of lytic enzymes (depolymerases) such as β-1,3 glucanases and chitinases that degrade the polymers of the pathogen cell wall and mycoparasitism.

Indirect antagonism implies mechanisms which are not a direct result of the activity of the BCA, but are the consequence of the response of the fruit tissue to the presence of these beneficial micro-organisms, ultimately resulting in induction of resistance. Competition for space and nutrients by the BCA mainly takes place when the BCA colonizes wounds of fruit, caused by bruising during handling and transportation. In order to successfully out-compete the pathogens, postharvest BCAs need to possess a sound wound colonization competence, *i.e.* the ability to rapidly colonize and thrive in fruit wounds. Since wounding in apple fruits causes the production of hydrogen peroxide (H_2O_2) and superoxide anion, it has been suggested that resistance to the oxidative stress caused by these reactive oxygen species (ROS) plays a role in wound competence of biocontrol yeasts and, as a consequence, in their competition with the pathogen for space (and nutrients). As with other climacteric fruits, the physiology of stored apples changes during maturation and senescence. The increased production of ROS in apple wounds negatively affects both wound colonization by BCAs and their antagonistic activity.

Table 17.1: Types of Antagonism Leading to Biological Control of Plant Pathogens

Type	Mechanism	Examples
Direct antagonism	Hyperparasitism/predation	Lytic/some nonlytic mycoviruses *Ampelomyces quisqualis* *Lysobacter enzymogenes* *Pasteuria penetrans* *Trichoderma virens*
Mixed-path antagonism	Antibiotics	2,4-diacetylphologlucinol Phenazines Cyclic lipopeptides
	Lytic enzymes	Chitinases Glucanases Proteases
	Unregulated waste products	Ammonia Carbon dioxide Hydrogen cyanide
	Physical/chemical interference	Blockage of soil pores Germination signals consumption Molecular cross-talk confused
Indirect antagonism	Competition	Contact with fungal cell walls Detection of pathogen-associated molecular patterns Phytohormone-mediated induction

Mycoparasitism

Mycoparasitism is a process by which biocontrol fungi may attack pathogenic fungi. Generally, mycoparasitism can be described as a four-step process.

1. Chemotropic Growth

The biocontrol fungi grow towards the target fungi that produce chemical stimuli. For example, a volatile or water soluble substance produced by the host fungus serves as a chemo attractant for parasites. However, the lack of available data for statistical comparison of different conditions or host parasite combinations is a limitation to understanding the phenomenon.

2. Recognition

Lectins of hosts (pathogens) and carbohydrate receptors on the surface of the biocontrol fungus may be involved in this specific interaction.

3. Attachment and Cell Wall Degradation

Mycoparasites can usually either coil around host hyphae or grow alongside it and produce cell wall degrading enzymes to attack the target fungus. These enzymes such as chitinases and β-1,3-glucanase may be involved in degradation of host cell walls and may be components of complex mixtures of synergistic proteins that act together against pathogenic fung.

4. Penetration

The biocontrol agent produces appressoria-like structures to penetrate the target fungus cell wall.

Evidence for these processes in *Trichoderma* spp. and other fungi have been presented. Most of these events have been described from in vitro studies even though mycoparasitic structures have been observed in situ on seeds. Recently, a lectin from *Sclerotium rolfsii* has been isolated and proven to play an important role in recognition. However, the biochemical basis for this phenomenon is not understood. Similarly, cell-wall-degrading enzymes have also been shown to be involved in the inhibition of pathogenic fungi.

Another mycoparasitic fungus, *Sporidesmium sclerotivorum*, is a biotrophic parasite and is often found only on sclerotia of plant pathogenic fungi such as *Sclerotinia minor* and *Sclerotium cepivorum* (the causal agents of lettuce drop).

Antibiosis

It plays an important role in plant disease suppression by certain bacteria and fungi. The process has been defined as the interactions that involve a low-molecular weight compound or an antibiotic produced by a micro-organism that has a direct effect on another micro-organism. The role of antibiotics in biocontrol has been studied by genetic analysis, *e.g.*, mutants that do not produce antibiotics to demonstrate a correlation between antibiotic productivity and biocontrol activity. A few systems have been thoroughly examined. For example, a phenazine antibiotic (Phz) produced by *Pseudomonas fluorescens* strain 2-79 has been implicated in control of takeall of wheat caused by *Gaeumannomyces graminis* var. *tritici.* Although antibiotic production plays major role in suppression of the take-all pathogen, it is not the only factor; some suppression of the pathogen is retained by the nonproducing mutants. Another example, chaetomin is produced by *Chaetomium globosum*, peptaibols are produced by *Trichoderma harzianum*, and pyrones are produced by *Trichoderma* spp. However, the roles of these antibiotics have not yet been demonstrated *in vivo* by genetic analysis.

Antibiotic-Mediated Suppression

Antibiotics are microbial toxins that can, at low concentrations, poison or kill other micro-organisms. Most microbes produce and secrete one or more compounds with antibiotic activity. In some instances, antibiotics produced by micro-organisms have been shown to be particularly effective at suppressing plant pathogens and the diseases they cause. In all cases, the antibiotics have been shown to be particularly effective at suppressing growth of the target pathogen in vitro and/or in situ. To be effective, antibiotics must be produced in sufficient quantities near the pathogen to result in a biocontrol effect. *In situ* production of antibiotics by several different biocontrol agents has been measured; however, the effective quantities are difficult to estimate because of the small quantities produced relative to the other, less toxic, organic compounds present in the phytosphere. And while methods have been developed to ascertain when and where biocontrol agents may produce antibiotics for detecting expression in the infection court is difficult because of the heterogenous distribution of plant-associated microbes and the potential sites of infection.

Mutant strains incapable of producing phenazines or phloroglucinols have shown to be equally capable of colonizing the rhizosphere but much less capable of suppressing soilborne root diseases than the corresponding wild-type and complemented mutant strains. Several biocontrol strains are known to produce multiple antibiotics which can suppress one or more pathogens. For example, *Bacilluscereus* strain UW85 is known to produce both zwittermycin and kanosamine. The ability to produce multiple antibiotics probably helps to suppress diverse microbial competitors, some of which are likely to be plant pathogens. The ability to produce multiple classes of antibiotics, that differentially inhibit different pathogens, is likely to enhance biological control and this is demonstrated in Table 17.2.

Antibiotic Resistance

An attractive feature of biocontrol strategies is that populations of pathogens resistant to antibiotics produced by biocontrol agents are likely to develop slowly. There are two reasons for this First: Most biocontrol agents produce more than one antibiotic, and resistance to multiple antibiotics should occur only at a very low frequency. Second: Total exposure of the pathogen population to the antibiotics is low because, in general, the populations of biocontrol agents are localized on the root; therefore, selection pressures are minimized. The use of antimicrobial agents in human, medicine and agriculture has shown that selection pressures drive the evolution of resistance faster than we expect. The development of multiple-drug resistance derived from spontaneous mutations was not predicted to occur, but now it is recognised as a common mechanism of antibiotic resistance in bacteria. Therefore, research is needed to understand the molecular bases of pathogen resistance to antibiotics produced by bio-control agents. Resistance should be studied before it occurs in the field so that fundamental knowledge can be applied to anticipate and prevent the breakdown of biocontrol. Examples of such approaches include inhibiting resistance proteins, combining antibiotics that select for different resistance genes, and avoiding use of biocontrol agents against pathogen populations in which a high frequency of resistance is predicted. Similarly, understanding resistance of insects to *B. thuringiensis* has led to innovative strategies to reduce the impact of resistance on insect biocontrol. Resistance to antibiotics usually arises in a sensitive population by spontaneous mutation or by horizontal gene transfer. Mutations conferring resistance may affect antibiotic uptake or target sensitivity. Self-resistance genes could be

Table 17.2: Antibiotics Secreted by Potential Biocontrol Agents against some Plant Pathogens

Antibiotic	Source	Target pathogen	Disease	Reference
2,4-diacetylphloroglucinol	*Pseudomonas fluorescens* F113	*Pythium* spp.	Damping off	Shanahan *et al.*, 1992
Agrocin 84	*Agrobacterium radiobacter*	*Agrobacterium tumefaciens*	Crown gall	Kerr, 1980
Bacillomycin D	*Bacillus subtilis* AU195	*Aspergillus flavus*	Aflatoxin contamination	Moyne *et al.*, 2001
Bacillomycin fengycin	*Bacillus amyloliquefaciens* FZB42	*Fusarium oxysporum*	Wilt	Koumoutsi *et al.*, 2004
Iturin	*B.subtilis* QST713	*Botrytis cinerea* and *R.solani*	Damping off	Paulitz and Belanger, 2001
Phenazines	*P.fluorescens* 2-79 an 30-84	*Gaeumannomyces graminis* var. *tritici*	Take-all	Thaomshow *et al.*, 1990
Pyoluterin, pyrrolnitrin	*P.fluorescens* Pf-5	*Pythium ultimum* and *R.solani*	Damping off	Howell and Stipanovic, 1980
Zwittermicin A	*Bacillus cereus* UW85	*Phytophthora medicaginis* and *P. aphanidermatum*	Damping off	Smith *et al.*, 1993

transferred to target pathogens in the soil; therefore, it is essential to understand mutations conferring resistance in target pathogens as well as self-resistance genes in producing organisms. The importance of self-resistance is illustrated in the crown gall biocontrol system. *Agrobacterium radiobacter* effectively controls crown gall, which is caused by *A.tumefaciens,* largely through the action of the antibiotic agrocin 84. Efficacy was threatened when agrocin 84-resistant strains of the pathogen were isolated from galls on *A. radiobacter*-treated plants. Little information is available concerning spontaneous mutations that confer antibiotic resistance on pathogens that are the targets of biocontrol strategies.

The wheat pathogen *Septoria tritici* acclimates to 1- hydroxyphenazine by inducing genes for catalase, superoxide dismutase, and melanin production; therefore, mutants of the pathogen that constitutively produce high levels of these protectants might not be suppressed by phenazine-producing biocontrol organisms. The widespread occurrence of cyanide-resistant respiratory pathways in micro-organisms suggests that prolonged application of hydrogen cyanide-producing biocontrol agents may select for pathogens containing cyanide-resistant oxidases. Recent evidence shows variation among strains of *G.graminis* for sensitivity to antibiotics produced by fluorescent pseudomonad biocontrol agents, and disease induction by these resistant strains is not suppressed effectively by the biocontrol agent.

Competition

From a microbial perspective, soils and living plant surfaces are frequently nutrient limited environments. To successfully colonize the phyllosphere, a microbe must effectively compete for the available nutrients. On plant surfaces, host-supplied nutrients include exudates, leachates, or senesced tissue. Additionally, nutrients can be obtained from waste products of other organisms such as insects like aphid honeydew on leaf surface and the soil. While difficult to prove directly, much indirect evidence suggests that competition between pathogens and non-pathogens for nutrient resources is important for limiting disease incidence and severity. In general, soilborne pathogens, such as species of *Fusarium* and *Pythium,* which infect through mycelial contact, are more susceptible to competition from other soil- and plant-associated microbes than those pathogens that germinate directly on plant surfaces and infect through appressoria and infection pegs. Genetic work of Anderson *et al.,* 1988 revealed that production of a particular plant glycoprotein called agglutinin was correlated with potential of *P. putida* to colonize the root system. *P. putida* mutants deficient in this ability exhibited reduced capacity to colonize the rhizosphere and a corresponding reduction in Fusarium wilt suppression in cucumber. The most abundant nonpathogenic plant-associated microbes are generally thought to protect the plant by rapid colonization and thereby exhausting the limited available substrates so that none are available for pathogens to grow. For example, effective catabolism of nutrients in the spermosphere has been identified as a mechanism contributing to the suppression of *Pythium ultimum* by *Enterobacter cloacae.* At the same time, these microbes produce metabolites that suppress pathogens. These microbes colonize the sites where water and carbon containing nutrients are most readily available, such as exit points of secondary roots, damaged epidermal cells, and nectaries and utilize the root mucilage. Biocontrol based on competition for rare but essential micronutrients, such as iron, has also been examined. Iron is extremely limited in the rhizosphere, depending on soil pH. In highly oxidized and aerated soil, iron is present in ferric form, which is insoluble in water (pH 7.4) and the concentration may be as low as 10-18 M. This concentration is too low to support the growth of micro-organisms, which generally need concentrations approaching 10-6 M. To survive in such an environment, organisms were found to secrete iron-binding ligands called siderophores having high affinity to sequester iron from the micro-environment. Almost all micro-organisms produce siderophores, of either the catechol type or hydroxamate type were the first to demonstrate the importance of siderophore production as a mechanism of biological control of *Erwinia carotovora* by several plant-growth promoting *Pseudomonas fluorescens* strains A1, BK1, TL3B1 and B10.

Nutrient competition has been believed to have an important role in disease suppression, although it is extremely difficult to obtain conclusive evidence. Biocontrol by nutrient competition can occur when the biocontrol agent decreases the availability of a particular substance thereby limiting the growth of the pathogen. Particularly, the biocontrol agents have a more efficient uptake or utilizing system for the substance than do the pathogens. For example, iron competition in alkaline soils may be a limiting factor for microbial growth in such soils. Some bacteria, especially fluorescent pseudomonads produce siderophores that have very high affinities for iron and can sequester this limited resource from other microflora thereby preventing their growth. A few studies have demonstrated that siderophore biosynthesis in *P. fluorescens* plays a role in pathogen suppression.

Cell-wall Degrading Enzymes

Extracellular hydrolytic enzymes produced by microbes may also play a role in suppression of plant pathogenic fungi. Chitin and b-1,3-glucans are major constituents of many fungal cell walls. Several studies have demonstrated in vitro lysis of fungal cell walls either by chitinase or β-1,3- glucanase alone or in combination. Diverse micro-organisms secrete and excrete other metabolites that can interfere with pathogen growth and/or activities. Many micro-organisms produce and release lytic enzymes that can hydrolyze a wide variety of polymeric compounds, including chitin, proteins, cellulose, hemicellulose, and DNA. Expression and secretion of these enzymes by different microbes can sometimes result in the suppression of plant pathogen activities directly. For example, control of *Sclerotium rolfsii* by *Serratia marcescens* appeared to be mediated by chitinase expression. While they may stress and/or lyse cell walls of living organisms, these enzymes generally act to decompose plant residues and nonliving organic matter. Currently, it is unclear how much of the lytic enzyme activity that can be detected in the natural environment represents specific responses to microbe-microbe interactions.

In postharvest disease control, addition of chitosan can stimulate microbial degradation of pathogens similar to that of an applied hyperparasite. Chitosan is a non-toxic and biodegradable polymer of β-1,4-glucosamine produced from chitin by alkaline deacylation. Amendment of plant growth substratum with chitosan suppressed the root rot caused by *Fusarium oxysporum* f. sp. *radicis-lycopersici* in tomato.

Although the exact mechanism of action of chitosan is not fully understood, it has been observed that treatment with chitosan increased resistance to pathogens. Other microbial byproducts also may contribute to pathogen suppression. Hydrogen cyanide (HCN) effectively blocks the cytochrome oxidase pathway and is highly toxic to all aerobic micro-organisms at picomolar concentrations. The production of HCN by certain fluorescent pseudomonads is believed to be involved in the suppression of root pathogens. *P.fluorescens* CHA$_0$ produces antibiotics, siderophores and HCN, but suppression of black rot of tobacco caused by *Thielaviopsis basicola* appeared to be due primarily to HCN production. It is clear that biocontrol microbes can release many different compounds into their surrounding environment, the types and amounts produced in natural systems in the presence and absence of plant disease have not been well documented and this remains a frontier for discovery.

Induction of Host Resistance

Plants actively respond to a variety of environmental stimuli, including gravity, light, temperature, physical stress, water and nutrient availability. Plants also respond to a variety of chemical stimuli produced by soil- and plant-associated microbes. Such stimuli can either induce or condition plant host defenses through biochemical changes that enhance resistance against subsequent infection by a variety of pathogens. Induction of host defenses can be local and/or systemic in nature, depending on the type, source, and amount of stimuli. Recently, phytopathologists have begun to characterize the determinants and pathways of induced resistance stimulated by biological control agents and other non-pathogenic microbes. The first of these pathways, termed systemic acquired resistance (SAR), is mediated by salicylic acid (SA), a compound which is frequently produced following pathogen infection and typically leads to the expression of pathogenesis-related (PR) proteins. These PR proteins include a variety of enzymes some of which may act directly to lyse invading cells, reinforce cell wall boundaries to resist infections, or induce localized cell death. A second phenotype, first referred to as induced systemic resistance (ISR), is mediated by jasmonic acid (JA) and/or ethylene, which are produced following applications of some nonpathogenic rhizobacteria. Interestingly, the SA- and JA- dependent defense pathways can be mutually antagonistic, and some bacterial pathogens take advantage of this to overcome the SAR. For example, pathogenic strains of *Pseudomonas syringae* produce coronatine, which is similar to JA, to overcome the SA-mediated pathway (He *et al.*, 2004). Because the various host-resistance pathways can be activated to varying degrees by different microbes and insect feeding, it is plausible that multiple stimuli are constantly being received and processed by the plant. Thus, the magnitude and duration of host defense induction will likely vary over time. Only if induction can be controlled, *i.e.* by synergistically interacting with endogenous signals, will host resistance be increased. A number of strains of root-colonizing microbes have been identified as potential elicitors of plant host defenses.

Some biocontrol strains of *Pseudomonas* sp. and *Trichoderma* sp. are known to strongly induce plant host defenses. In several instances, inoculations with plant-growth-promoting rhizobacteria (PGPR) were effective in controlling multiple diseases caused by different pathogens, including anthracnose (*Colletotrichum lagenarium*), angular leaf spot (*Pseudomonas syringae* pv. *lachrymans* and bacterial wilt (*Erwinia tracheiphila*). A number of chemical elicitors of SAR and ISR may be produced by the PGPR strains upon inoculation, including salicylic acid, siderophore, lipopolysaccharides, and 2,3-butanediol, and other volatile substances. Again, there may be multiple functions to such molecules blurring the lines between direct and indirect antagonisms. More generally, a substantial number of microbial products have been identified as elicitors of host defenses, indicating that host defenses are likely stimulated continually over the course of a plant's lifecycle. Excluding the components directly related to pathogenesis, these inducers include lipopolysaccharides and flagellin from Gram-negative bacteria; cold shock proteins of diverse bacteria; transglutaminase, elicitins, and β-glucans in Oomycetes; invertase in yeast; chitin and ergosterol in all fungi; and xylanase in *Trichoderma*. These data suggest that plants would detect the composition of their plant-associated microbial communities and respond to changes in the abundance, types, and localization of many different signals. The importance of such interactions is indicated by the fact that further induction of host resistance pathways, by chemical and microbiological inducers, is not always effective at improving plant health or productivity in the field.

Microbial Diversity and Disease Suppression

Plants are surrounded by diverse types of mesofauna and microbial organisms, some of which can contribute to biological control of plant diseases. Microbes that contribute most to disease control are most likely those that could be classified competitive saprophytes, facultative plant symbionts and facultative hyperparasites. These can generally survive on dead plant material, but they are able to colonize and express biocontrol activities while growing on plant tissues. A few, like avirulent *Fusarium oxysporum* and binucleate *Rhizoctonia*-like fungi, are phylogenetically very similar to plant pathogens but lack active virulence determinants for many of the plant hosts from which they can be recovered. Others, like *Pythium oligandrum* are currently classified as distinct species. However, most are phylogenetically distinct from pathogens and, most often, they are subspecies variants of the same microbial groups. Due to the ease with which they can be cultured, most biocontrol research has focused on a limited number of bacterial genera like *Bacillus, Burkholderia, Lysobacter, Pantoea, Pseudomonas,* and *Streptomyces* and fungal genera like *Ampelomyces, Coniothyrium, Dactylella, Gliocladium, Paecilomyces,* and *Trichoderma*.

Lastly, there are many general micro- and meso-fauna predators, such as protists, collembola, mites, nematodes, annelids, and insect larvae whose activities can reduce pathogen biomass, but may also facilitate infection and/or stimulate plant host defenses by virtue of their own herbivorous activities. While various epiphytes and endophytes may contribute to biological control, the ubiquity of mycorrhizae deserves special consideration. Mycorrhizae are formed as the result of mutualist symbioses between fungi and plants and occur on most plant species. Because they are formed early in the development of the plants, they represent nearly ubiquitous root colonists that assist plants with the uptake of nutrients especially phosphorus and micronutrients. The vesicular arbuscular mycorrhizal fungi (VAM), also known as arbuscular mycorrhizal or endomycorrhizal fungi are all members of the zygomycota and the current classification contains one order, the Glomales, encompassing six genera into which 149 species have been classified. Arbuscular mycorrhizae involve aseptate fungi and are named for characteristic structures like arbuscles and vesicles found in the root cortex. Arbuscules start to form by repeated dichotomous branching of fungal hyphae approximately two days after root penetration inside the root cortical cell. Arbuscules are believed to be the site of communication between the host and the fungus. Vesicles are basically hyphal swellings in the root cortex that contain lipids and cytoplasm and act as storage organ of VAM. These structures may present intra- and inter- cellular and can often develop thick walls in older roots. These thick walled structures may function as propagules. During colonization, VAM fungi can prevent root infections by reducing the access sites and stimulating host defense. VAM fungi have been found to reduce the incidence of root-knot nematode. Various mechanisms also allow VAM fungi to increase a plant's stress tolerance. This includes the intricate network of fungal hyphae around the roots which block pathogen infections. Inoculation of apple-tree seedlings with the VAM fungi *Glomus fasciculatum* and *G. macrocarpum* suppressed apple replant disease caused by phytotoxic myxomycetes. VAM fungi protect the host plant against root-infecting pathogenic bacteria. The mechanisms involved in these interactions include physical protection, chemical interactions and indirect effects. The other mechanisms employed by VAM fungi to indirectly suppress plant pathogens include enhanced nutrition to plants; morphological changes in the root by increased lignification; changes in the chemical composition of the plant tissues like antifungal chitinase, isoflavonoids, etc. In contrast to VAM fungi, ectomycorrhizae proliferate outside the root surface and form a sheath around the root by the combination of mass of root and hyphae called a mantle. Disease protection by ectomycorrhizal fungi may involve multiple mechanisms including antibiosis, synthesis of fungistatic compounds by plant roots in response to mycorrhizal infection and a physical barrier of the fungal mantle around the plant root (Duchesne, 1994). Ectomycorrhizal fungi like *Paxillus involutus* effectively controll root rot caused by *Fusarium oxysporum* and *Fusarium moniliforme* in red pine. Inoculation of sand pine with *Pisolithus tinctorius*, another ectomycorrhizal fungus, controlled disease caused by *Phytophthora cinnamomi*.

Plant diseases may be suppressed by the activities of one or more plant associated microbes, researchers have attempted to characterize the organisms involved in biological control. The introduction of *Pseudomonas fluorescens* that produce the antibiotic 2,4-diacetylphloroglucinol can result in the suppression of various soilborne pathogens. However, specific agents must compete with other soil- and root-associated microbes to survive, propagate, and express their antagonistic potential during those times when the targeted pathogens pose an active threat to plant health. In contrast, general suppression is more frequently invoked to explain the reduced incidence or severity of plant diseases because the activities of multiple organisms can contribute to a reduction in disease pressure. High soil organic matter supports a large and diverse mass of microbes resulting in the availability of fewer ecological niches for which a pathogen competes. The extent of general suppression will vary substantially depending on the quantity and quality of organic matter present in a soil. Functional redundancy within different microbial communities allows for rapid depletion of the available soil nutrient pool under a large variety of conditions, before the pathogens can utilize them to proliferate and cause disease. For example, diverse seed-colonizing bacteria can consume nutrients that are released into the soil during germination thereby suppressing pathogen germination and growth. Manipulation of agricultural systems, through additions of composts, green manures and cover crops is aimed at improving endogenous levels of general suppression.

Current Scenario and BCA Products Available in Market

Most pathogens will be susceptible to one or more biocontrol strategies, but practical implementation on a commercial scale has been constrained by a number of factors. Cost, convenience, efficacy, and reliability of biological controls are important considerations, but only in relation to the alternative disease control strategies. Cultural practices like good sanitation, soil preparation, and water management and host resistance can go a long way towards controlling many diseases, so biocontrol should be applied only when such agronomic practices are insufficient for effective disease control. The use of pest- and disease-resistant cultivars, developed through conventional breeding or genetic engineering, provides the next line of defense. However, such measures are not always sufficient to be productive or economically sustainable. In such cases, the next step would be to deploy bio-rational controls of insect pests and diseases. These include BCAs, introduced as inoculants or amendments, as well as active ingredients directly derived from natural origins and having a low impact on the environment and non-target organisms. If these foundational options are not sufficient to ensure plant health and/or economically sustainable production, then less specific and more harmful synthetic chemical toxins can be used to ensure productivity and profitability. With the growing interest in reducing chemical inputs, companies involved in the manufacturing and marketing of BCAs should experience continued growth. However, stringent quality control measures must be adopted so that farmers get quality products. New, more effective and stable formulations also will need to be developed. However, if the infection court or target pathogen can be effectively colonized using inoculation, the ability of the living organism to reproduce could greatly reduce application costs. In general, though, regulatory and cultural concerns about the health and safety of specific classes of pesticides are the primary economic drivers promoting the adoption of biological control strategies in urban and rural landscapes. Self-perpetuating biological controls like hypo-virulence of the chestnut blight pathogen are also needed for control of diseases in forested and rangeland ecosystems where high application rates over larger land areas are not economically-feasible.

During the past decade, more emphasis has been put on the safety of fruits and vegetables as an increasing number of outbreaks with food borne human diseases are reported each year following consumption of various fruit and vegetables. The situation is worsened by growing consumption of fresh-cut fruits and vegetables, which provide a conducive environment for the growth of various bacterial human pathogens. This also creates an opportunity for the use of microbial antagonists to combat these food-borne pathogens. The discovery of a very effective antagonist producing antifungal volatiles indicates that a single MBC can be sufficient to provide adequate control of fruit decay. This also justifies screening for a single MBC, if an effective mechanism can be identified, and makes in vitro screening more meaningful and effective because it allows the screening of vast numbers of organisms in a short period of time before resorting to more expensive and time consuming tests on fruit. Biological control of post harvest diseases now encompasses the use of antagonists to control postharvest diseases of fruits, vegetables and grains. Postharvest biocontrol of flower diseases is another area awaiting exploration. Early attempts to use a metabolite (pyrrolnitrin) of a bacterial antagonist to control cut rose flower infections were very encouraging but little has been done in this field. Biocontrol of grain spoilage during storage in silos has made significant advances and appears to be on the way to commercialization. Examples from commercially used biocontrol products indicate that biocontrol agents developed for BCPD on one commodity could also be effective against the same or different pathogens on other commodities. The activity of biocontrol agents is not as universal as fungicides but is less specialized than originally anticipated. Broadening the application of biocontrol agents is a good strategy for commercial success. It not only makes the product more profitable and allows for a quicker return on the investment made in the commercialization of the product, but it also allows a buffer in the fluctuation in the market due to registration of new fungicides or other alternative products. As postharvest biocontrol products are coming to the market, their anticipated limitations are elucidated and much of the current research is focused on addressing these limitations. The most commonly used approach is combining antagonists with various substances Generally Regarded as Safe (GRASS), sodium bicarbonate (SBC), calcium chloride, or ethanol. These combinations both reduced the fluctuation and increased the level of decay control. Combining antagonists with other alternatives to fungicide treatments also showed good results, but its implementation often requires modifying currently used postharvest practices or adds substantially to the cost of the treatment. For example, a heat treatment of apples with hot air at 38°C for 4 days or oranges at 30°C for 1 day after harvest in combination with antagonists gave superior control including eradicative activity of blue mold and gray mold, respectively, compared to the individual treatments. However, this approach will require adding heating equipment and changing the temperature in of storage rooms. Some packinghouses use high temperatures to sanitize empty bins and storage rooms, but even this practice is limited. It is also well established that the proper selection of the combination of two antagonists can provide superior decay control to either antagonists applied individually. Although using this approach is very attractive, it doubles the cost of registration because each antagonist must be registered separately. This problem could be eliminated if biocontrol products currently on the market could be combined resulting in additive or synergistic effects.

BioSave™–two strains of *P. syringae* was originally registered for postharvest application to pome and citrus fruits, and this was later extended to cherries, potatoes, and more recently to sweet potatoes. YieldPlus™ was developed in South Africa for postharvest application to pome fruits but the success of this product is largely unknown and there is no published literature or information available to determine extent of its use. Avogreen™ has been used for control of postharvest disease of avocado. Its use has been limited, possibly due to inconsistent results. More recently, Shemer™ was registered in Israel for both pre- and postharvest application on various fruits and vegetables including apricot, citrus, grapes, peach, pepper, strawberry and sweet potato. There are three more products coming to the market: Candifruit™ based on *Candida sake*, developed in Spain; Boni-Protect®, based on *Aureobasidium pullulans*, developed in Germany and NEXY, based on *Candida oleophila*, developed in Belgium. All of these products have been registered for control of postharvest diseases of pome fruits. These new products are increasing interest in BCPD and is not just a matter of scientific curiosity but have resulted in diligent efforts to implement this approach. Gradual removal of the major regulatory barriers to registration of antagonists for BCPD in different countries is also very encouraging. All biocontrol agents currently registered for postharvest application control fruit decays originating from wound infections made during or after harvest. Although for some fruits, such as pome or citrus (depending on the region) fruits, wounds are the main court of entry for postharvest decay causing fungi, many postharvest decays of stone fruits and subtropical fruits develop in storage from latent infections occurring in the orchard. These infections are difficult to control because the intimate relationship of the pathogen with the host has been already established, and melanized appressoria often formed by these fungi on fruit surface are very resistant to environmental factors and penetration by fungicides.

Much has been learned from the biological control research conducted over the past forty years. A successful biocontrol requires considerable understanding of cropping system; disease epidemiology; the biology, ecology, and population dynamics of biocontrol organisms; and the interactions among these variables. Understanding the mechanisms or activities for antagonist-pathogen interactions will be one of important steps because it may provide a reasonable basis for selection and construction of more effective biocontrol agents. Over the past few years, the novel applications of molecular techniques have broadened our insight into the basis of biological control of plant diseases. New molecular approaches have been available for assessment of interaction between the antagonist and pathogen, ecological traits of antagonists in rhizosphere and improving the efficacy of bacterial, fungal and viral biocontrol agent. Consequently, there has been a significant increase in the number of biological disease control agents registered or on the market worldwide in the last few years. For example, there are approximately 30 bacterial and fungal products for control of foliar, soil-borne and postharvest diseases. Currently, fundamental advances in computing, molecular biology, analytical chemistry, and statistics have led to new research aimed at characterizing the structure and functions of biocontrol agents, pathogens, and host plants at the molecular, cellular, and ecological levels. The greatest successes in biological control have been achieved in situations where environmental conditions are most controlled or predictable and where biocontrol agents can preemptively colonize the infection court. Monocyclic, soil-borne and postharvest diseases have been controlled effectively by biological control agents that act as bioprotectants that prevent infections. Specific applications for high value crops targeting specific diseases like fireblight, downy mildew, and several nematode diseases have also been adopted. As research unravels the various conditions needed for successful biocontrol of different diseases, the adoption of BCAs in IPM systems is bound to increase in the years ahead. The market share of biopesticides of the total pesticide market is less than three percent. The challenge is to develop a formulation and application method that can be implemented on a commercial scale, that must be effective, reliable, consistent, economically feasible, and with a wider spectrum. Continual laboratory works followed by field experiments are needed to establish excellent biocontrol agents particularly against plant fungal pathogens.

Conclusion

In this chapter, an outline of plant diseases and plant pathogens that affect crop plants and their biological control is being discussed. Chemicals that are used as pesticides and insecticides have prolonged negative effects on the environment and living organisms. Biological control is recommended over chemical control which has no side effects but takes a longer time period. Different approaches of biological control of plant pathogens, the products currently available in market have been presented in this chapter. Further lot of research has to be undertaken in this field to improve the control conditions and reduce the pathogenic effects on plants.

References

J. Usall, N. Teixidó, M. Abadias, R. Torres, T. Cañamas, and I. Viñas 2010. Improving Formulation of Biocontrol Agents Manipulating Production Process In: *Postharvest Pathology* Dov Prusky M. Lodovica Gullino (Eds) Springer Science+Business Media B.V. 211pp.

M.A. Manning, H.A. Pak, and R.M. Beresford 2010. Non-fungicidal Control of Botrytis Storage Rot in New Zealand Kiwifruit Through Pre- and Postharvest Crop Management In: *Postharvest Pathology* Dov Prusky M. Lodovica Gullino (Eds) Springer Science+Business Media B.V. 211pp.

M. Dickinson 2003. *Molecular Plant Pathology* School of Biosciences, University of Nottingham, Nottingham, UK BIOS Scientific Publishers. 273pp.

Michael E. Hochberg and Robert D. Holt 2004. The uniformity and density of pest exploitation as guides to success in biological control; Pest exploitation and success in biological control In: *Theoretical Approaches to Biological Control* Bradford A. Hawkins and Howard V. Cornell (Eds) Cambridge University Press. 412pp.

P. Narayanasamy 2011. *Microbial Plant Pathogens-Detection and Disease Diagnosis Bacterial and Phytoplasmal Pathogens*, Volume 2 Springer. 256pp.

Raffaello Castoria and Sandra A.I. Wright 2010. Host Responses to Biological Control Agents In: *Postharvest Pathology* Dov Prusky M. Lodovica Gullino (Eds) Springer Science+Business Media B.V. 211pp.

Rob Jenkins, C.K. Jain 2010. *Advances in Soil-borne Plant Diseases* Oxford Book Company Mehra Offset Press, Delhi. 276pp.

Roger Hull 2009. *Comparative plant virology* second edition, Elsevier, Academic Press. 376pp.

Wojciech J. Janisiewicz 2010. Quo Vadis of Biological Control of Postharvest Diseases In: *Postharvest Pathology* Dov Prusky M. Lodovica Gullino (Eds) Springer Science+Business Media B.V. 211pp.

2015, Recent Trends in Microbiology, Mycology and Plant Pathology   *Pages* 283–295
Editor: **Dr. H.C. Lakshman**
Published by: **DAYA PUBLISHING HOUSE, NEW DELHI**

Chapter 18

Eco-friendly Approaches in the Management of Post-harvest Fruit Diseases of Ivy Gourd (*Coccinia inidica* Wight and Arn.)

*V.S. Chatage and U.N. Bhale**

*Research Laboratory,
Department of Botany, Arts, Science and Commerce College,
Naldurg, Tq. Tuljapur, Dist. Osmanabad – 413 602, M.S.
E-mail: unbhale2007@redffmail.com

ABSTRACT

Post-harvest diseases cause considerable losses to harvested fruits and vegetables during transportation and storage. Synthetic fungicides are primarily used to control postharvest decay loss. However, the recent trend is shifting toward safer and more eco-friendly alternatives for the control of postharvest decays. Ivy gourd fruits are attacked by *Alternaria pluriseptata* and *Bipolaris tetramera* which was severe in Marathwada region of Maharashtra after harvesting. Pathogencity test of the organisms were confirmed as Koch's postulates and fully satisfied. Therefore, the present study based on botanical pesticides for controlling these diseases. Leaf extracts, *Adhatoda vassica* leaf extract showed reduction of radial growth of *A.pluriseptata* (96.66 per cent) and *B. tetramera* (97.77) at 100 per cent conc respectively. *Azadirachta indica* also showed significantly results at 100 per cent conc. *Jatropa curcus* latex extract showed 100 per cent reduction of radial growth of *A. pluriseptata* and *B. tetramera* at 75 and 100 per cent conc respectively.

Castor oil was completely inhibited the radial growth of both pathogen at 100 per cent conc. *T. pseudokiningii* (77.70 per cent) was found antagonism in case of *A. pluriseptata*. All *Trichoderma* species were found suitable for antagonistic activity against *B.tetramera* but *T. pseudokiningii* (74.44 per cent) was better than that of others.

Keywords: *Coccinia indica, Botanical pesticides, Trichoderma spp.*

Introduction

Plants have always been an consummate source of drugs and many drugs currently available have been derived directly or indirectly from them. A vast majority of population particularly those living in villages depends largely on medicinal plants for treating and curing diseases. One such medicinal plant, ivy gourd (*Coccina indica* Wight and Arn.) of the family *Cucurbitaceae* is most important vegetable and medicinal plant, distributed in Tropical Asia, Africa, Pakistan, India and Sri Lanka (Cooke, 1903; Sastri1,950). It is a climber and trailer (Nasir and Ali, 1973). Different names of ivy gourd like the parwal, kundru, tondli are in market. It is native to Africa and has been growing in the Indo-Malayan region of Asia for many centuries (Singh, 1990). It has white flowers and small cucumber like fruits which turn bright scarlet red when ripened. Ivy gourd has vitamin A, β–carotene and is a good source of protein. The fruit of *Coccinia* is used as vegetable when green and eaten fresh when ripened into bright scarlet color. Every part of this plant is valuable in medicine and various preparations have been mentioned in indigenous system of medicine for skin diseases bronchial catarrh, bronchitis and unani systems of medicine. It shows also hypoglycemic activities (Mukerjee *et al.,* 1972 and Nahar *et al.,* 1998). The juice of the roots and leaves are considered to be a useful in treatment of diabetes (Chopra *et al.,* 1925). A post and pre-harvest food loss constitutes a vast complex of physical and biological changes due to micro-organisms like fungi and bacteria. Diseases are very important in reducing market quality of ivy gourd fruit and are primarily responsible for the post and pre harvest losses up to 10-35 per cent.

However, ivy gourd fruits are attacked by *Alternaria pluriseptata* (Karst and Har) and *Bipolaris tetramera* (Mc Kinney) which was severe in Marathwada region of Maharashtra. Since, biocontrol agents for protection of seeds and control of seed borne diseases offers farmers an alternative source for chemical fungicides which is highly effective (Callan *et al.,* 1997). It is therefore necessary to develop alternative ways of control. Various disease management methods have been implemented to combat and eradicate pathogenic fungi which include cultural, regulatory, physical, chemical and biological methods. All these methods are effective only when employed well in advance as precautionary measure (Kata, 2000). Therefore an investigation was made to evaluate the biopesticides such as plant (leaf) extract, plant latex extract, essential oils and *Trichoderma* species against *A. pluriseptata* and *B. tetramera* inciting fruit rot of ivy gourd.

Materials and Methods

a) Symptomatology of Pathogens

Fruit showing whitish green and dark green spots symptoms were collected from different localities of Marathwada region of Maharashtra state. It causes post-harvest diseases of *Coccinia indica* fruit when ivy gourd contacts with soil particles during processing.

b) Isolation and Identification of Test Pathogens

Post-harvest fruit rot samples were collected and infected fruit parts were cut into small pieces by sterilized blade then surface sterilized with Mercuric Chloride (0.1 per cent) for 1 min. The pieces were then washed thrice with sterilized distilled water and dried by sterilized blotting paper. These pieces were placed on potato dextrose agar (PDA) medium and incubated at 28±2°C. The fungus was isolated and identified by using different manuals (Ellis, 1971; Barnett, 1960; Subramanim, 1971) and confirmed by Agharkar Research Institiute (ARI), Pune and fungal culture were deposited at Research Laboratory, Department of Botany, Arts, Science and Commerce College, Naldurg.

c) Pathogencity Test

Pathogenicity was tested on healthy fruits of uniform sizes were collected. The fruits were surface sterilized with 0.1 per cent Mercuric Chloride, washed thrice with sterile distilled water, wounded by sterile needle and inoculated with fungal spore suspension. The inoculated fruits were placed in pre- sterilized polythene bags. These fruits were injured by pin prick method (Thomson, 1996). Some fruits were peeled by sterile blade at certain place. The inoculum was prepared from seven day old culture of fungi grown on PDA slants. The spore suspension was inoculated on the injured peeled portions of fruits aseptically. These fruits were then placed in separate sterile plastic bags and were incubated at $27 \pm 2°C$ for 8th days. Symptoms were recorded and isolates were compared with original.

d) Eco-friendly Approaches

i) Plant (Leaf) Extract

Fungi toxicity of leaf extracts was studied by food poisoning technique described by (Mishra and Tiwari, 1992). Fresh plant material of *Azadirachta indica, Ocimum gratissimum, Adhatoda vassica, Aegle mormelos* and *Santalum album* were collected, washed with sterilized distilled water, oven dried and pulverized to obtain dry powder. Plant extract of each prepared with water *i.e.* 100gm powder dissolved in 100 ml distilled water. Mixed well and filter through double filter muslin cloth, it served as stock. This stock was used against tested fungi in four different concentrations of (25, 50, 75 and 100 per cent). Petri plates containing CZA supplemented with different leaf extract at four different concentrations with three replications were inoculated with fresh 8th days old culture of test fungi and (8mm) cork borer disc and kept upside down and incubated in BOD incubators at 27± 2°C. Plates without leaf extracts were served as control. Radial growth of the tested pathogens was measured at regular intervals.

ii) Plant Latex Extract

Plant Material and Latex Collection

The fresh latex of *Jatropha curcas, Calotropis gigantea, Ficus bengalensis* and *Ficus glumerata* were aseptically collected from the aerial parts of the healthy plants as described by Aworh *et al.* (1994) in clean glass tubes containing distilled water to yield a dilution rate of 5:5 (v/v). The latex mixture was gently handled to maintain homogeneity during transport to the laboratory where it was stored at (4°C) until further use.

Preparation of Latex Extract

The fresh latex was selectively decanted and centrifuged at 5000 rpm for 5 min. The precipitated material showing rubber aspect (poly-isoprene) was pooled apart and the supernatant was decanted carefully. Finally the samples were centrifuged as previously described and the clear soluble supernatant was collected and lyophilized The stock solutions of latex extract was diluted suitably as required from stock solution (Juncker *et al.,* 2009).

Determination of Antifungal Activity

Plant latex aqueous extracts of each prepared with distilled water and condensed to serve as stock extract was determined by food poisoning technique (Mishra and Tiwari 1992) against tested pathogens in five different concentrations. Petriplates containing Czapek Dox Agar (CZA) medium, supplemented with different plant latex extracts at four concentrations (25, 50, 75 and 100 per cent) with three replications were inoculated with fresh 7 days old culture of test fungi in 8 mm discs and kept upside down. The plates were incubated in BOD incubator at 27 ± 2°C Plates without plant latex extracts served as control. Starting two days after inoculation (DAI), radial growth was recorded daily for 8 days or until the plates were overgrown. The growth inhibition was calculated by using the formula: $100 \times C-T/C$, Where C = growth in control and T = growth in treatment (Vincent, 1947). The lowest concentration of the extracts that inhibited the growth of the test pathogens was recorded as the minimum inhibitory concentration (MIC).

iii) Essential Oils

Antifungal activity of four essential oils such as Groundnut, Sesame, Neem and Castor oils were collected. Hundred ml of oil directly mixed with sterilized hot 100ml PDA medium *i.e.* considered as 100 per cent conc. and tested against pathogens. For this experiment four different concentrations (25, 50, 75 and 100 per cent) were evaluated. Petri plates containing PDA supplemented with different oils at four different concentrations with three replications were inoculated with fresh 7th days old culture of test fungi and (8mm) cork borer disc and kept upside down and incubated in BOD incubators at 27 ± 1°C. Plates without oils were served as control. Radial growth of the tested pathogens was measured at regular intervals.

iv) *Trichoderma* Species

Dual Culture Experiment

Antagonistic efficacy of different isolates of *T. viride, T. harzianum, T. virens, T. koningii* and *T. pseudokoningii,* were tested against the isolated pathogenic fungus by dual culture experiment (Morton and Stroube, 1955). *Trichoderma* species and test fungus was inoculated at 6 cm apart. Three replicates were maintained for each treatment and incubated at 27 ± 2°C for 9 days. Monoculture plates of both served as control. Nine days after incubation (DAI), radial growth of test fungus and *Trichoderma* isolates were measured. Colony diameter of test fungus in dual culture plate was observed and compared with control. The growth inhibition was calculated by using the formula: 100 X C–T/C, Where C = growth in control and T = growth in treatment (Vincent, 1947).

Statistical analyses were performed by the book of Biometry (Mungikar, 1997).

Results

Symptomatology of Pathogens

Alternaria pluriseptata

Symptoms are green spots; colony a characters is whitish green color with prominent concentric rings were studied under natural and artificial conditions. Microscopic slides in cotton blue and lactophenol were prepared from the infected fruit to observe mycelium, conidiophores and conidia of the pathogen.

Bipolaris tetramera

Symptoms are whitish green spots, colony a character is cottony growth with concentric rings having green color. Studied under natural and artificial conditions Microscopic slides in cotton blue and lactophenol were prepared from the infected fruit to observe mycelium, conidiophores and conidia of the pathogen.

Pathogencity test of the organisms were confirmed as Koch's postulates and fully satisfied.

Plant (Leaf) Extracts

Plant leaf extracts was used in this study against two pathogenic fungi to determine their antifungal activity. Different concentrations of plant leaf extracts (25, 50, 75 and 100 per cent) were tested against pat hogenic fungi. Minimum Inhibitory Concentration (MIC) was measured to determine the antifungal activity. The inhibition effects of the plant extracts on pathogenic fungi were represented in Table 18.1. Among tested leaf extracts, *Adhatoda vassica* leaf extract showed reduction of radial growth of *Alternaria pluriseptata* (96.66 per cent) and *Bipolaris tetramera* (97.77) at 100 per cent conc respectively. *Azadirachta indica* also showed significantly results at 100 per cent conc.

Plant Latex

Plant latex extracts was used in this study against two pathogenic fungi to determine their antifungal activity. Different concentrations of plant latex (25, 50, 75 and 100 per cent) were tested against pathogenic fungi. Minimum Inhibitory Concentration (MIC) was measured to determine the antifungal activity. The inhibition effects of the plant latex extracts on pathogenic fungi were represented in Table 18.2. *Jatropa curcus* latex extract showed 100 per cent reduction of radial growth of *Alternaria pluriseptata* and *Bipolaris tetramer* at 75 and 100 per cent conc respectively. In some

Table 18.1: Antifungal Activity of Plant Extracts against Pathogenic Fungi of *Coccinia indica*

Plant Species	Family	Conc. (Per cent)	Radial Growth of A. pluriseptata (mm)	Radial Growth of B. tetramera (mm)
Azadirachta indica	Meliaceae	25	17(81.11)	22 (75.55)
		50	15 (83.33)	14 (84.44)
		75	11 (87.77)	10 (88.88)
		100	09 (90.00)	07 (92.22)
Ocimum gratissimum	Lamiaceae	25	30 (66.66)	47 (47.77)
		50	26 (71.11)	42 (53.33)
		75	22 (75.55)	33 (63.33)
		100	15 (83.33)	26 (71.11)
Adhatoda vassica	Acanthaceae	25	12 (86.66)	10 (88.88)
		50	11 (87.77)	09 (90.00)
		75	04 (95.55)	03 (96.66)
		100	03 (96.66)	02 (97.77)
Aegle mormelos	Rutaceae	25	45 (50.00)	20 (77.77)
		50	37 (58.88)	17 (81.11)
		75	33 (63.33)	11 (87.77)
		100	25 (72.22)	09 (90.00)
Santalum album	Santalaceae	25	50 (44.44)	38 (57.77)
		50	44 (51.11)	27 (70.00)
		75	35 (61.11)	20 (77.77)
		100	28 (68.88)	15 (83.33)
Control			90.00	90.00

Figures in parentheses are value of per cent inhibition.

Table 18.2: Antifungal Activity of Plant Latex Extracts against Pathogenic Fungi of *Coccinia indica*

Plant Species	Family	Conc. (Per cent)	Radial Growth of A. pluriseptata (mm)	Radial Growth of B. tetramera (mm)
Jatropa curcas	Euphorbiaceae	25	04 (95.55)	10 (88.88)
		50	05 (94.44)	03 (96.66)
		75	00 (100.00)	06 (93.33)
		100	00 (100.00)	00 (100.00)
Calotropis gigantean	Asclepiadaceae	25	40 (55.55)	50 (44.44)
		50	35 (61.11)	45 (50.00)
		75	28 (68.88)	30 (66.66)
		100	25 (72.22)	22 (75.55)

Contd...

Table 18.2–Contd...

Plant Species	Family	Conc. (Per cent)	Radial Growth of A. pluriseptata (mm)	Radial Growth of B. tetramera (mm)
Ficus bengalensis	Moraceae	25	24 (73.33)	20 (77.77)
		50	15 (83.33)	16 (82.22)
		75	06 (93.33)	14 (84.44)
		100	05 (94.44)	10 (88.88)
Ficus glumerata	Moraceae	25	45 (50.00)	35 (61.11)
		50	35 (61.11)	25 (72.22)
		75	30 (66.66)	15 (83.33)
		100	22 (75.55)	08 (91.11)
Control			90.00	90.00

Figures in parentheses are value of per cent inhibition.

Table 18.3: Antifungal Activity of Essential Oils against Pathogenic Fungi of *Coccinia indica*

Essential Oils	Family	Conc. (Per cent)	Radial Growth of A. pluriseptata (mm)	Radial Growth of B. tetramera (mm)
Groundnut oil	Fabaceae	25	60 (31.81)	75 (15.73)
		50	52 (40.90)	50 (43.82)
		75	45 (48.86)	45 (49.43)
		100	29 (67.04)	30 (66.29)
Sessame oil	Pedaliaceae	25	55 (37.50)	40 (55.05)
		50	38 (56.81)	33 (62.92)
		75	25 (71.59)	25 (71.91)
		100	20 (77.27)	18 (79.77)
Neem oil	Meliaceae	25	15 (82.95)	17 (80.89)
		50	14 (84.09)	15 (82.95)
		75	10 (88.63)	14 (84.09)
		100	09 (89.77)	11 (87.64)
Castor oil	Euphorbiaceae	25	10 (88.63)	20 (77.27)
		50	03 (96.59)	18 (79.77)
		75	02 (97.72)	04 (95.50)
		100	00 (100.00)	05 (94.38)
Control			88.00	89.00

Figures in parentheses are value of per cent inhibition.

extent, *F. bengalensis* also showed significant reduction of *A. pluriseptata* at 100 per cent conc. The inhibitory effect of *F. glomerata* was also shown in case of *B. tetramera* at 100 per cent conc. However, there was no significant reduction of radial growth in case of *C. gigantea, F. bengalensis* and *F.glomerata*.

Essential Oils

Essential oil was used in this study against two pathogenic fungi to determine their antifungal activity. Different concentrations of essential oil (25, 50, 75 and 100 per cent) were tested against pathogenic fungi. Minimum Inhibitory Concentration (MIC) was measured to determine the antifungal activity. The inhibition effects of the essential oils on pathogenic fungi were represented in Table 18.3. Castor oil was completely inhibited the radial growth of *A. pluriseptata* and *B. tetramera* at 100 per cent conc.

Antagonistic Activity of *Trichoderma* Species

Table 18.4 illustrated that, all *Trichoderma* species were found antagonism over 50 per cent but *T. pseudokiningii* (77.70 per cent) was found antagonism in case of *Alternaria pluriseptata*. All *Trichoderma* species were found suitable for antagonistic activity against *Bipolaris tetramera* but *T. pseudokiningii* (74.44 per cent) was better than that of others (Table 18.5).

Table 18.4: Antagoistic Activity of *Trichoderma* Species against *Alternaria pluriseptata*

Sl.No.	Trichoderma Species	Radial Growth of Alternaria pluriseptata (mm)	Radial Growth of Trichoderma Species	Inhibition per cent	Duration Required for Point of Contact (d)	Per cent Zone Overlapped by Antagonist after 7d of Incubation
1.	*T. viride*	38	52	57.77	7	100
2.	*T. koningii*	37	53	58.88	6	100
3.	*T. virens*	35	55	61.11	3	80
4.	*T. harzianum*	31	59	77.77	5	100
5.	*T. pseudokiningii*	39	51	56.66	–	–
	Control	100.00		4.65		
	CD (p=0.05)					

In *A. pluriseptata, T. viride* showed less duration (2d) for point of contact and percent zone overlapped by *T. viride, T. harzianum, T. koningii* and *T. pseudokoningii* after 7d of incubation. In case of *B. tetramera, T. koningii* showed less duration (2d) for point of contact and per cent zone overlapped by *T. virens* only after 7d of incubation.

Discussion

Earlier workers, Akhter *et al.* (2006) was tested eight ethanolic plant extracts and ten aqueous plant extracts in combination with cow urine to inhibition of conidial

Table 18.5: Antagonistic Activity of *Trichoderma* Species against *B. tetramera*

Sl.No.	Trichoderma Species	Radial Growth of Bioplaris tetramera (mm)	Radial Growth of Trichoderma Species	Inhibition per cent	Duration Required for Point of Contact (d)	Per cent Zone Overlapped by Antagonist after 7d of Incubation
1.	*T. viride*	40	50	55.55	3.4	–
2.	*T. koningii*	36	54	60.00	2	–
3.	*T. virens*	35	55	61.11	4	100
4.	*T. harzianum*	37	53	58.88	4	70
5.	*T. pseudoiningii*	23	67	74.44	–	–
	Control	90		4.65		
	CD (p=0.05)					

germination of *Bipolaris sorokiniana* causing leaf blight disease of wheat and recorded that *Adhatoda vasica* (leaf) and *Zingiber officinalis* (rhizome) extracts were most effective in inhibition of conidial germination at 2.5 per cent concentration where, most cases *Ocimum sactum* extracts exhibited less inhibitory effect. Shekhawat and Prasad (1971) reported that out of nine plant extracts tested *viz.*, *Allium cepa* L., *Allium sativum* L., *Ocimum sanctum* L., *Mentha piperita* L. and *Beta vulgaris* L. showed strong inhibitory action against *Alternaria tenuis* Nees. From bean, *Helminthosporium* sp. from watermelon and *Curvularia penniseti* (Mitra) Boed. from bajra used as test fungi. Nargis *et al.* (2006) reported that the extracts of *Adhatoda vasica* Nees., *Zingiber officinale*.L., *Vinca rosea* and *Azadirachta indica* Juss. in combination with cow dung, *Calotropis procera* (Aiton) W.T. Aiton and cow urine posses high ability to inhibit conidial germination of *Bipolaris sorokiniana* which might be used for controlling phytopathogens of crop plants. Antifungal activity of crude extracts of plants, Hussein *et al.* (2002) reported that leaf extracts of *Datura stramoniam* reduced the development of rust pustules on the leaves of wheat. Mughal *et al.* (1996) observed that aqueous leaf extracts of *Allium sativum, Datura alba* and *Withania somnifera* inhibited the growth of *Alternaria alternata, A. brassicola* and *Myrothecium roridam*.

Leaf extracts, chopped leaves and latex of *C. procera* have shown great promise as a nematicide *in vitro* and *in vivo* (Khirstova and Tissot, 1995). The mycelia growth, percentage spores germination and germ- tube extension in *Fusarium oxysporum* and *Aspergillus carbonaris* decreased when *Calotropis procera* extract concentration increases, where as growth of *Humicola brevis* and *Penicillium lanosum* were not affected (Rizk, 2008). The antifungal potency of *C. gigantea* latex extract on the *C. albicans* showed a larger diameter of clearance than that of other fungal strains (Venkatesan and Subramanian, 2010).Recently Ambuse and Bhale (2012) reported *Jatropa curcas* latex showed 75 per cent reduction of radial growth of *Alternaria tenuissima* and *Fusarium proliferatum* at 50 per cent conc and *Pythium* sp. at 75 per cent conc.

The general antifungal activity of essential oils is well documented (Reuveni *et al.*, 1984; Deans and Ritchie, 1987; Alankararao *et al.*, 1991; Baruah *et al.*, 1996; Gogoi *et al.*, 1997; Pitarokili *et al.*, 1999; Meepagala *et al.*, 2002) and there have been some studies on the effects of essential oils on post-harvest pathogens (Bishop and Thornton, 1997). Examined the *in vitro* effect of extracts of different neem (*Azardirachta indica* A. Juss) plant parts such as leaf, bark, oil cake and neem oil on the growth, mycelial yield and sclerotial survival of *Macrophomina phaseolina* (Dubey, 2009).

Trichoderma viride T112 and T. viride (MO), *T. harzinmum* (M) and *T.harzianum* T194 were used as potential biological agent for control of common root rot caused by *Bipolaris sorokiniana*,(Salehpour *et al.*, 2005). Kumar *et al.* (2007) reported *T. viride* as the most effective antagonist for *A. alternata* while Sempere and Santamarina (2007) found *T. harzianum* as the potential antagonist for *A. alternata*. Potphode (2004) studied the efficacy of *T. viride* and *T. harzianum* against *C. gloeosporioides* causing anthracnose of Jasmine and found that maximum inhibition of the pathogen was achieved by *T. harzianum* when placed at the center of the test fungus.

Kumar (2006) reported strong antagonistic effect of *B. subtilis, A. niger* and *T. viride* against *M. phaseolina in vitro* whereas, *A. flavus, T. harzianum, T. longibrachyatum, G. virens* and *P. fluorescens* appeared as potent antagonists. Different species of *Trichoderma* gained considerable success against pathogenic fungi. *T. harzianum* protects the root system against *F. solani, R. solani* and *M. phasoelina* infection on a number of crops (Malik and Dawar, 2003). Sankar and Sharma (2001) stated that 2 out of 9 isolates of *T. viride* evaluated in preliminary tests showed superior performance against *Macrophomina phaseolina* (the causal of charcoal rot in maize) in the laboratory. The efficacy of four fungal bioagents *viz., Trichoderma hamatum, T. harzianum, T. polysporum* and *T. viride* were evaluated *in vitro* condition against the Eggplant root-rot pathogen, *Macrophomina phaseolina* (Hesamedin Ramezani, 2008). Swami and Mukadam (2004) observerd the efficacy the efficacy of *T. viride* against the tomato fungi *Geotrichum candidum, Aspergillus niger, Alternaria solani, fusarium oxysporum and Rhizopus stolonifer.*

References

Akhter N, Begum MF, Alam S and Alam MS. 2006. Inhibitory effect of different plant extracts, cow dung and cow urine on conidial germination of *Bipolaris sorokiniana*. *J. Bio.Sci.* 14: 87–92.

Alankararao GSJG, Baby P, Prasad RY. 1991. Leaf oil of *Coleus amboinicus* Lour: *in vitro* antimicrobial studies. *Perfumerie Kosmetics,* 72: 744–745.

Ambuse MG and Bhale UN. 2012. Evaluation of antifungal activity of plant latex extracts against resistant isolates of pathogens associated in *Rumex acetosa* L. *International Journal of Ayurvedic and Herbal Medicine,* 2(2): 389–393.

Aworh OC, Kasche V, Apampa OO. 1994. Purification and properties of Sodom apple latex proteinases. *Food Chem.,* 50: 359–362.

Barnett H L. 1960. Illustrated Genera of imperfect fungi. Burgress Publishing Company, IInd edition, West Virginia, pp.127.

Baruah P, Sharma RK, Singh RS and Ghosh AC. 1996. Fungicidal activity of some naturally occurring essential oils against *Fusarium monoliforme. J. essential Oils Res.,* **8,** 411–441.

Bishop CD, Thornton IB. 1997. Evaluation of the antifungal activity of the essential oils of Monarda citriodora var. citriodora and Melaleuca alternifolia on the post-harvest pathogens. *J. Essential Oil Res.,* 9 (1)**:** 77–82.

Callan NW, Mathre DE, Miller JB and Vavrina CS. 1997. Biological seed treatments: factors involved in efficacy. *Hort Science,* 32: 179 –183.

Chopra RN and Bose JP. 1925. *Cephalandra indica* (Telakucha) in diabetes. *Indian J. Med.Res.,* 13: 11–16.

Cooke CIET. 1903. Flora of Presidency of Bombay, Published under the Authority of Secretary of State for Council. Vol 1: 571–572.

Deans SG, Ritchie G. 1987. Antimicrobial properties of plant essential oils dry root caused by *Macrophomina phaseolina. Karnataka J Agric Sci.,* 20**:** 54–56.

Dubey RC, Kumar H, Pandey RR. 2009. Fungitoxic Effect of Neem Extracts on Growth and Sclerotial Survival of *Macrophomina phaseolina in vitro. Journal of American Science,* 5(5): 17–24.

Ellis MB.1971.Dematiaceous Hypomycetes. Commonwealth Mycological. Institute, Kew, Surrey, England. Pp.608.

Gogoi R, Baruah P, Nath SC. 1997. Antifungal activity of the essential oils of Litsea cubeba *Pers. J. Essential Oils Res.,* 9**:** 213–215.

Hesamedin R. 2008. Biological Control of Root–Rot of Eggplant caused by *Macrophomina phaseolina* American–Eurasian. *J. Agric. and Environ. Sci.,* 4 (2): 218–220.

Hussein MMA, El–Feki MA, Iatai KI and Yamamoto KA. 2002. Inhibitory effects of Thymoquinone from Nigella sativa onpathological Saprolegnia in fish. *Bio. Control Sci.,* 7: 31–35.

Juncker T, Schumacher M, Dicato M. and Diederich M. 2009. UNBS1450 from *Calotropis procera* as a regulator of signaling pathways involved in proliferation and cell death. *Biochem Pharmacol.,* 78(1): 1–10.

Kata J. 2000. Physical and cultural methods for the management of soil borne pathogens. *Crop Protection,* 19**:** 725–731.

Khirstova P and Tissot M. 1995. Soda –Anthroquinone pulping of *Hibiscus Sabdariffa* (Karkadeh) and *Calotropis procera* from Sudan.*Bioresource Technology,* 53**:** 677–72.

Kumar A. 2006. Investigation on leaf spot disease (*Macrophomina phaseolina*) (Tassi.) Goid. of cowpea (*Vigna unguiculata* L.) under south Gujarat condition. M.Sc. (Agri.) thesis submitted to N.M. College of Agriculture, NAU, Navsari.

Kumar S., Upadhyay JP and Kumar S. 2007. Bio–control of *Alternaria* leaf spot of *Vicia faba* using antagonistic fungi. *J. Bio. Control,* 20 (2): 247–250.

Malik G and Dawar S. 2003. Biological control of root infecting fungi with *Trichoderma harzianum. Pak.J.Bot.*, 35(5): 971–975.

Meepagala KM, Sturtz G and Wedge DE. 2002. Antifungal constituents of the essential oils fraction of *Artemisia drancunculus* L. var. *dracunculus. J Agric Food Chem.,* 50: 6989–6992.

Mishra M and Tiwari SN. 1992. Toxicity of *Polyalthia longifola* against fungal pathogens of rice. *Indian Phytopathol.*, 45: 59–61.

Morton DJ and Stroube WH. 1955. Antagonistic and stimulating effects of soil microorganisms upon sclerotium. *Phytopathol.,* 45: 417–420.

Mughal MA, Khan TZ and Nasir MA. 1996. Antifungal activity of some plant extracts. *Pak. J. Phytopathol.,* 8: 46–48.

Mukerjee K, Ghosh NC, and Datta T. 1972. *Coccinia indica* as a potential hypoglycemic agent. *Indian J. Exp. Biol.*, 5: 347–349.

Mungikar AM. 1997. An Introduction to Biometry. Saraswati Printing Press, Aurangabad. pp. 57–63.

Nahar Nilufar, Mosihuzzaman M, Khan M and Haque S. (1998) Determination of free sugars in plant materials having antidiabetic activity. *Dhaka Univ. J. Sci.,* 46: 167–170.

Nargis A, Ferdousi M, Alam BS and Alam MS. 2006. Inhibitory effect of different plant extracts, cow dung and cow urine on conidial germination of *Bipolaris sorokiniana. J. Bio–Sci.,* 14: 87–92.

Nasir E and Ali SI. 1973. Flora of West Pakistan, Cucurbitaceae, No. 154. *Botany Department,University of Karachi.,* pp.154.

Pitarokili D, Tzakou O, Couladis M and Verykokidou E. 1999. Composition and antifungal activity of the essential oils of Salvia pomifera subsp. calycina growing wild in Greece. *J Esential Oils Res.,* 11: 655–659.

Potphode PD. 2004. Studies on anthracnose of Jasmine (*Jasminum sambac* L.) and its management. M.Sc. (Agri.) thesis submitted to Dr. B.S.K.K.V., Dapoli, and M.S. R.W. Robinson and C. Jeffrey (eds), Biology and utilization of the Cucurbitaceae.

Reuveni R, Fleischer A and Putievski E. 1984. Fungistatic activity of essential oils from *Ocimum basilicum* Chemotypes. *Phytopathol.,* 10: 20–22.

Rizk MA. 2008. Phytotoxic effect of *Calotropis procera* extract on seedling development and rhizosphere microflora of tomato plants in soil infested with *Fusarium oxysporum f. sp. lycopersici.World Applied Sciences Journal*, 3(3): 391–397.

Salehpour M, Etebarian HR, Roustaei A, Khodakaramian G and Aminian H. 2005. Biological Control of Common Root Rot of Wheat (*Bipolaris sorokiniana*) by *Trichoderma* isolates. *Plant Pathology Journal,* 4 (1): 85–90.

Sankar P and Sharma RC. 2001. Management of charcoal rot of maize with *Trichoderma viride. Indian Phytopathology,* 54(3): 390–391.

Sastri BN. 1950. The Wealth of India, A Dictionary of Raw Material and Industrial Products. Publication and Information Directorate, CSIR, New Delhi, 2 (8): 257 and 285–293.

Sempere F and Santamarina M. 2007. *In Vitro* biocontrol analysis of *Alternaria alternate* (Fr.) Keissler under different environmental conditions. *Mycopathologia,* 163(3): 183–190.

Shekhawat PS and Prasad R. 1971. Antifungal properties of some plant extracts. Inhibition of Spore Germination. *Indian Phytopath.*, 59: 883.

Singh A K. 1990. Cytogenetics and evolution in the Cucurbitaceae. In: D.M. Bates, *Cornell.*

Subramanian CV. 1971. Hypomycetes: an account of Indian species except cercosporaceae. Indian Agriculturral Research Institute (IARI), New Delhi.pp.811 and 848.

Swami CS and Mukadam DS. 2004. Biological control of tomato fungi using *Trichoderma viride.* Abs. National Semi. on Biotechnological approaches towards the integrated management of crop disease.,pp.47.

Thompson,A.K. 1996. Post –harvest technology of fruits and vegetables. *Blackwell Science Ltd.London. Univ. Press, Lthaca and London.*

Venkatesan S and Subramanian SP. 2010. Evaluation of antifungal activity of *Calotropis Gigantea* latex extract: An *in vitro* study. *International Journal of Pharmaceutical Sciences and Research,* 1(9): 88–96.

Vincent JM. 1947. Distortion of fungal hyphae in the presence of certain inhibitors. *Nature* 150, 850.

2015, Recent Trends in Microbiology, Mycology and Plant Pathology   *Pages* 297–312
Editor: **Dr. H.C. Lakshman**
Published by: **DAYA PUBLISHING HOUSE, NEW DELHI**

Chapter 19

Diseases on some Important Common Leafy Vegetables and their Control Management

Shwetha C. Madgaonkar* and H.C. Lakshman

Microbiology Laboratory, Post Graduate Department of Studies in Botany, Karnatak University, Dharwad – 580 003 Karnataka, India
E-mail: shweta.madgaonkar@gmail.com

Introduction

Vegetables constitute a major part of daily food intakes by human population all over the world. Vegetables play an important role in well-balanced diet (Kawashima, 2003). Vegetables are the edible parts of plant that are consumed wholly or in parts, raw or cooked as part of main dish or salad. Leafy vegetables called potherbs, greens, vegetable greens, leafy greens or salad greens, are plant leaves eaten as a vegetable. The leafy vegetables are said to tone up the energy and vigour in human being. Though leafy vegetables are low in calories and fat but high in dietary fibre, iron, Phosphorous, calcium and magnesium content and very high in phytochemicals such as vitamin (A, B, C, K), carotenoids, lutein, folate, magnesium. As the green leafy vegetables are excellent sources of micronutrients, consumption of green leafy vegetables may play an important role to overcome the micronutrient deficiencies as well as to prevent the degenerative diseases (Martins *et al.,* 2011; Khader and Sarma, 1998). It has been found that the person who consumes less amount of vegetables suffer from malnutrition, which in turn hampers the immune system. Several chemical compounds from leafy vegetables having therapeutic effect against several ailments have been identified. It could potentially prevent chronic disease due to their antioxidant content. But, broadly reporting leafy vegetables should form a regular component in our day to day diet to suffice the vital protective nutrients that are required for healthy living. In India, leafy vegetables are greatly preferred by the countrymen, either cooked to different delicacies or sometimes adorn the table in the form of fresh salads. Leaf vegetables most often come from short-lived herbaceous plants.

India has achieved adequate source and good degree of stability of vegetable crop production. Among them leafy vegetables are most essential component of our diet which nourishes with nutrients, minerals and vitamins. For healthy diet, daily minimum consumption should be required. Therefore, there is an urgent need to explore and cultivate leafy vegetables in India. Leafy vegetables account for around 60 per cent of the total vegetable production in the country. In India, out of total production, leafy vegetables are prone to several fungal diseases most commonly causing leaf spot and wilting. Annually billions of rupees loss occurs throughout the country due to these diseases, though 74 per cent of Indian population is engaged in agriculture. So plant diseases control or elimination is paramountly important to all concerns. In India, the leaves of a large number of wild and cultivated plants are used as vegetables. They have a very high protective food value and are very easy to grow. Almost all the leafy vegetable crops are propagated from the seed which is sown directly in the field.

Diseases can occur at any stage during the course of plant growth. The rapid, accurate diagnosis of the cause of a disease, along with the implementation of a rapid treatment, is essential to ensure the protection of the crop. Certain infectious diseases caused by living, microscopic organisms have the potential to rapidly ruin a crop. However, for any particular vegetable, these diseases are not that numerous and, so, it would not be difficult for a grower to become familiar with them and take proper preventative action. Diseases caused by nonliving things (*i.e.* not infectious) can be much more difficult to diagnose. Usually, it is easier to rule out an infectious agent as the cause of a disease before investigating possible nonliving (abiotic) causes. This stresses the need for the grower to become familiar with the more common infectious diseases that can occur on the crop.

Disease is the outcome of an interaction between the host, the disease agent, and their environment. If the cause of infectious disease, the pathogen, is next to the host, nothing will happen unless environmental factors are favorable for its infection and development within the plant. With foliar pathogens, there is usually a minimal period of leaf wetness required to stimulate spore germination and infection. For some soil-borne pathogens, infection occurs in combination with high soil moisture and certain critical soil temperatures. Not all diseases are caused by pathogenic organisms. Determining whether a disease is caused by a pathogen, or has nonliving (abiotic) causes requires not only the examination of individual plants, but also, noting the pattern of symptom occurrence in a field. Examine individual plants for unusual symptoms, such as leaf spots, wilts, stunting, fruit rots, misshapen leaves, cankers and stem blight. Roots should be examined for galls, root rot and necrosis (dead areas). Fields should be observed to determine if the problem is widespread and whether different plants species in and around the field are affected, which could indicate an abiotic cause. Symptoms with a nutritional or physiological cause have a more widespread occurrence within a field than infectious diseases. Initially, most of the disease causing pathogens will be isolated in areas and spread outward from those areas. Also, weeds or nonrelated crops are not typically affected. Soil borne pathogens are even more restricted within a field than foliar pathogens.

Plant diseases were known to man when organized agriculture came into existence and when man had to move to other areas in search of food. Inspite of many thousand years of progress in human civilization, based on great scientific advances and also the opening of new frontiers in scientific agriculture, it remains stark truth even today that a very large proportion of world population still remains underfed, undernourished or even starved. Guided by International Agency like FAO, there had been sustained effort all over the world to boost crop production, particularly the food crops, through the adoption of improved technology. In a few advanced countries, the estimation of losses due to the plant diseases, which is decidedly a complex issue, has not been made in many agricultural countries.

Basically, plant diseases are placed in two categories: parasitic or non-parasitic depending on their mode of existence as a parasite or not. Some prefer to keep viral diseases in a third group, as there is doubt about the living and non-living nature of viruses. Since virus can live actively and replicate only within living cells, diseases they cause should come under parasitic diseases. On this basis, plant diseases are classified as follows: the disease of parasitic origin share the potential to spread extensively over a large area, through transmission of their inoculum and thus to cause heavy damage to the crops. It is of interest, however, to note that symptom expression in the non-parasitic diseases and also in many parasitic diseases can directly be traced to one or other kind of physiological disorder. Plant leaves showing a few minute dead spots in response to infection do not come in the way of normal growth or reproduction of the plant, one may take the plant as a whole as healthy. The plant becomes diseased, however, if large number of spots develop on the leaves to affect its growth or reproduction.

Diseases Caused by Different Types of Pathogens on Leafy Vegetables

Many species of micro-organisms, including fungi, bacteria, viruses, phytoplasmas and nematodes, cause diseases of vegetable crops. For disease to occur, plant pathogens must come in contact with a susceptible host plant. Pathogens can be carried to the plants by various means, including transplants, soil, humans, animals, insects, infested seed, and wind or water, alone or in combination. Favorable environmental conditions must be present for the plant pathogen to infect and thrive on the plants.

Fungi

Fungi are multicellular microscopic organisms that can grow to their food, usually in the form of filamentous strands. Their growth pattern is radial, so on surfaces such as plant leaves, the effects of their growth may be seen as circular spots. However, fungal infections of other plant parts, such as roots, may produce no visible structures. Some symptoms can indicate these infections. For example, browning of the water-conducting tissues of the stem, in combination with wilt, can indicate infection by the *Fusarium* wilt fungus. Other disease symptoms, such as blight (a general death of tissue), which can have a variety of causes, may require laboratory testing to confirm fungi as a cause. Fungi can produce specialized structures, such as spores, which are used for reproduction, dissemination through space and time, and survival. *Sclerotia* are structures that function in the long term survival of many soil borne pathogens. Most fungi that infect leaves require free moisture to initiate infection, with the exception of powdery mildew fungi, which need only high humidity to initiate infection.

Bacteria

Bacteria are single celled microscopic organisms, which survive by becoming dormant. The most notable exception to this is the pathogen that causes common scab which is a filamentous, multicelled bacterium that produces spores. They can be transported by wind driven rain, by insects, or the movement of infected plant parts, including seed. Bacteria that infect leaves may cause circular spots, but irregular shaped lesions that don't extend beyond veins are more characteristic for bacterial infections. They can also cause soft rots of vegetative parts that are usually characterized by a foul smell.

Mycoplasma

Mycoplasmas are bacteria-like, only structurally simpler and smaller than bacteria. They are transmitted by leafhoppers. The most notable disease of vegetables caused by a mycloplasma is aster yellows. The symptoms of aster yellows are distinctive: leaves have a bronzed appearance and flowers are abnormal (leaf-like tissue grows from them, instead of petals).

Viruses

Viruses are submicroscopic entities that can only replicate inside living plant cells. They require agents such as insects to transmit them to plants. A typical virus symptom is a mosaic pattern on leaves, but viruses may cause other symptoms, such as necrotic lesions and stunting, which can have other causes. Insect vectors of viruses include by aphids, leafhoppers, white flies and thrips. Viruses often have wide host ranges, including weeds that are not botanically related to the crop, and symptoms are not always produced in these plants.

Nematodes

Nematodes are microscopic roundworms that feed on the roots of plants. The root knot nematode produces root galls and deformed roots on a wide range of crops. Heavy infestation can lead to wilting and death of plants. Nematodes can also cause stubbiness, necrosis and stunting of roots, but these symptoms are not distinctive. Soil or root analysis is required to confirm diagnosis.

Diseases on some Leafy Vegetables

1. Fenugreek

Fenugreek commonly known as "Methi" (*Trigonella foenum graecum* L.) belongs to Leguminosae–Family. The species name *"foenum-graecum "* means "Greek hay" indicating its use as a forage crop in the past. Fenugreek is believed to be native to the Mediterranean region (Petropoulos 2002), but now is grown as a spice in most parts of the world. It is reported as a cultivated crop in parts of Europe, northern Africa, west and south Asia, Argentina, Canada, United States of America (USA) and Australia (AAFRD 1998; Edison 1995; Fazli and Hardman 1968; Petropoulos 2002). India is the leading fenugreek producing country in the world (Edison 1995).Major producing states are Rajasthan, Madhya Pradesh, Gujarat, Uttar Pradesh, and Tamil Nadu. It is used as a condiment for flavouring of foods. It has got medicinal value, hence used as medicine. Its seed and leaves have medicinal value, and have been used to reduce blood sugar and lower blood cholesterol in humans and animals (Dahanukar *et al.*, 2000).

Disease Organisms Affecting Fenugreek

Fungal, bacterial, viral and insect mediated diseases are reported to be associated with considerable lowering of forage and seed yield in fenugreek and hence is a serious agronomic concern (Fogg *et al.*, 2000; Jongebloed, 2004; Petropoulos, 1973; Prakash and Sharma, 2000). Fenugreek diseases are broadly classified on the basis of their pathogenicity into the following groups:

A. Fungal Diseases

Leaf Spot Disease
Causal organism: *Cercospora traversinia.*

Symptoms: The disease is characterized by the presence of numerous small, brown circular spots on the leaves. In the beginning, the spots are small, roundish with concentric rings but later on these spots increase in size and sometimes they coalece forming bigger spots.

Life cycle: primary infection occurs through conidia, carried over the plant debris or by seeds. The fungus spreads through air borne conidia. The role of perfect stage in perpetuating the disease is not fully known.

Control measures: It may be controlled by spraying Bordeaux mixture (5:5:50) or 0.3 per cent Blitox three times at an interval of 15 days and follow crop rotation

Powdery Mildew
Causal organisms: *Erysiphe polygoni* D.C.

Symptoms: In this disease, white powdery patches appear on the lower and upper surface of leaves and other parts of plant.

Life cycle: The infection starts at flowering stage of the plant. The disease perennates through cleistothecia on the plant debris or in soil. The conidiophores arise vertically from leaf surface each bears conidia in chain. The Conidia serves as secondary inoculums and it disseminated by wind and other agencies.

Control measures: Crop should be dusted with 300 mesh Sulphur dust @ 25 kg/ha to control this disease as soon as the symptoms are noticed. Spraying of wettable Sulphur or Dinocap (Thiowet) can also be used to control the disease @ 20-25 g per 10 liter of water at the initial stage of this disease. If needed two more sprays should be given at an interval of 15 days after first spray. Application of fungicides like Cosan, Kerathan etc.

Downy Mildew

Causal organisms: *Perenospora trigonellae* Gaum.

Symptoms: Yellow patches on the upper surface of leaves appear in the infected plants and white cottony mycelium on the lower surface of leaves.

Life cycle: This is soil borne disease. The oospores survive in the soil along with plant debris. Secondary infection occurs through wind borne conidia.

Control measures: This disease can be controlled by spraying of 0.2 per cent solution of Difoltan or any other copper fungicide.

Wilt

Causal organisms: *Fusarium oxsysporum* Synder and Hansen.

Symptoms: The first symptoms are slight yellowing and dropping of the lower leaves, followed by those of the next leaves in order up the stem.

Life cycle: The pathogen is soil born and survives on the dead host roots saprophytically in the soil. Penetration in the host occurs through the fine rootlets. Optimum temperature for the development of the disease is between 20°C to 27°C. it has the ability to survive in the soil for a long time (8-20 years).

Control measures: Preventative biological fungicide (Companion Liquid Biological Fungicide^OG) for control and suppression of soil and foliar diseases. Activates ISR (induced systemic resistance).

B. Bacterial Diseases

Bacterial leaf spot in fenugreek which was caused by *Pseudomonas syringae* pv. *syringae* (Fogg *et al.*, 2000). It also has been suggested that the bacterium *Xanthomonas alfalfa* can infect fenugreek (Petropoulos, 2002).

C. Viral Diseases

Bean Yellow Mosaic Virus, Alfalfa Mosaic Virus, Cow Pea Mosaic Virus, Soybean Mosaic Virus, Pea Mosaic Virus, Potato Virus A and Y, and Clover Vein Mosaic Virus are common viral infections of fenugreek. These viral diseases have been associated with moderate loss of seed and forage yield (Petropoulos, 2002).

D. Insect Diseases

insects such as thrips, pod-borers and heliothis can cause serious damage to forage yield in fenugreek (Lucy, 2004). Root rot by the soil borne nematode *Meloidogyne incognita,* which causes the death of immature plants has also been reported (Jongebloed, 2004).

2. Spinach

Distribution: Spinach (*Spinacia oleracea* L.) belongs to the family Chenopodiaceae. It is a herbacious plant which produces edible leaves as an annual and seed as a biennial crop in the plains. The edible part of spinach is a compact rosette of leaves. Spinach was probably originated in South-west Asia and was used in Iran over 2000 years ago. Now it is commercially grown in the United States, Europe, India, etc. In India it is popular all over the country. Spinach is a wonderful green-leafy vegetable often recognized as one of the functional foods for its nutritional, antioxidants and anti-cancer constituents. Its tender, crispy, dark-green leaves are favorite ingredients of chefs all around the planet. *Spinacia* plant grows about 1 foot in height. Although, it can be grown year round, fresh greens are best available just after the winter season in the Northern hemisphere from March through May and from September until November, in the South of the equatorial line.

Diseases on Spinach

Leaf Spot Disease
Causal organism: *Cladosporium variabile* (Cooke) G.A. de Vries

Symptoms: Early symptoms of infection occur as numerous small, circular white to yellow spots, beginning on the older leaves and progressing to the younger ones as the plant grows during the season. Spots often join together and become irregular in shape. When the fungus produces spores, the leaf spots become olive black in color from the spore masses. Generally, older leaves are killed, but in the case of severe infection all of the leaves on the plant may be affected and perish. Ripening seed may also develop lesions and become shriveled.

Life cycle the disease is favored by high humidity and cool temperatures, the fungus can grow under a wide range of temperatures, ranging from 41° to 86°F. Spores of *Cladosporium* can germinate and penetrate leaf stomata within 48 hours of infection in the presence of free moisture. Symptoms of the disease generally follow 5 to 10 days later. Once infection is established, the fungus grows in the leaf tissue; spores produced within leaf lesions start new infection cycles. Spores can be spread by wind, rain splash, or carried on equipment.

Control measures: It can be controlled by the application of Bordeaux mixture (5:5:50) or Cupravit. Cultural control is achieved through sanitation measures to reduce sources of the fungus and spread.

White Rust
Causal organism: *Albugo occidentalis* G.W. Wilson

Symptoms: Symptoms of white rust begin as chlorotic lesions on the upper leaf surface. As the lesions develop, small white pustules (sori) containing sporangia are produced on the underside of the leaf and occasionally on the upper leaf surface (Raabe, 1951; Raabe and Pound, 1952). The pustules often are so abundant that nearly the entire leaf surface is covered.

Life cycle: Primary infection is thought to occur from soilborne oospores that are splashed on plants by rainfall or overhead irrigation, or by airborne sporangia

((Dainello *et al.,* 1990; Thomas, 1970; Dainello and Jones, 1986). It is speculated that oospores germinate and infect plants through open stomata. Secondary infection results from airborne sporangia discharged from pustules. Sporangia can germinate directly, but usually germinate indirectly to produce six to nine biflagellate zoospores Specific temperatures and durations of free surface moisture are required for infection and subsequent disease development (Byrne *et al.,* 1998; Gross *et al.,* 1998; Hong *et al.,* 1996; Sirjusingh and Sutton, 1996).

Control measures: The disease can be controlled by keeping the field and surrounding areas clean and by spraying Bordeaux mixture (5:5:50).

Stemphylium Leaf Spot
Causal organisms: *Stemphylium botryosum* Wallr.

Symptoms:, Initial symptoms consist of circular, gray-green leaf spots. In advanced stages of the disease, leaf spots are light tan and papery in texture and coalesce. Sporulation of the pathogen is generally absent. Young spinach foliage can result in significant leaf death.

Life cycle: Spore germination increased with time in free water, and the relative susceptibility of host plants to infection was proportional to the duration of water retention on leaves.

Control Measures: Apply fungicides as needed Crop rotation Burial of crop residue.

Downy Mildew
Causal organisms: *Peronospora farinosa* f. sp.*spinaciae* Byford.

Symptoms: Chlorosis on upper leaf surface, and grayish brown felt-like mildew growth on leaf undersurface.

Life cycle: Pathogen mostly penetrates through oospores. Oospores are produced abundantly and retained within the host tissues and released after the decomposition of infected debris. The oospores after germination infect the healthy plants. Secondary spreads occurs through conidia.

Control Measures: Crop rotation Soil treatment with metalaxyl or mefenoxam Resistant cultivars, Burial of crop residue.

Anthracnose
Causal organism: *Colletotrichum spinaciae* Ellis and Halst.

Symptoms: the spots at first are small and dark olivaceous or water-soaked. As they enlarge, they become irregular in shape, variable in size and give a scorched appearance. Eventually the spots coalesce and cause the death of the entire leaf. In a wet weather the crop may look so much blighted that it could be considered as lost but if weather changes, new leaves develop and the crop grows normally. Elongated grayish spots are abundantly formed on the seed stocks. Acervuli of the fungus frequently appear as black dots on the surface of the seed.

Life cycle: the pathogen perpetuates through plant debris in the field. It is also presumed that fugus is externally seed borne. Secondary spread through air borne conidia.

Control measures: For the control of this disease Dithane Z-78 (0.3 per cent) may be sprayed at 15 days interval. Since the leaves are consumed frequently, insecticides application should be avoided. Precautions such as clean seed, destruction of crop debris and a three year crop rotation should be followed.

Viral Diseases

Spinach Blight

Cucumber mosaic virus (CMV) infection of spinach has long been called "spinach blight" by growers, but should not be confused with fungal diseases for which the term b*light is* more commonly used. The disease is caused by CMV, which is covered in detail in other sections of this fact sheet. Infected spinach plants may show a variety of symptoms including stunting, yellowing, and mottling of the older leaves and malformation of the younger leaves. Good resistance to CMV is available, but this resistance is temperature dependent; at temperatures above 80° F crown necrosis will develop which is similar to infection with broad bean wilt virus (BBWV) without the need for high temperatures. With the exception of the high temperature response for CMV infection, resistance for CMV in spinach has provided an effective control measure for over 60 years.

Broadbean wilt virus (BBWV) has previously been mentioned in this report. This virus does cause a major disease of spinach, particularly in the autumn crop when aphid vectors are most plentiful and much inoculum is present from earlier plantings.

3. Dill

Anethum graveolens L. (dill) has been used in ayurvedic medicines since ancient times and it is a popular herb widely used as a spice and also yields essential oil. It is an aromatic and annual herb of apiaceae family. Dill is an annual that is best grown in rich, light, well-drained soils in full sun. Plants are more apt to fall over in part shade. Shelter plants from strong winds. Close-planting, stakes or cages may be used to provide support. Best growth occurs in cool summer climates. Plants appreciate consistent soil moisture. Soils should not be allowed to dry out. Plants tend to bolt when conditions remain dry. The Ayurvedic uses of dill seeds are carminative, stomachic and diuretic. *Anethum* grows up to 90 cm tall, with slender stems and alternate leaves finally divided three or four times into pinnate sections slightly broader than similar leaves of fennel.

Diseases

Powdery Mildew
Causal organisms: *Erysiphe polygoni* D.C.

Symptoms: Fungal growth appeared as typical white, dense, persistent powdery mildew colonies on leaves, inflorescences, and stems. Hyphae were 6 to 10 μm wide. Conidia were produced singly on unbranched three-celled conidiophores, were cylindrical to ovate. No fibrosin bodies were observed. Germ tubes were formed from the ends of conidia. Appressoria from mycelia were lobed.

Life cycle: the fungus obligate parasite. It is known that clestothecia survives from season to season in soil and they liberate ascospores, which causes infection to lower leaves.

Control measures: apply sulfur fungicides weekly to prevent infection of susceptible plants. Destroy plants that are heavily infected.

4. *Basella alba*

Distribution: *Basella alba* is a wildly cultivated, cool season vegetable with climbing growth habit. It is a succulent, branched, smooth, twining herbaceous vine, several meters in length. *Basella alba* is a perennial vine belonging to the family *Basellaceae*. The plant is a perennial vine and grown as annual or biennial pot-herb. It is known as Malabar spinach, Red vine spinach, creeping spinach. It is commonly found in the tropical regions of the world, and is widely used as a leaf vegetable. *Basella alba* is a vigorous soft-stemmed climbing vine and grows up to 10m long. It has broad, fleshy heart-shaped, thick semi-succulent green leaves, 5-12cm wide. The plant is a good source of vitamins A, and C, Iron and Calcium. Stems are ready for harvesting about 35 to 45 days after planting (about 50 days after seeding). It prefers hot humid climate and moist, fertile, well-drained soil to flourish. Although its seeds can be sown directly for planting, usually thick cuttings about the length of 20 cm preferred for easy propagation and fast growth. It bears white or white-pink color tiny flowers depending upon the species and deep-purple to black color berries.

Diseases

Infections on Stems

Die Black

Causal organisms:, *Colletotrichum capsici* (Syd.) Butl. and Bisby

Symptoms: Discoloured streaks appear first which extend linearly aquiring ash-colour at maturity dotted by small black fructifications. The stems and leaves dry premature losing chlorophyll and are thus devalued economically.

Life cycle: the pathogen perpetuates through plant debris in the field. It is also presumed that fungus is externally seed borne. Secondary spreads occurs through air borne conidia.

Control measure: field sanitation, seed treatment with Thiram, Brassicol and fungicidal spray- 3to 4 spray of 0.2 per cent Perenox have been recommended for effective control of the disease.

Anthracnose

Causal organisms: *Glomerella cingulata* (Stonem.) Spauld. and Sehr.

Symptoms: On *Basella alba* stems the initial light-brown patches turn into ash-colour at maturity bounded by irregular dark-brown, upraised halo. Usually lesions are elongated which coalesce.

Life cycle: The pathogen perpetuates on the leaves and twigs, which fall on the ground. The fungus can also survive saprophytically for long duration on these debris. Infection occurs through the young fruits, just after blossoming period.

Control measures: fungicidal spray of 6:6:100 bordeaux mixture or other fungicides like Fytolon or blitox-50 at the time of flowering is recommended and sanitation is important to destroy the infected twigs and leaves fallen on the ground.

5. *Amaranthus*

Amaranthus paniculatus L. is commonly called as "Rajgira" (Amaranth) in India. *Amaranthus* can be upright or spreading herbaceous, fast growing, cereal-like annuals or short-lived perennial. *Amaranthus paniculatus* L. is the world's most nutritious plant. *Amaranthus paniculatus* is recommended as good food with medicinal properties. The vegetable form of *Amaranthus paniculatus* L. was probably introduced in the tropics and subtropics of the Old World during colonial times. It is more popular in humid lowland than in highland or arid areas. It is also an important vegetable in many tropical areas outside Africa *e.g.* in India, Bangladesh, Sri Lanka and the Caribbean.

Diseases

Leaf Spot
Causal organisms: *Phyllosticta* species.

Symptoms: In early stages the infection appears as light yellow specks which later increase into spots and coalesce resulting in shrivelling and drying up of the leaves.

Life cycle: Two types of spores of this fungus germinate when moisture is present: conidia and ascospores. Conidia can quickly be carried from diseased plants to healthy ones by splashing rainwater, sprinklers or watering. In addition, the ascospores are discharged into the air and can travel between plants on a breeze or current. If they land on a moist leaf, ascospores germinate, infect the host, and begin the cycle anew.

Control measure: The effective control measure is the spraying of fungicide like shell copper, BIitox or Cupramar (0.3 per cent) at 15-20 days interval.

Leaf Spot
Causal organism: *Cercospora brachiata* Ellis and Everh.,

Life cycle: Development of cercospora leaf spot disease begins when the fungus spores are dispersed by rain, irrigation water and wind. Germination occurs in humid conditions, usually during late spring and summer, and fungus growth is encouraged by frequently damp leaves. Plants that mature in the fall may escape acute infection. In general, the more rain, the worse the disease spreads. Primary infection occurs through the conidia, carried over the plant debris or by seeds.

Symptoms: The disease is characterized by the presence of numerous small, brown circular spots on the leaves.

Control measures: It may be controlled by spraying Bordeaux mixture (5:5:50) three times at an interval of 15 days.

White Rust
Causal organism: *Albugo bliti,* (Biv.) Kuntze

Symptoms: The white rust is characterized by the white, blister like, circular or irregular pustules on the lower surface of the leaf and opposite to each pustule on the upper surface a yellow patch develops.

Life cycle: The thick-walled oospores are the main overwintering structures, but the mycelium can also survive in conditions where all the plant material is not destroyed during the winter. In the spring the oospores germinate and produce sporangia on short stalks called sporangiophores that become so tightly packed within the leaf that they rupture the epidermis and are consequently spread by the wind. The liberated sporangia in turn can either germinate directly with a germ tube or begin to produce biflagellate motile zoospores. These zoospores then swim in a film of water to a suitable site and each one produces a germ tube–like that of the sporangium–that penetrates the stoma. When the oomycete has successfully invaded the host plant, it grows and continues to reproduce.

Control measures: The disease can be controlled by keeping the field and surrounding areas clean and by spraying Bordeaux mixture (5:5:50).

Damping-off

Causal organisms: *Pythium aphanidermatum* (Eds.) Fitz.

Symptoms: This disease causes seeds to rot in the ground, or developing seedlings to die before they emerge through the soil. Although poor seed is usually blamed and can be involved, even good seed can be so badly affected that very few seedlings emerge. This disease affects seedlings which have already emerged. A rot develops and constricts the stems near the soil surface. The affected stems are weakened and the young plants collapse. This disease usually occurs when seedlings are overwatered, are planted too thickly, or both.

Life cycle: the disease is soil-borne. The disease causing fungi are habitually soil-dwellers. At sufficient moisture and optimum temperature (10°C to 17°C), the fungus germinates producing coenocytic mycelium. Mycelial stage of fungus is capable of infecting the host plant and producing asexual zoospores very rapidly.

Control measures: It is controlled by good drainage. Over-dense sowing should be avoided. Fungicides such as dithiocarbamates have some effect.

Leaf Spot

Causal organisms: *Alternaria tenuissima* (Kunze) Wiltshire.

Symptoms: Brown to black, circular to oval, necrotic lesions were observed on leaves.

Life cycle: Disease is both soil and air borne. The mycelium and conidia can survive in the soil or on the diseased plant debris.

Control measures: Broad-spectrum fungicide with preventative, curative and locally systemic activity.

Wet Rot or Stem Rot

Causal organisms: *Choanephora cucurbitarum* (Berk. and Ravenel) Thaxt.

Symptoms: So many long sporangiophores are produced on diseased stem that a whisker like growth is evident. Water soaked lesions appear on the leaves and the margins and leaf tips are blighted. Older lesions turn necrotic and appear dried out. The entire plant may wilt. Flowers and flower buds turn dark and wilt. Young fruit can be infected.

Life cycle: Outbreaks of this disease occur during extended rainy periods and high temperatures. Wet weather and high humidity favor disease development from inoculum that is typically soil-borne. The pathogen grows profusely on diseased plants and the host may kill in few days.

Control measures: Chemical control by repeated spraying with fungicides such as maneb or carbatene reduces the losses, but is seldom applied.

Preventing Vegetable Diseases

Below are a few general rules that can reduce vegetable disease problems:

Avoid Common Gardening Mistakes: Avoid common mistakes like planting too early or overcrowding vegetables.

Choose the right vegetables for location: Most vegetables require at least 6 hours of sun a day to thrive, and 8-10 hours is better in the summer. If vegetables of sun-loving or heat-loving try to grow in a shady, cool, coastal climate, it will have a lot more disease problems instead of plants grow more suited to the conditions, like leafy green vegetables.

Grow disease resistant varieties: This is one of the most important weapons gardeners have against vegetable diseases. Many plants, though more expensive to buy at the garden stores and nurseries, may be disease resistant varieties which are able to battle diseases better than their non-resistant varieties. At times, climatic conditions are such that the onset of a particular disease is inevitable; season to season. In this case, the best option would be to grow a disease resistant version of that plant. Plant breeders are constantly selecting for disease resistance, so choose varieties that are resistant to diseases prevalent in selected area.

Grow robust vegetables: Plant vigor is the first and best defense against pests and diseases. Prepare the soil well, space your vegetables correctly, and make sure they have enough water. Adding organic soil amendments like kelp meal and rock dust at the time of planting can boost plant immunity and make them more resilient in the face of extreme temperatures.

Indulge in crop rotation: Many plant diseases build up in garden soil over a period of time, even over years. Crop rotation is a means of breaking this cycle so that the habit of the disease agents is disturbed and affects the disease agent population. Avoid planting the same crop or a member of the same plant family in the same place year after year. Pest and disease populations build up over time, and rotating a non-susceptible crop into the plot disrupts this buildup. Plants in the same plant family also tend to strip the same nutrients from the soil, so the next generation of plants is less vigorous. When a different crop is grown in an area, at times, these disease agents are unable to thrive as the plant may be unfavourable or repelling.

Resort to soil inoculation: Soil inoculation which includes beneficial micro-organisms like Arbuscular mycorrhiza, *Rhizobium*, PSB is a method of adding 'good' agents to the soil that fight against the 'bad diseases causing agents' (Lakshman *et al.*, 2000). It is simply adding good microbes that fight against or feed on the disease microbes. It acts as a defence mechanism for the soil where natural and organic methods are adopted without the use of harmful chemicals (Lakshman *et al.*, 2006; Santosh *et al.*, 2006).

Practice Good Sanitation

Good sanitation is critical in preventing vegetable diseases: Diseases and pests overwinter in plant debris, and spores will infect the next crop planted there. Remove and compost plant debris and mulches, or till them into the soil before winter. Diseased plants should be placed in the trash, not the compost pile. While a hot compost pile will kill vegetable diseases, all diseased materials have to be in the hottest core of the pile. Spores outside this hot zone can survive to infect your garden again.

Avoid working in the garden when foliage is wet: Wet foliage is heavier and more easily damaged by movement through the garden. Most diseases enter leaves through injuries, and fungal diseases often require water on the leaves for infection.

Use drip irrigation for summer vegetables: Water on the foliage of summer vegetables is an invitation to fungal diseases. It's okay to use overhead watering for seedlings, but once the plants start flowering and setting fruit, their energy shifts from defending leaves against diseases to setting seeds.

Use Mulch

Mulch the soil under vegetables: Mulching is one of the least-used, but most important organic gardening practices. A layer of mulch reduces stress on plants. It protects the soil surface from baking sun, pounding rain, and scouring by the wind. It keeps roots cooler, holds in moisture, and reduces watering by 20 per cent or more. Rain striking bare soil causes soil and disease spores and bacteria to splash up onto the undersides of leaves, where it can infect the plant. Mulch acts as a barrier that helps prevent vegetable diseases. Mulch and compost are rich in nutrients and nutrient releasing micro-organisms, that when they are added to the soil, they enrich the soil and make sick plants healthy.

Use pesticides/fungicides: Most plant diseases are caused by fungi. Chemicals which are organic in nature and do not cause drastic damage to soil nutrient levels may be sprayed on plants and in the soil. Some sulphur based mixtures may be used, as sulphur is a naturally obtainable pesticide from the earth and is good in pesticide combinations. The most common are Bordeaux Mixture and Lime-sulphur Mixture. They can be prepared at home and have to be used immediately after preparation.

Acknowledgement

First author is very thankful to U.G.C. New Delhi, for awarding Rajiv Gandhi National Fellowship for doing Ph.D. Second author is also indebted for sanctioning the major research project on rare millets of North Karnataka.

References

Agriose G. N., 1997. *Plant Pathology* (4 ed.), Academic Press, London.

Alberta Agriculture, Food and Rural Development (AAFRD). 1998. *Fenugreek,* agri-fax. Agdex. 147/20–5.

Byrne, J. M., Hausbeck, M. K., Meloche, C., and Jarosz, A. M. 1998. Influence of dew period and temperature on foliar infection of greenhouse–grown tomato by Colletotrichum coccodes. *Plant Dis.* 82: 639–641.

Dahanukar, S.A., Kulkarni, R.A. and Rege, N.N. 2000. Pharmacology of medicinal plants and natural products. *Indian J. Pharmacol.* 32: S81–S118.

Dainello, F. J. and Jones, R. K. 1986. Evaluation of use–pattern alternatives with metalaxyl to control foliar diseases of spinach. *Plant Dis.* 70: 240–242.

Dainello, F. J., Black, M. C., Kunkel, T. E. 1990. Control of white rust of spinach with partial resistance and multiple soil applications of metalaxyl granules. *Plant Dis.* 74: 913–916.

Edison, S. 1995. Spices–Research support to productivity. *In* N. Ravi (ed.) *The Hindu Survey of Indian Agriculture*, Kasturi and Sons Ltd., National Press, Madras. pp. 101–105.

Fazli, F.R.Y. and Hardman, R. 1968. The spice fenugreek (*Trigonella foenum–graecum* L.). Its commercial varieties of seed as a source of diosgenin. *Trop. Sci.* 10: 66–78.

Fogg, M.L., Kobayashi D.Y., Johnston S.A. and Kline, W.L. 2000. Bacterial leaf spot of fenugreek: A new disease in New Jersey caused by *Pseudomonas syringae* pv. *syringae.* Publication No. P–2001–0012–NEA. *In* Northeastern Division Meeting Abstracts, 1–3 Cape Cod, North Falmouth, MA, USA.

Gross, M. K., Santini, J. B. Tikhonova, L., and Latin, R. 1998. The influence of temperature and leaf wetness duration on infection of perennial ryegrass by *Rhizoctonia solani. Plant Dis.* 82: 1012–1016.

Hong, C. X., Fitt, B. D. L., and Welham, S. J. 1996. Effects of wetness period and temperature on development of dark pod spot (*Alternaria brassicae*) on oilseed rape (*Brassica napus*). *Plant Pathol.* 45: 1077–1089.

Jongebloed, M. 2004. Coriander and Fenugreek, *In* S. Salvin *et al.* (ed.) *The New Crop Industries Handbook*, Rural Industries Research and Development Corporation (RIRDC), Australian Government, Australia. p. 229–235

Kawashima, L.M. 2003. LMV Soares. *Journal of Food Composition and Analysis*, 16: 605–611.

Khader, V and S. Sarma., 1998. *Plant Food for Human Nutrition.* 53: 71–81.

Lakshman, H.C. Mulla, F.I., Inchal R.F., Rajanna, L. and Waghamore B. P. 2000. Some coastal potential plants with VAM fungi and their use in land reclamation programmes. S.D.M.C.E.T. Dharwad. Spl. Pub. *Coastal Zone Manage.* 2(1): 189–196.

Lakshman, H.C., H. Ramesh, N. Ratageri, M. Rolli and M. G. Nadagouda. 2006. Interaction between *Azospirillum* and Phosphate solublising bacteria (*Bacillus polymyxa*) on the yield of *Capsicum annum* L. (chilli). *J. Theo. Expt. Bio.* 2(3and 4): 133–135.

Lucy, M. 2004. Fenugreek [Online] Available at *www.dpi.qld.gov.au/fieldcrops/9050.html* [Accessed September18, 2005].

Martins, D., Barros, L and A.M. Carvalho, 2011. ICRF Ferreira. *Food Chem*, 125, 488–494.

Petropoulos, G. A. 1973. Agronomic, genetic and chemical studies of *Trigonella foenumgraecum* L. *Ph.D. Diss.* Bath University, England.

Petropoulos, G. A. 2002. Fenugreek –The genus *Trigonella,* 1st ed. Taylor and Francis, London and New York, pp. 1–127.

Prakash, S., and Sharma G.S. 2000. Conidial germination of *Erysiphe polygoni* causing powdery mildew of fenugreek. *Indian Phytopathol.* 53(3): 318–320.

Raabe, R. D. 1951. The effect of certain environal [sic] factors on initiation and development of the white rust disease of spinach. *Ph.D. dissertation.* University of Wisconsin, Madison.

Raabe, R. D. and Pound, G. S. 1952. Relation of certain environal [sic] factors to initiation and development of the white rust disease of spinach. *Phytopathology* 42: 448–452.

Santosh G. Hiremath, Bheemareddy V.S. and Lakshman,H.C. 2006. Microbial inoculations for *Capsicum annum*, Linn. GB–4 var. *J. Micro. World.* 8(2): 184–186.

Sirjusingh, C. and Sutton, J. C. 1996. Effects of wetness duration and temperature on infection of geranium by *Botrytis cinerea. Plant Dis.* 80: 160–165.

Thomas, C. E. 1970. Epidemiology of spinach white rust in South Texas (Abstr.). *Phytopathology.* 60: 588.

2015, Recent Trends in Microbiology, Mycology and Plant Pathology   *Pages* **313–324**
Editor: **Dr. H.C. Lakshman**
Published by: **DAYA PUBLISHING HOUSE, NEW DELHI**

Chapter 20

Plant Diseases Caused by Viruses and Mycoplasma and their Control Measures

Pushpa K. Kavatagi and H.C. Lakshman*

P.G. Department of Studies in Botany,
Karnatak University, Dharwad – 580 003, Karnataka, India
*E-mail: pushpak.k2@gmail.com

Introduction

The term disease is coined by combining the words Dis + Ease = Disease. The prefix Dis means negative, reverse, or opposite, and the word Ease means comfort, or freedom from pain or discomfort. Dis-Ease therefore means not well, and the cause can be many. A plant disease may therefore be defined as: Any harmful deviation or alteration from the normal functioning of physiological processes. It is also defined by some as: Disease is a malfunctioning process that is caused by continuous irritation which results in suffering. A more practical definition of a disease would be: A plant is diseased when its systems are not normal and, therefore, it is not producing as well as it should according to normal expectations of the farmers. Since the beginning of agriculture, farmers have had to develop means for managing weeds, insect pests, and diseases. Even today, with all the scientific research on crop protection, it is estimated that insects, diseases, weeds, and animal pests eliminate half the food produced in the world during the growing, transporting, and storing of crops.

Landscape and turf plant damage can result from many causes such as insect feeding, equipment injury, heavy use, growth disorders and infection by disease organisms. Growth disorders occur when plants are grown with incorrect amounts of water, nutrients, or light. Temperature extremes can also cause growth disorders.

Plant diseases caused by pathogens are called infectious diseases. These diseases can be hard to diagnose and treat. They are caused by organisms that are too small to see with the naked eye. Micro-organisms that infect plants are called plant pathogens. Pathogens include fungi, bacteria, viruses, and nematodes. When they infect a plant, they change the way in which it functions. They can kill or stunt the plant. A disease is passed from plant to plant as the pathogen multiplies and spreads. Plant disease is often caused by a combination of environmental conditions and pests. Plants already weakened by environmental stress are more likely to be attacked by pathogens or insects. For example, wood boring insects often attack weakened trees. Stressed plants sustain more damage and heal more slowly from any type of damage. Removing the source of the stress improves plant appearance in landscapes. It also helps them to resist pest attacks.

Viruses are very small obligate parasites that are so simply constructed that they do not consist of cells, but of particles. The particles of a simple virus consist of a protein coat surrounding a genome of either ribonucleic acid (RNA) or deoxyribonucleic acid (DNA). They have been recognized as agents of disease in plants and other living organisms since the late 1880s. In 1886, Adolf Mayer, a German agricultural chemist working at Wageningenin the Netherlands, found that a mosaic disease of tobacco could be transmittedto healthy plants by rubbing them with sap extracted from plants showing disease symptoms. In 1892, the Russian botanist, Dmitrii Ivanowski. showed that sap from diseased plants retained its infectivity after passing through a filter that eliminated bacteria. Virus genomes are composed of either RNA or DNA. Most plant viruses contain single stranded RNA (ssRNA). Some however, contain double stranded DNA (dsDNA) with the same helical structure as in host plant cells. Others contain single stranded DNA (ssDNA) or double stranded RNA (dsRNA) which occur either in linear strands or in rings (circular nucleic acids).

Mycoplasmas are smallest and wall-less prokaryotes capable of self-replication, taxonomically, belong to class Mollicutes (meaning soft skin). The term 'Mollicutes' or 'Mycoplasma' is now used interchangeably to denote any species included in the four orders, five families, eight genera and about 200 species of mollicutes. However, in this article the term ' mycoplasma' and 'acholeplasma' are used to describe members of the genera Mycoplasma and *Acholeplasma*, respectively. Many mollicutes cause significant diseases in man, animals and plants. The identity of the evolutionary ancestors of mycoplasmas is unclear. The smallest genome of mycoplasmas is little more than twice the genome size of certain large viruses. Mycoplasmas are the smallest organisms that can be free-living in nature and self-replicating on laboratory media.

Importance of the Plant Diseases

Globally, enormous losses of the crops are caused by the plant diseases. The loss can occur from the time of seed sowing in the field to harvesting and storage. Important historical evidences of plant disease epidemics are Irish Famine due to late blight of potato (Ireland, 1845), Bengal famine due to brown spot of rice (India, 1942) and Coffee rust (Sri Lanka, 1967). Such epidemics had left their effect on the economy of the affected countries.

Concept of Plant Disease

The normal physiological functions of plants are disturbed when they are affected by pathogenic living organisms or by some environmental factors. Initially plants react to the disease causing agents, particularly in the site of infection. Later, the reaction becomes more widespread and histological changes take place. Such changes are expressed as different types of symptoms of the disease which can be visualized macroscopically. As a result of the disease, plant growth in reduced, deformed or even the plant dies.

Classification of Plant Disease

To facilitate the study of plant diseases they are needed to be grouped in some orderly fashion. Plant diseases can be grouped in various ways based on the symptoms or signs (rust, smut, blight etc.), nature of infection (systemic or localized), habitat of the pathogens, mode of perpetuation and spread (soil-, seed- and air-borne etc.), affected parts of the host (aerial, root disease etc.), types of the plants (cereals, pulses, oilseed, ornamental, vegetable, forest diseases etc.). But the most useful classification has been made based on the type of pathogens that cause plant diseases. Since this type of classification indicates not only the cause of the disease, but also the knowledge and information that suggest the probable development and spread of disease along with their possible control measures. The classification is as follows:

1. Infectious Plant Diseases
   a. Disease caused by parasitic organisms: The organisms included in animate or biotic causes can incite diseases in plants.
   b. Diseases caused by viruses and viroids.

2. Non-Infectious or Non-Parasitic or Physiological diseases
The factors included in inanimate or abiotic causes can incite such diseases in plants under a set of suitable environmental conditions.

Plant Diseases

In the case of disease, the source of continual irritation may be abiotic (non-living) or biotic (caused by a pathogen). Abiotic diseases are also referred to as non infectious diseases as they do not spread from plant to plant. In lay terms, they are "not contagious." Abiotic diseases are very common and should be considered the likely suspect when attempting to diagnose the cause of decreased plant vigor or death. Biotic diseases are caused by pathogens and are often referred to as infectious diseases because they can move within and spread between plants. Plant pathogens are very similar to those that cause disease in humans and animals. Pathogens may infect all types of plant tissues to include leaves, shoots, stems, crowns, roots, tubers, fruit, seeds, and vascular tissue and can cause a wide variety of disease types ranging from root rots and rusts to cankers, blights, and wilts.

The Disease Cycle of Infected Plants

The development of visual disease symptoms on a plant requires that the pathogen must (a) come into contact with a susceptible host (referred to as inoculation);

(b) gain entrance or penetrate the host through either a wound, a natural opening (stomates, lenticels, hydathodes) or via direct penetration of the host; (c) establish itself within the host; (d) grow and reproduce within or on the host; and ultimately, (e) be able to spread to other susceptible plants (referred to as dissemination). Successful pathogens must also be able to survive prolonged periods of unfavorable environmental conditions in the absence of a susceptible plant host. Collectively, these steps are referred to as the disease cycle. If this cycle is disrupted, either naturally or via the concerted efforts of a grower, the disease will be less intense or fail to develop. In general, there are five methods used to manage plant diseases. These include use of genetically resistant plants, cultural practices, chemical application, beneficial micro-organisms to suppress or counter the activity of the pathogen (known as biological control), and the use of quarantines and other regulatory practices.

Viruses

Matthew (1981) defined a virus as "a set of one or more nucleic acid template molecules, normally encased in a protective coat, or coats of protein or lipoprotein, which is able to organize its own replication within suitable host cells. Within such cells, virus production is (a) dependent on the host's protein synthesizing machinery, (b) organized from pools of the required materials rather than by binary fission, and (c) located at sites which are non separated from the host cell contents by a lipoprotein, bilayer membrane". About 10 per cent of all viruses are transmitted through seed but many can be spread vegetatively in tubers, bulbs, cuttings and other plant parts used for propagation.

Viruses and Viriods

Morphologically, virus particles are (i) isometric (spherical, polyhedral) and (ii) anisometric (rigid or flexuous rods, bacilliform or bullet-shaped). Many isometric viruses have symmetric polyhedra which are either of three cubic symmetry *i.e.* tetrahedral, octahedral or icosahedral. The rod shaped particles of tobacco mosaic virus (TMV) consist of protein sub-units (capsomeres) built up in a regular, helical array, with the RNA chain compactly coiled in a corresponding helix on the inside of the protein sub-units.

Chemical Composition

Plant virus particles consist of infectious nucleic acid (the genome), which is encapsidated within a protective protein coat or shell. The genome, essential for virus replication, is composed of ribonucleic acid (RNA in most groups of viruses) and deoxyribonucleic acid (DNA in the caulimovirus and geminivirus groups). The RNA and DNA may be single stranded (ss) or double stranded (ds) Besides these two basic components, an envelop of lipid or lipoprotein membrane is present in some plant viruses. In some RNA viruses, the genetic information is divided into two or more parts. They are called multi-component viruses and the individual components are not infectious alone. Hence two or more genomic elements are needed to cause infection and replication.

Multiplication and Infection Nature of Plant Viruses

The events in virus infection involve three steps: adsorption, penetration or entry and uncoating or disassembly. The initial contact between virus particle and host cell is referred to as adsorption or entry. The process during which the virion or its nucleic acid passes into the cytoplasm of the cell is known as penetration or entry. Uncoating is the removal of various components of the mature virion and subsequent release of viral genome and other constituents that plays a major role in establishing infection.

Disease Caused by Viruses

Common Tobacco (Tomato) Mosaic

Light and dark green mottled areas, which are usually somewhat raised and puckered, develop in the leaves. The young leaves at the tips of the growing shoots tend to bunch and unfold unevenly. Plants infected when young are commonly stunted and have a yellowish cast. Leaflets on young plants growing under glass or plastic are often long, pointed, and narrow. Symptoms appear about 10 days after the plants become infected.

Sources and Transmission

TMV is extremely infectious, being spread easily by persons touching a healthy tomato plant after touching an infected plant. Plant-to-plant contact during pulling or pricking –out of young seedlings, transplanting, cultivating, weeding, and wind-whipping of the leaves are other very common ways of spreading the virus. Hand operations, such as pruning, tying, pollinating, spraying, watering, and picking fruit are also important in transmission, especially in greenhouse-grown tomatoes. TMV has remained infective in dried leaves and stems for over 50 years, longer than any known virus. It may persist at least 12 months in moist heavy soils, probably in the debris from the preceding crop, and 24 months or longer in dry soils. The use of tobacco products by persons handling plants may result in TMV infection, since this virus is often found in tobacco tissue.

Aucuba or Yellow Mosaic

Leaves develop a striking, bright yellow mottling. The foliage may be curled downward, distorted or crinkled, and dwarfed. Fruit may be mottled, light and dark green when immature, and yellow and red when ripe, and are generally not acceptable for marketing. A new method of classifying strains of TMV is based on the ability of the virus to overcome specific tomato genes for TMV resistance.

Sources and Transmission

This strain of TMV is introduced and spread among tomato plants in much the same way as the common tobacco mosaic virus.

Cucumber Mosaic Virus (CMV)

Cucumber mosaic is the most destructive and widespread disease of cucumber and muskmelons worldwide. All vine crops as well as a wide range of annual, biennial, and perennial crop and weed plants in about 40 families are attacked.

Plants infected when young are yellowed, bushy, and stunted. The leaves may be mildly mottled, suggestive of common mosaic. The most pronounced symptom is the extremely distorted, shoestring-like leaves. Sometimes the leaves are so distorted that little remains but the midrib. In tomatoes, cucumber mosaic is much less prevalent than tobacco (tomato) mosaic.

Sources and Transmission

This virus, unlike TMV, is not easily transmitted to tomato by persons brushing against or handling plants. Also, it does not persist in the soil or crop refuse and does not remain active for any great length of time on the hands or clothing. Most of the infection in tomatoes occurs through transmission by insects, particularly aphids. CMV symptoms in cucumber are more severe on plants exposed to short days or reduced light than on plants exposed to long days and bright light. Cucurbit plants rarely become infected in the seedling stage. When this happens, the cotyledons may turn yellow and wilt. If CMV-affected plants also have a root rot, caused by soil borne fungi including species of *Pythium* and *Fusarium*, they wilt, collapse, and die within seven to ten days of showing the first symptoms. This suggests that these pathogens have a synergistic effect.

Squash Mosaic Virus (SQMV)

Squash mosaic virus is much less common than cucumber mosaic virus. SqMV affects most cucurbits; however, most isolates or strains do not affect watermelon. Other plants infected by SqMV include garden and sweet peas, coriander, and salad chervil. SqMV may cause considerable loss in late-season squash and muskmelon crops.

Sources and Transmission

A slight vein yellowing followed by mottling, yellow spotting, and a green banding along the veins is common on infected muskmelon plants. Other muskmelon leaves develop a yellow streaking, spotting, or general yellowing along the veins. A few leaves become slightly distorted with the veins extending beyond the margins of the leaves. SqMV survives between vine crop plantings in infected cucurbit weed hosts, in infected seed (up to 4.5 percent depending on the isolate and crop), and in over wintering beetles. Long distance spread, of course, is possible with seed. Insect vectors of SqMV include the 12-spotted, western striped and banded cucumber beetles as well as the beetles *Acalyma thiemei thiemei* and *Epilechna chryssomelina*.

Tobacco Ringspot Virus (TRSV)

It affects many cultivated and weed plants. A cucumber strain differs from the tobacco strain and induces more severe symptoms on cucumber than does the common tobacco strain.

Sources and Transmission

Watermelon plants are dwarfed and yellowish. The tips of infected vines are often upright instead of the normal prostrate position on healthy plants. The leaves are coarsely mottled and speckled with irregular black spots that somewhat resemble anthracnose lesions. Severely affected leaves become tattered and brittle. Plants tend to recover slightly as they mature with the symptoms being more persistent on watermelon than on muskmelon. Tiny yellow spots appear on cucumber leaves. New leaves that form are mottled and appear like those infected with cucumber mosaic virus. TRSV survives between cucurbit plantings in numerous crop, weed, and wild host plants, in infected seed, and possibly in a common dagger nematode (*Xiphinema americanum*) vector. Seed transmission is rare in cantaloupe, cucumber and muskmelon. It has been detected in up to 2.5 percent of butternut squash seed from infected plants, but has not been detected in other cucurbit seed. Seed transmission, however, occurs in dandelion, globe-amaranth, lettuce, petunia, soybean, and tobacco.

Alfalfa Mosaic (Alfalfa Mosaic Virus)

Alfalfa mosaic is a destructive disease of tomato because of the severe damage it causes to fruit. However, it is currently of minor importance because it occurs at low levels, mostly in tomatoes situated near old alfalfa fields. Symptoms of alfalfa mosaic are a bright yellow mosaic of newly expanded leaves and an extensive browning and splitting of fruit.

Sources and Transmission

The virus has a wide host range, but infection in tomato is thought to arise from old alfalfa fields that harbor the virus. At least 14 species of aphids transmit the virus from infected to healthy plants in sap that clings to their mouthparts. The virus can also be spread mechanically.

Curly Top (Beet Curly Top Virus)

Curly top or western yellows disease is caused by one or more strains of the beet curly top virus.

Sources and Transmission

All leaves eventually become curled and plants appear pale green and stunted as new growth is stopped. Leaves of affected plants become thickened and often have purple colored veins. Leaf stems and branches become brittle and are easily snapped. Plants affected early in the season are usually killed. On plants that develop symptoms after fruit set, fruit prematurely ripens and becomes dull red and wrinkled. Curly top or western yellows disease is caused by one Beet curly top virus is spread long distances and transmitted to plants by the beet leafhopper. The virus is rapidly acquired by leafhoppers while they sample or feed on infected plants. The virus has a wide host range that includes over 300 species of broadleaf plants. The level of curly top in tomatoes is thought to correspond with leafhopper migration patterns and the proportion of leafhoppers within a population that are carrying the virus. Sources of the leafhoppers and the virus which affect tomatoes in Oklahoma are not currently known. In arid and semiarid areas of the western U.S. where curly top is a chronic problem, leafhoppers over winter in dessert areas on winter vegetation.

Control measures

1. Eradicate all biennial and perennial weeds and wild reservoir hosts in and around greenhouses, seedbeds, gardens, and fields. It is especially

important to eradicate bur- and wild-cucumbers, catnip, chickweeds, clovers, curly dock, dandelions, fleabane, flowering spurge, groundcherries, horsenettle, Jimsonweed, milkweed, motherwort, nightshades, pokeweeds, and white cockle.

2. Apply insecticides regularly in and around the greenhouse, garden, or field to eliminate aphids, cucumber beetles, and other insects.

3. Avoid touching healthy plants after handling mosaic-affected plants. If this is necessary, first wash hands thoroughly with hot running water and strong soap.

4. Maintain good cultural conditions and ventilation. Eliminate all plant debris and unsterilized soil that could act as a source of contamination.

5. Proper use of fertilizer can reduce disease. Some diseases are suppressed by reduced amounts of nitrogen. Others are likely to be suppressed by increased amounts of nitrogen.

6. The selection of disease-free stock is another good control measure. This includes not only healthy ornamental plants but also healthy transplants for such crops as tobacco, tomatoes, peppers, strawberries, and fruit trees.

Mycoplasma

Mycoplasma are a cross between bacteria and viruses. They hide inside cells, often in the most secret places of our bodies. Members of the genus Mycoplasma are the smallest organisms lacking cell walls that are capable of self-replication taxonomically, belong to class Mollicutes (meaning soft skin) and cause various diseases in humans, animals, and plants. Diseases in man, animals and plants. Aster yellows, corn stunt, and other plant diseases appear to be caused by mycoplasmas. They are transmitted by insects and can be suppressed by tetracyclines. The importance of mycoplasmas in human and animal infections has accentuated with recent reporting of instances of human infections with animal mycoplasmas and *vice-versa* leading to recognition of zoonotic mycoplasmoses. The slow reporting of zoonotic mycoplasmoses can be attributed mainly to difficulty in measuring zoonotic potential of mycoplasmoses as in some other diseases *viz.* multiple sclerosis, rheumatoid arthritis, leukemia, atherosclerosis, chronic fatigue syndrome, CJD, Crohn's colitis *etc.*

Phytoplasmas (previously referred to as mycoplasma-like-organisms, MLO) are cellular parasites recognized as causing diseases in many crops. Although phytoplasmas are graft transmissible and cause symptoms similar to those typically associated with infection by a virus, they are actually bacteria-like micro-organisms; similar to bacteria, Phytoplasmas typically cause abnormalities in growth (rosetting, willowy growth, yellowing) that may be confused with symptoms of infection by a virus, and tests to confirm the presence of a phytoplasma are therefore necessary. Unlike typical bacteria, phytoplasmas cannot be cultured on artificial media in the laboratory. In plants, 'yellow' diseases were thought to be caused by some viral pathogens. Nullifying the role of viruses from Japan reported the involvement of mycoplasma like organisms (MLO) in causing 'yellow' diseases in plants. Little leaf of brinjal, grassy shoot of sugarcane, sandal spike are some of the important plant disease caused by MLOs (=Phytoplasma *i.e.* plant mycoplasma).

Before 1967 most MLO were grouped together with viruses and known as yellows type and witches broom viruses. Although the original report (Doi *et al.,* 1967). Left the identification open specified that the MLO might belong to either of several groups such as the Mycoplasmataceae the Bedsonia or the Chlamydia, subsequent descriptions from other laboratories sometimes used the term "Plant mycoplasma" or "Mycoplasma like agents" indiscriminately for all MLO. Temporary remission of MLO-caused plant diseases by application of antibiotics, first reported (Ishiie *et al.,* 1967) has been confirmed in many laboratories. Morphologically MLO destruction by antibiotics can be distinguished from natural degeneration (Hirumi and Maramorosch, 1972). The MLO observed by electron microscopy in thin sections of diseases tissues of higher plants and of insect vectors comprise many diverse forms (Maramorosch *et al.,* 1968, 1970; Davis and Whitcomb, 1971). In addition to MLO observed in phloem tissues of diseased plants, xylem invading micro-organisms have also been discovered (Plavsic-Banjac and Maramorosch, 1972; Maramorosch *et al.,* 1973; Goheen *et al.,* 1973; Hopkins and Mollen hauer, 1973; Hopkins *et al.,* 1973).

Reproduction of Mycoplasmas

MLOs can reproduce by binary fission, by formation of spores (elementary bodies), by filamentous growth and budding. The chromosome replication starts at a fixed site followed by bidirectional progression. Further, the outline of chromosome replication of mycoplasmas is somewhat similar to that of *E. coli.* However, the process of mycoplasmas cell reproduction has not yet been well classified.

Mycoplasma pneumoniae (Type Strain: ATCC 15531, NCTC 10119)

Mycoplasma pneumoniae was first identified and described in the early 1960s. Primarysite of colonization of *M. pneumoniae* is oropharynx and is pathogenic to humans. Thepneumonia is designated 'primary atypical' to distinguish it from 'typical' pneumonia due to *Steptococcus pneumoniae,* the pneumococcus. A detailed definition of epidemiology of *M. pneumoniae* disease has emerged (Lind and Bentzon, 1988).

Transmission

Transmission by droplets require a rather high inoculum unless the host is immunocompromised.

Mycoplasma salivarium (Type Strain: ATCC 23064, NCTC 10113)

Mycoplasma salivarium exists as commensal organisms in the oropharynx of human and nonhuman primates. Pathogenicity of *M. salivarium* is not known. However, isolation of the organism with a significantly higher incidence from the gingival sulci of individuals with periodontal disease (87 per cent) than in persons with healthy periodentium (32 per cent) has stimulated interest in its possible role in periodontal pathology (Engel and Kenny, 1970; Forest, 1979).

Mycoplasma canis (Type Strain: ATCC 19525, NCTC 10146)

Mycoplasma canis is found as a common parasitic inhabitant of the mucous membrane of upper respiratory tract, conjunctivae and genitals of dogs. It has also been isolated occasionally from nonhuman primates. Although one member of the family was taking immunosuppressant, the pathogenic role of *M. canis* in human disease was neither established nor the dynamics of zoonotic transfer of *M. canis* studied. Several strains of *M. canis* were isolated from the oropharynx of members of a family at the time when their dog had an acute respiratory disease infection. (Armstrong *et al.,*1971).

Mycoplasma felis (Type Strain: ATCC 23391, NCTC 10160)

Mycoplasma felis is a common parasitic inhabitant of the upper respiratory and lower genital tract of asymptomatic and diseased cats. Strains related to *M. felis* have also been isolated frequently from tonsils and other regions of respiratory tract of healthy and diseased horses. Bonilla *et al.* (1997) described first documented case of *M. felis* infection in a women who had common variable immune deficiency and who presented with septic arthritis of the left hip and right knee.

Disease Caused by Mycoplasma

Little Leaf of Brinjal or Eggplant

Little leaf of brinjal (*Solanum melongena*) is found throughout India and other neighbouring countries. This graft transmissible disease was first reported from Coimbatore by Thomas and Krishnaswami, 1939. Almost all the brinjal varieties are susceptible to this disease. This disease has become a serious threat to the cultivation of this vegetable in most of the states.

The main symptoms of the disease are production of very short small leaves by the affected plant. The petioles are so much reduced in size that the leaves are narrow, soft, smooth and yellowish in colour, newly formed leaves are further reduced in size.

Little Peach

Little peach symptoms may resemble some aspects of the symptoms of yellows and rosette and, like rosette and yellows, only one or two branches of whole trees may be affected. Apical dominance is lost and all buds on an infected limb develop thin, willowy shoots with small leaves. The shoots are longer than on rosette-affected trees but shorter than on yellows-affected trees.

Red Suture

Red suture may appear as a dark red suture on fruit that ripens prematurely, with the flesh in the suture soft and watery. In other instances, a prominent red suture that softens is observed on an otherwise green fruit. The tree, except for fruit, is symptom less at this point. After several years, an infected tree or branch may have symptoms similar to little peach-affected trees. An infected tree may survive for a number of years, but the fruit is inedible.

Witches' Broom of Withania (Ashwagandha)

A witches broom diseased of *W. somanifera* was reported from Lucknow, India. The disease first appeared in the experimental plantations. The disease has spread to commercial field causing considerable damage. Typical symptoms of the disease are small leaves, shortening of internodes, excessive branching resulting in witches broom appearance, premature drying and death of infected twigs and leaves. Infected plants remain stunted.

Greening Disease of Citrus

This disease has been reported to be widely prevalent in India. The disease causes various kinds of leaf symptoms and defoliation. It is very widespread and has been found in various *Citrus* species. The disease is not mechanically transmissible but it is transmitted by bud or wedge grafting as also by dodder (*Cuscuta reflexa* Roxb.). The chief symptoms of the disease on most *Citrus* spp. Are yellowing of the midrib and lateral veins of the old mature leaves.

Control Measures

1. Eradicate known diseased trees as soon as they occur. Removal of trees eliminates sources of infection within the orchard.

2. The eradication of known alternate hosts has been effective when the alternate hosts are limited in species and number.

3. Immediate slaughter of infected breeding flocks to prevent transmission of mycoplasma to the progeny.

4. Diseases caused by living agents are controlled by different methods. For some diseases, resistant varieties are the best means of control. The use of resistant varieties is not new.

5. Avoid susceptible host plants, Choose disease-resistant cultivars or plants not affected by diseases. This should be considered for all new plants and replacement plants in landscapes.

6. Make the environment less favorable to the pathogen. Grow seedlings in warm conditions to slow the spread of damping-off diseases. Pruning plants promotes good air circulation among branches and allows leaves to dry more quickly.

7. Certain other control measures can be taken to protect against the introduction and spread of virus and mycoplasma diseases of trees. Only virus-free plants should be bought and planted. This in most cases can be achieved only by a concerted effort on the part of nurserymen to propagate trees from healthy mother stock trees which are indexed periodically to ascertain their freedom from virus or mycoplasma.

Acknowledgement

First Author is thankful to UGC New Delhi for awarding RFSMS fellowship to pursue Ph.D. and for the financial support.

References

Armstrong, D. Yu, B.H., Yagoda, A. and Kagnoff, M.F. (1971). Colonization of humans by *Mycoplasma canis. J. Infect. Dis.* 124: 607–609.i). pp. 164–174.

Bonilla, H.F., Chenoweth, C.E., Tully, J.G., Blythe, L.K., Robertson, J.A. and Kauffman, C.A.(1997). *Mycoplasma felis* septic arthritis in a patient with hypogamma-globulinemia. *Clin. Infect. Dis.* 24: 222–225.

Engel, L. D. and Kenny, G. E. (1970). *Mycoplasma salivarium* in human gingival sulci. *J. Periodon. Res.* 5: 1–9.

Forest, N. (1979). Caracterisation de *Mycoplasma salivarium* dans les parodontopathies. *J. Biol. Buccale.* 7: 321–330.

Doi, y., Teranaka, M., Yora, K., and Asuyama,H (1967). Mycoplasma or PLT group–like micro–organisms found in the phloem elements of plants infected with mulberry dwarf, potato witches broom, aster yellows or pautownia witches broom (In Japanesse, Engl abstr.) *Ann. Phytopath.Soc.*Jap., 33, 259–266.

Davis, R.E and Whitcomb, R.F(1971). Mycoplasma, rickettsiae and chlamydiae; possible relation to yellows diseases and other disorders of plants and insects. *A.Rev.Phytopath.*, 9, 119–154.

Goheen, A.C., Nyland, G and Lowe, S,K (1973). Association of a rickettsia–like organisms with pierce's disease of grapevines and alfalfa dwarf and heat therapy of the disease in grapevines. *Phytopathology*, 63, 341–345.

Hirumi, H., and Maramorosch, K(1972). Natural degeneration of Mycoplasma–like bodies in an aster yellows infected host plant. *Phytopathol.*Z., 75,9–26.

Hopkins, D.L., French, W.J and Mollenhauer, H.H (1973). Association of a rickettsia–like bacterium with Phonypeach disease. *Phytopathology*, 63, 443, (Abstr.).

Ishii, T., Doi, Y., Yora, K., and Asuyama, H (1967). Suppressive effects of antibiotics of tetracycline group of symptom development of mulberry dwarf disease (in Japanese, Engl, abstr). *Ann. Phytopath. Soc.Jap.*, 33,267–275.

Lakshman, H.C(2009). The role of AM fungi on the agricultural crops. In: ICAR proceedings.14–16 April. National Aerospace Labrotary. Bangalore.Eds. Venkataswamy, R.Rajendran, H.Sridhara and K.Venkateshwarlu. P.268–278.

Lind, K. and Bentzon, M. W. (1988). Changes in the epidemiological pattern of *Mycoplasma pneumoniae* infections in Denmark: a 30 years survey. *Epidemiol. Infect.* 101: 377–386.

Maramorosch, K., Shikata, E and Grandos, R.R (1968). Structures resembling Mycoplasma in diseased plants and in insects vectors. Trans. N.Y. Acad. Sci, 30, 841–855.

Maramorosch, K., Grandos, R.R and Hirumi, H (1970). Mycoplasma diseases plants and insects. *Adv. Virus Res.*, 16, 135–193.

Maramorosch, K., Palvsic–Banjac., Biljana, Bird, J and Liu, L. J (1973). Electron microscopy of ratoon–stunted sugar cane micro-organisms on xylem. *Phytopathol.* Z., 77, 270–273.

Palvsic–Banjac, B and Maramorosch, K (1972). Electron microscopy of ratoon–stunted sugar cane micro–organisms on xylem. *Phytopathol. Z.*, 62, 498–499.

2015, Recent Trends in Microbiology, Mycology and Plant Pathology   *Pages* 325–338
Editor: **Dr. H.C. Lakshman**
Published by: **DAYA PUBLISHING HOUSE, NEW DELHI**

Chapter 21

Induction of Defense Responses in Tomato against Tolcv by Consortium of Plant Growth Promoting Rhizobacteria Formulated with Chitosan

Shefali Mishra[1], K.S. Jagadeesh[1], P.U. Krishnaraj[2],
G. Jyothi[1], A.S. Byadagi[3] and A.S. Vastrad[4]

[1]*Department of Agricultural Microbiology*
[2]*Department of Biotechnology*
[3]*Department of Plant Pathology*
[4]*Department of Entomology*
College of Agriculture, University of Agricultural Sciences,
Dharwad – 580 005, Karnataka India

Introduction

Tomato (*Lycopersicon esculentum* Mill.) is a herbaceous fruiting plant. It was originated in Latin America and became one of the most widely grown vegetables with ability to survive in diverse environmental conditions (Rice *et al.,* 1987). It has high cash value with potential for value-added processing. Recently, there has been more emphasis on tomato production not only as a source of vitamins, but also as a source of income and food security. Apart from this, lycopene present in this fruit is valued for its anti-cancer property, since it acts as an antioxidant and scavenger of free radicals.

India is the third largest tomato producer in the world after China and USA, accounting for about 8 per cent of the World tomato production. In terms of area, it occupies the third largest place after potato and onion at the national level (Anon., 2010). During 2008–09, the area of tomato, in India, was about 5,99,000 ha with a production of 11,149,000 MT (Anon., 2010). In Karnataka, it occupies an area of 46,000 hectares with a production of 17.65 lakh tonnes (Anon., 2001). Bangalore and Kolar districts produce nearly 35 per cent of total tomato production in Karnataka (Anon., 1998). Though, the area under tomato cultivation is high, the productivity (15 t/ha) is rather low. This is attributed to the potential loss in yield due to a number of diseases.

Besides fungal, bacterial and phytoplasmal infections, it is also affected by a large number of viral diseases (Anon., 1983). Of all the viral diseases reported on tomato, tomato leaf curl virus (ToLCV) is the most important and destructive viral pathogen in many parts of India (Sastry and Singh, 1973; Saikia and Muniyappa, 1989; Harrison *et al.*, 1991). The incidence of ToLCV in tomato growing areas of Karnataka ranged from 17-100 per cent in different seasons and 50 to 70 per cent yield loss was observed in tomato Cv. Pusa Ruby grown in February – May (Saikia and Muniyappa, 1989). Devaraja *et al.* (2005) reported cent per cent infection in summer month which caused yield losses ranging from 27 to 90 per cent in Karnataka.

Tomato leaf curl virus disease is characterized by the curling and twisting of leaves followed by marked reduction in leaf size (Figure 21.1). The diseased plants look pale and stunted due to shortening of internodal length with more lateral branches resulting in a bushy appearance (Vasudeva and Samraj, 1948). Whitefly *Bemisia tabaci* (Gennadius) (Homoptera: Aleyrodidae) is the vector of the virus (Butter and Rataul, 1973; Muniyappa and Veeresh, 1984). Tomato plants of all ages are vulnerable to ToLCV and generally exhibit symptoms after two to three weeks of infection. Development of strategies for integrated disease management would significantly reduce yield loss. Use of genetically resistant varieties, integration of selected cultural practices, application of insecticides to control insects that might serve as vectors and their combinations (Hull, 1994) have been attempted. The use of genetically resistant varieties is clearly the most economically and environmentally sound choice. However, commercially acceptable varieties which resist virus infection are not always available. The application of insecticides, however, also has associated environmental concerns. When the doses exceed, they become phytotoxic and pollute the environment. Further, they are also toxic to beneficial micro-organisms in the soil (Newell *et al.*, 1981).

Under the above circumstances, it becomes inevitable to develop a bio-based, ecofriendly, biodegradable, plant derived or microbial derived method in order to control plant pathogens. Hence, in this context, management of tomato leaf curl virus through biocontrol agents is an alternative strategy, which is also ecologically sound and environmentally safe. Bio-control agents, mainly bacterial inoculants are believed to induce systemic defense responses in the plants besides other mechanisms including direct antagonism, antibiosis and siderophore production. Recent investigation on the mechanism of biological control by plant growth promoting fluorescent *Pseudomonas* revealed that these strains protect plants from various

Stunted plant growth with prominent yellow margins

Upward curling of leaves

Withering of flowers

Reduced Leaflets size with erect shoots

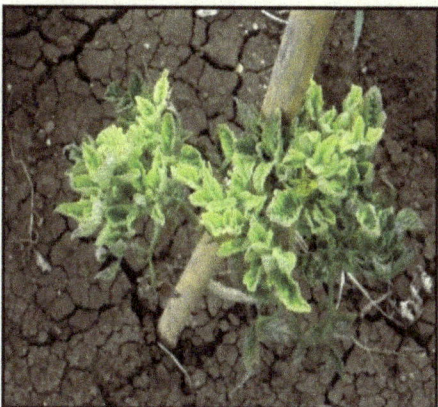

Reduced Leaflets size with puckered leaves

Distorted leaves with yellow margins

Figure 21.1: Typical Symptoms of ToLCV.

pathogens in several crops, by activating defense genes encoding chitinase, beta-1,3 glucanase, peroxidase, phenylalanine ammonia lyase and other enzymes (Maurhofer *et al.,* 1994). Induced systemic resistance once expressed, activates multiple potential defense mechanisms that enhance the increased activity of chitinase and peroxidase, which showed resistance to various pathogens (Xue *et al.,* 1998).

The elicitiors stimulate the contacts of plant-phyto-pathogens and thereby trigger defensive mechanisms that constrain the invasion of pathogenic fungi, bacteria and viruses. Chitosan is one of the most studied elicitiors. It regulates the expression of resistance genes and induces jasmonate synthesis (Doares *et al.,* 1995). Fragments from chitin and chitosan are known to have eliciting activities leading to a variety of defense responses in host plants in response to microbial infection, inducing the accumulation of phytoalexins, pathogen-related (PR) proteins and proteinase inhibitors, lignin synthesis and callose formation. These molecules were shown to display toxicity and inhibit fungal growth and development. They were reported to be active against viruses, bacteria and other pests (Abdelbasset *et al.,* 2010). Based on these proprieties, interest has been growing in using them in agricultural systems to reduce the negative impact of diseases on yield and quality of crops.

Materials and Methods

Field Study

A field experiment was conducted to assess the effect of selected PGPR strains and chitosan on the disease severity control as well as on growth and yield of tomato (Figure 21.2). It was carried out at Main Agricultural Research Station, U A S, Dharwad, during Kharif season (June-Dec. 2011). Five weeks old seedlings of variety Pusa Ruby, raised in a glasshouse were transplanted in the main field with plot size 25m x 12 m, following a spacing of 75cm x 60 cm. In the chemical control treatment, confidor @2 ml/L was sprayed at weekly intervals to control the vector, as per the package of practices for tomato crop.

Rhizobacterial Treatment

Fluorescent bacteria were grown in King's B broth and non –fluorescent bacteria in nutrient broth medium on a shaker (150 rpm) for two days and centrifuged at 10,000 rpm for 5 min and the pellet mixed with sterile carboxy methyl cellulose(CMC) ('Hi Media') suspension(1 per cent).Tomato seeds of the variety, Pusa Ruby (susceptible to ToLCV) were used in the experiment. They were surface sterilized with sodium hypochlorite solution, placed in CMC cell suspension, air dried inside a laminar flow chamber (Jagadeesh, 2000) and the biocoated seeds sown in pots. For soil application, the lignite based culture (1:3) was applied to soil @ 5kg/ha before sowing seeds and mixed well. For foliar application, the lignite based culture was filtered through a muslin cloth and sprayed @ 1 per cent (w/v) at 10 days after sowing (DAS) and 20 DAS. Control plants in pots without application of rhizobacteria were also maintained. All treatments were replicated five times and arranged in a randomized complete block design (RCBD).

Figure 21.2: General View of the Field Experiment Conducted during Kharif' 2011.

Chitosan Treatment

Bacteria were grown in nutrient broth medium on a shaker(150 rpm) for 2 days and centrifuged at 10,000 rpm for 5 min. Chitosan was dissolved in 100 mM acetate buffer (pH 4.5) and the pH adjusted to 6.5 using 1 N NaOH. The cell pellet was mixed with chitosan solution(5 per cent). Tomato seeds surface sterilized with sodium hypochlorite solution were soaked in chitosan cell suspension and kept at shaking condition for 3 hrs at 28°C. The seeds were shaken in chitosan solution until they became fully coated. The biocoated seeds were dried inside a laminar flow chamber. At 25 DAS,both upper and lower surfaces of the leaves were sprayed with the Chitosan solution (1 mg/ml) prepared in 100 mM acetate buffer (pH 4.5) and adjusted with 1 N NaOH to pH 6.5.

ToLCV Inoculation

Whiteflies were collected from cotton and tobacco plants from fields by sucking with the help of an aspirator by slowly turning the leaves slightly upwards. Whiteflies were released on to the ToLCV diseased tomato plants grown in insect proof rearing cages and continuously maintained by introducing younger plants in to the rearing cage,thus, making insects viruliferous. The viruliferous insects were sucked from the diseased plants with the help of an aspirator and released on to healthy, rhizobacteria treated tomato seedlings on the top leaves. Immediately, the seedlings were placed in an insect proof rearing cage and allowed insects for a week to feed on them and bring about infection by the virus. Twenty five days old seedlings were used for release of the viruliferous insects.Thus, it was ensured that all seedlings were infected with ToLCV (Figure 21.3).

Sample Collection, Enzyme and Phenol Estimation

Leaf samples were collected at 45 DAS and 75 DAS from both inoculated and uninoculated tomato plants. They were frozen immediately in liquid nitrogen, ground to a powder and stored at -80°C until determination of phenylalanine ammonia lyase (PALase), chitinase, polyphenol oxidase and peroxidase activities.

The Peroxidase activity was assayed spectrophotometrically following the method described by Mahadevan and Sridhar (1986). The PALase activity was determined using the method described by Ross and Sederoff (1992). The polyphenol oxidase activity in leaves was estimated at 45 and 75 DAS following the method of Mayer *et al.* (1965). The chitinase activity was estimated at 45 and 75 DAS, following the method described by Miller (1959). The total phenol content in leaves was estimated at 45 and 75 DAS by following Folin Cio-calteau method (Sadasivam and Manickam, 1991).

Statistical Analysis

The data obtained from field experiments were subjected to Randomized Complete Block Design analysis as described by Gomez and Gomez (1984). The level of significance used in the 'F' test was P=0.05. The critical difference values were calculated whenever the F test values were significant.

Results and Discussion

All the three promising rhizobacterial strains were tested for their ability to induce peroxidase, polyphenol oxidase, PALase, chitinase and production of phenol in tomato plants. It was interesting to note that all the strains triggered the defence related enzymes in plants at varied levels. The rhizobacteria are well known inducers of defense mechanisms and their application has often resulted in increased rates of plant growth and reduced disease incidence in many crops (Leeman *et al.*, 1995; Liu *et al.*, 1995; Nandakumar *et al.*, 2001; Viswanathan and Samiyappan, 2002).

One of the major biological properties of phenolic compounds is their antimicrobial activity (Saini *et al.*, 1988) and it is, often, assumed that their main role in plants is to act as protective compounds against disease causing agents such as

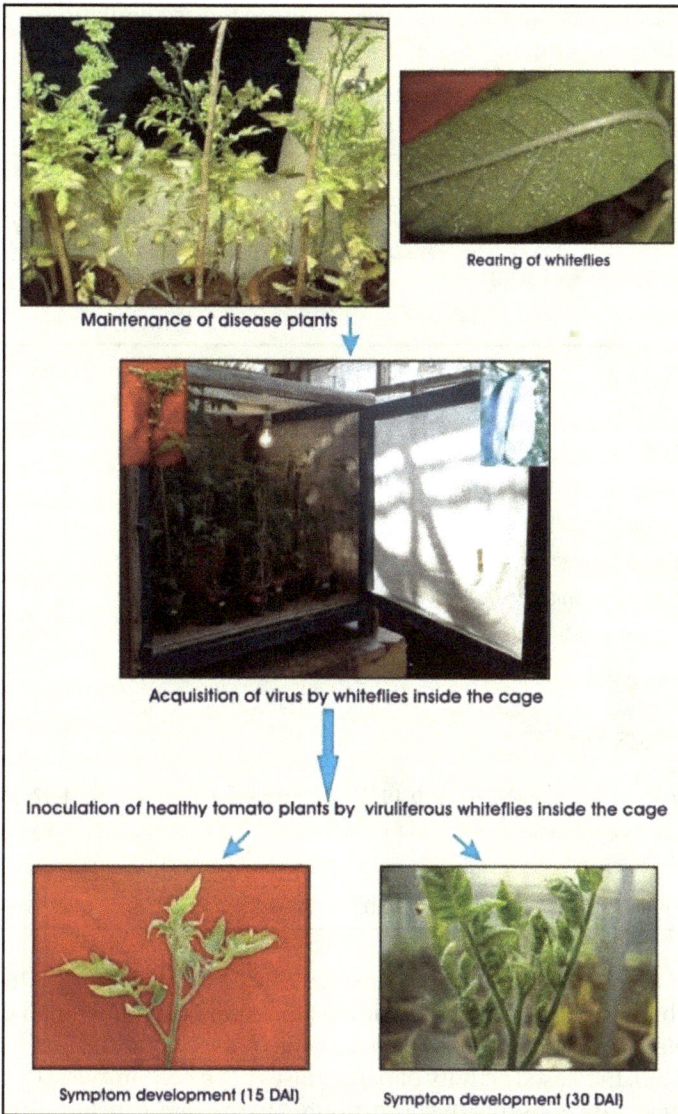

Figure 21.3: Sequential Events in the Development of ToLCV Disease in Tomato.

fungi, bacteria and viruses. Phenolic compounds are known to enhance the mechanical strength of the host cell wall and also to inhibit the invading pathogens. In the present study, higher level of accumulation of phenolics was observed in plants inoculated with *Pseudomonas* sp. 206(4) + B-15+ JK-16+ Chitosan compared to diseased control plants. This treatment recorded 41.66 per cent higher phenol content than the disease control (Table 21.1). The present findings are in agreement with Kandan *et al.* (2003), who also reported increased production of phenolic compounds in cowpea due to *P. fluorescens* inoculation which, in turn, protected plants from spotted wilt virus. Seed treatment with *P. fluorescens* 63 induced the accumulation of phenolics in tomato root tissue (M'piga *et al.*, 1997).

Table 21.1: Effect of the Selected Rhizobacterial Strains and Chitosan on Defense Molecules Activity (At 45 DAS)

Treatments	Phenol (mg/g Dry Weight)	Peroxidase (ΔOD/g Protein/ min)	Chitinase (µgGlc NAc/µg Protein/ min)	Polyphenol Oxidase (ΔOD/g Protein/ min)	Phenyl-ammonia Lyase (Changes in Cinnamic Acid/ min/g)
Pseudomonas 206(4)	0.26	1.42	2.41	1.29	0.176
Pseudomonas B-15	0.25	1.38	2.39	1.24	0.171
Pseudomonas JK-16	0.24	1.30	2.36	1.23	0.170
Pseudomonas (206(4) +B-15+JK-16)	0.28	1.49	2.46	1.33	0.184
Pseudomonas 206(4)+ Chitosan	0.32	1.76	2.51	1.42	0.189
Pseudomonas B-15 + Chitosan	0.30	1.58	2.48	1.39	0.187
Pseudomonas JK-16 + Chitosan	0.30	1.53	2.47	1.35	0.185
Pseudomonas (206(4) +B-15+JK-16) + Chitosan	0.36	2.05	3.10	1.66	0.216
Chitosan	0.23	1.19	2.15	1.22	0.164
Reference strain (*P. fluorscencs* NCIM 2099)	0.25	1.39	2.40	1.26	0.175
Chemical control (confidor 2 ml/L)	0.19	1.08	1.91	1.07	0.138
Diseased control (only ToLCV)	0.21	1.12	2.05	1.10	0.145
SEm+	0.01	0.10	0.10	0.06	0.02
LSD @ 5 per cent	0.02	0.31	0.29	.17	.05

Tomato plants treated with *Pseudomonas* sp. 206(4) + B-15+ JK-16+ Chitosan showed the highest induction of peroxidase (45.36 per cent higher than the diseased control, Table 21.1). Peroxidase is a key enzyme in the biosynthesis of lignin (Bruce and West, 1989). Increased activity of peroxidases have been implicated in a number of physiological functions that may contributed to resistance including exudation of hydroxyl cinnamyl alcohol into free radical intermediates (Gross, 1980) and lignification (Walter, 1992). Peroxidase is also associated with deposition of phenolic compounds into plant cell walls during resistance interactions (Graham and Graham, 1991).

Inoculation of tomato plants with *Pseudomonas* sp. 206(4) + B-15+ JK-16+ Chitosan also resulted in the higher synthesis of polyphenol oxidase (33.73 per cent higher than the diseased control),whereas the reference strain showed 12.69 per cent higher PPO activity than the diseased control (Table 21.1). The role of polyphenol oxidase (PPO) in disease resistance is to oxidize phenolic compounds to quinones,which are often more toxic to micro-organisms than the original phenols and the enzyme itself is inhibitory to viruses by inactivating the RNA of the virus (Vidhyasekaran, 1988).Similar kind of induced higher PPO activity was noticed by Kandan *et al.* (2005), when *P. fluorescens* strain CHAO was applied alone or in combination with other strains to tomato seeds for controlling tomato spotted wilt virus in tomato.

Inoculation of tomato plants with *Pseudomonas* sp. 206(4) + B-15+ JK-16+ Chitosan resulted in the highest synthesis of phenylalanine ammonia lyase (32.87 per cent higher than the diseased control, Table 21.1). Phenylalanine ammonia lyase (PALase) plays an important role in the biosynthesis of various defence chemicals in phenyl propanoid metabolism (Daayf *et al.*, 1997). PALase activity was induced during plant-pathogen and plant-pest interactions (Bhaarathi *et al.*, 2004 and Harish, 2005). Thus, higher induction of peroxidase and PALase and phenols might have reduced disease incidence and increased disease control in all the rhizobacteria treated plants.

Similar results were obtained by Kavino *et al.* (2003) who observed control of banana bunchy top virus (BBTV) due to induction of increased peroxidase, PALase and phenol content in banana plants when treated with *Pseudomonas fluorescens viz.*, Pf1 and CHAO alone as well as when amended with chitin. Rajinimala *et al.* (2003) also reported that *Pseudomonas chlororaphis* and *P. fluorescens* treated bittergourd plants, when challenge inoculated with bittergourd yellow mosaic virus (BGYMV), significantly increased peroxidase activity (140.37 per cent more than the unioculated control) which led to reduction in disease incidence.

Another fascinating observation was the enhanced chitinase activity in rhizobacterial mixture and chitosan treated plants (33.87 per cent higher than the diseased control, Table 21.1). This increase in chitinase activity might have prevented the damage caused by viral pathogen and, thus, increased the per cent disease control in all the rhizobacteria treated plants. Synthesis and accumulation of PR proteins have been reported to play an important role in plant defence mechanisms. Chitinases, which are classified under PR-3 have been reported to associate with resistance in plants against pests and diseases (Maurhofer *et al.*, 1994 and Van Loon, 1998).

Chitin and chitosan are known to have eliciting activities leading to a variety of defense responses in host plants in response to microbial infections, Chitosan was shown to inhibit the systemic propagation of viruses and viroids throughout the plant and to enhance the host's hypersensitive response to infection. They are reported to control a number of plant viruses like potato virus X, tobacco mosaic and necrosis virus, alfalfa mosaic virus, peanut stunt virus and cucumber mosaic virus (Chirkov., 2002).

All the plants treated with the combination of rhizobacteria and chitosan showed elicitation of enzyme activities and phenolics compounds. A significant increase in phenol content, peroxidase, polyphenol oxidase, PALase and chitinase activity was observed in all the treatments when compared to healthy control and diseased control. Liana *et al.* (2011) in an earlier study observed control of tobacco mosaic virus by *Bacillus* sp. through induced systemic resistance in tobacco as evidenced by increased levels of defense enzymes such as phenylalanine ammonia-lyase, peroxidase, polyphenol oxidase and pathogenesis-related (PR) proteins in tobacco.

Nandeeshkumar *et al.* (2008) observed enhanced activation of catalase, PALase, peroxidase, PPO and chitinase level in sunflower when seeds were treated with 5 per cent chitosan for controlling downy mildew. Ting *et al.* (2007) found that combination of chitosan and *Cryptococcus laurentii* resulted in a synergistic inhibition of the blue mold rot caused by *Penicillium expansum* in apple. Liu *et al.* (2007) found that chitosan treatment induced a significant increase in the activities of PPO, peroxidase and enhanced content of phenolic compounds which controlled the grey mold and blue mold in tomato. Chirkov *et al.* (2001) reported callose, ribonuclease and β-1,3 glucanase induction in potato plants as defense response against potato virus X (PVX) when plants were sprayed with chitosan solution (1 mg/ml). Similar kind of inhibition was reported on tomato leaves when treated with chitosan and challenge inoculated with potato spindle tuber viroid (Pospiezny, 1997). Thus, it has been proved that PGPR treatment along with chitosan significantly inhibited virus accumulation and systemic propagation of virus inside the treated plants.

Though all the treatments tested in this investigation induced biosynthesis of defense molecules, rhizobacterial mixture + chitosan showed the highest induction of defense molecules. Thus, the study has clearly brought out that the rhizobacterial mixture + chitosan was effective in reducing the disease severity of ToLCV through induced systemic resistance.

References

Abdelbasset EH Adam LR Hadrami IE and Daayf F. 2010. Chitosan in Plant Protection. *Marine Drugs* 8: 968–987.

Anonymous 1983. *Pest control in Tropical Tomatoes*.Centre for overseas pest research, London.130 pp.

Anonymous 1998. Indian Horticulture Database *National Horticulture Board Progress Report*. 247 pp.

Anonymous 2001. *Horticultural crops statistics of Karnataka state at a glance*. Department of Horticulture Government of Karnataka. 109 pp.

Anonymous 2010. *Indian Horticulture database 2009*. National Horticulture Board Gurgaon.

Baker B Zambryski P Staskaawicz B and Dineshkumar SP. 1997 Signaling in plant microbe interaction. *Sci* 276: 726–733.

Bhaarthi R Vivakananthan R Harish S Ramananthan A and Samiyappan R. 2004. Rhizobacteria based bioformulation for the management of fruit infection in chillies. *Crop Prot* 23: 835–843.

Boller T. 1991. *Ethylene in pathogenesis and disease resistance*. In: Mattoo AK Suttle JC (eds) The plant hormone ethylene. CRC Press Boca Raton FL. 293–314 pp.

Bruce RJ and West CA. 1989. Elicitation of lignin biosynthesis and isoperoxidase activity by pectic fragments in suspension culture of castor bean. *Plant Physiol* 91: 889–897.

Butter NS and Rataul HS. 1973. Control of tomato leaf curl virus (ToLCV) in tomatoes by controlling whitefly *Bemisia tabaci* Genn. by mineral oil sprays. *Curr. Sci* 42: 846–865.

Chen N Goodwin PH Hsiang T. 2003. The role of ethylene during the infection of *Nicotiana tabacum* by *Colletotrichum destructivum. J Exp Bot* 54: 2449–2456.

Chirkov SN. 2002. The antiviral activity of chitosan (review). *Appl. Biochem. Microbiol* 3: 81–89.

Chirkov SN Ii'ina AV Surgucheva NA Letunova EV Varitsev YA Tatarinova NY and Varlamov VP. 2001. Effect of chitosan on systemic viral infection and some defense responses in potato plants. *Russian J. Plant Physiol* 48: 774–779.

Compant S Duffy B Nowak J Clement C Barka EA. 2005. Use of plant growth promoting bacteria for biocontrol of plant diseases: principles mechanisms of action and future prospects. *Appl Environ Microbiol* 71: 4951–4959

Daayf F Bel– Rhlid R and Belanger RR. 1997. Methyl easter of P– coumaric acid: A phytoalexin like compound from long English cucumber leaves. *J. Chem. Ecol* 23: 1517–1526.

Devaraja Narayanaswamy K Savithri HS and Muniyappa V. 2005. Purification of *Tomato leaf curl Bangalore virus* and production of polyclonal antibodies. *Curr. Sci* 89(1): 15–20.

Doares SH Syrovets T Weiler EW and Yan CA. 1995. Oligogalacturonides and chitosan activate plant defensive genes through the octadecanoid pathway. *Proc National Academy Science* USA 92: 4095–4098.

Fitchen JH and Beachy RN. 1993. Genetically engineered protection against viruses in transgenic plants. *Annu. Rev. Microbiol* 47: 739–763.

Gomez KA and Gomez AA. 1984. Statistical Procedures for Agricultural Research. An International Rice Research Institute Book Wiley–Inter Science Publication New York, UAS. 680 pp.

Graham MY and Graham TL. 1991. Rapid accumulation of anionic peroxidases and phenolics polymers in soybean cotyledon tissue following treatment with *Phytopthora megasperma* f. sp. glycinea wall glucan. *Plant Physiol* 97: 1445–1455.

Gross GG. 1980. The biochemistry of lignifications. *Adv. Bot. Res* 8: 25–63.

Harish S. 2005. *Molecular biology and diagnosis of banana bunchy top virus and its management through induced systemic resistance.* Ph.D. Thesis Tamil Nadu Agric. Univ. Coimbatore India.

Harrison BD Muniyappa V Swanson MM Roberts IM and Robinson DJ. 1991. Recognition and differentiation of seven whitefly transmitted Gemini viruses from India and their relationships to African cassava mosaic and Thailand mungbean yellow mosaic viruses. *Ann. Appl. Biol* 118: 299–308.

Hull R. 1994. The movement of plant virus. *Ann. Rev. Phytopathol* 27: 213–240.

Jagadeesh KS. 2000. *Selection of rhizobacteria antagonistic to Ralstonia solanacearum causing bacterial wilt in tomato and their biocontrol mechanisms*. Ph. D. Thesis, Univ. Agric. Sci. Dharwad (India).

Kandan A Radjacommare R Nandakumar R Raghuchander T Ramiah M and Samiyappan R. 2002. Induction of phenyl propanoid metabolism by *Pseudomonas fluorescens* against tomato spotted wilt virus in tomato. *Folia Microbiologia* 47(2): 121–129.

Kandan A Radjacommare R Ramiah M Ramanathan A and Samiyappan R. 2003. PGPR induced systemc resistance in cowpea against tomato spotted wilt virus by activating defense against tomato spotted wilt virus by activating defense related enzymes and compound. *6th Int. PGPR Workshop* pp. 480–486.

Kandan A Ramiah M Vasanthi VJ Radjacommare R Nandakumar R Ramanathan A and Samiyappan R. 2005. Use of *Pseudomonas fluorescens*-based formulations for management of tomato spotted wilt virus (TSWV) and enhanced yield in tomato. *Biocontrol Sci. and Technol* 15(6): 553–569.

Kavino M Harish S Kumar N Soorianathasundaram K Ramanathan A and Samiyappan R. 2003. PGPR induced systemic resistance against banana bunchy top virus in banana. *6th Int PGPR Workshop*.pp 486–492.

Kirankumar R. 2007. *Evaluation of Plant growth promoting rhizobacterial stains against TMV on tomato*. M. Sc (Agri.) Thesis Univ. Agric. Sci. Dharwad (India).

Lecoq H. 1998. Control of plant virus diseases by cross protection. *Plant Virus Dis. Control* pp. 33–40.

Leeman M Van Pelt JA Ouden FM Heinbroek M Bakker PA and Schippers B. 1995. Induction of systemic resistance against Fusarium wilt of radish by lipopolysacchaides of *Pseudomonas fluorescens*. *Phytopathol* 85: 1021–1027.

Liana LC Liyan XB Luping ZB and Qiying LB. 2011. Induction of systemic resistance in tobacco against Tobacco mosaic virus by *Bacillus* spp. *Biocontrol Sci Technol*. 21: 281–292.

Liu J Tian S Meng X and Xu Y. 2007. Effects of chitosan on control of postharvest diseases and physiological responses of tomato fruit. *Postharvest Biol. Technol.* 44: 300–306.

Liu L Kloepper JW and Tuzun S. 1995. Induction of systemic resistance in cucumber against bacterial leaf spot by plant growth promoting rhizobacteria. *Phytopathol* 85: 843–847.

M'Piga Belanger RR Paulitz TC and Benhamli N. 1997. Increaesd resistance to tomato plants treated with endophytic bacterium *Pseudomonas fluorescens* 63–28 *Physiol. Molecul. Plant Pathol* 50: 301–320.

Mahadevan A and Sridhar R. 1986. *Methods in Physiological Plant Pathology*. Sivakami Publishers, Madras. 103 pp.

Maurhofer M Hase C Meuwly P Metraux JP and Defago G. 1994. Induction of systemic resistance of tobacco to tobacco necrosis virus by the root colonizing *Pseudomonas fluorescens* strain CHAO: Influence of the gacA gene and of pyoverdine production. *Phytopathol* 84: 139–146.

Mayer AM Harel E and Shaul RB. 1965. Assay of Catechol oxidase: a critical comparison of methods. *Phytochem* 5: 783.

Miller GL. 1959. Use of dinitrosalicyclic acid reagent for determination of reducing sugar. *Analytical Chem* 31: 426–428.

Mohanty AK and Basu AN. 1987. Biology of whitefly vector *Bemisia tabaci* Genn. on four host plants throughout the year. *J. Ent. Res* 11: 15–18

Muniyappa V and Veeresh GK. 1984. Plant virus diseases transmitted by whiteflies in Karnataka. *Proc. Indian Acad. Sci* 93: 397–406.

Nandakumar R Babu S Viswanathan R Sheela J Raghuchander T and Samiyappan R. 2001. A new bioformulation of containing plant growth promoting rhizobacterial mixture for the management of sheath blight and enhanced grain yield in rice. *Biocontrol* 46: 493–510.

Nandeeshkumar P Sudisha J Kini K Ramachandra HS Prakash SR Niranjana and Shetty H Shekar. 2008. Chitosan induced resistance to downy mildew in sunflower caused by *Plasmopara halstedii. Physiol. and Mol. Plant Pathol* 72: 188–194

Newell SY Cooksey KE Fell JW Master IM Miller C and Walter MA. 1981. Acute impact of an organophosphorus insecticides on microbes and small invertebrates of a mangrove estuary. *Arch. Environ. Cont. Toxicol* 10(4): 427–435.

Pospieszny H. 1997. Antiviroid activity of chitosan. *Crop Prot* 16: 105–106.

Rajinimala P Rabindran R Ramaiah M Nagarajan P and Varanavasiappan S. 2003. PGPR mediated disease resistance in bitter gourd against bitter gourd yellow mosaic virus. *6th Int. PGPR Workshop* Calicut, India.

Rice RP Rice LW and Tindall HD. 1987. Fruit and Vegetable Production in Africa. Macmillan Publishers U.K. 371 pp. FAO 1990 Production Year Book 1989 Vol. 43

Ross WW and Sederoff RR. 1992. Phenylalanine ammonia lyase activity from lobally pine: Purification of the enzyme and isolation of complementary DNA clones. *Plant Physiol* 98: 380–386.

Sadasivam S and Manikam A. 1991. *Biochemical Methods for Agriculture Science* Wiley Eastern Limited, New Delhi. 106–108 pp.

Saikia AK and Muniyappa V. 1989. Epidemiology and control of tomato leaf curl virus in Southern India. *Trop. Agric* 66: 350–354.

Saini RS Arora YK Chawla HKL and Wagle DS. 1988. Total phenols and sugar content in wheat cultivars resistant and susceptible to *Vgtilago nuda* (Jens) Rostrup. *Biochem. Physiologie de Pglanzon* 183: 89–93.

Sastry KSM and Singh SJ. 1973 Assessment of losses in tomato leaf curl virus. *J Mycol Plant Path* 3: 50–54.

Sastry KSM and Singh SJ. 1974. Effect of yellow vein mosaic virus infection on growth and yield of okra crop. *Indian Phytopathol* 27: 294–297.

Ting Y Hong YL and Xiao D Z. 2007. Synergistic effect of chitosan and *Cryptococcus laurentii* on inhibition of *Penicillium expansum* infections. *Int. J. Food Microbiol* 114: 261–266.

Van Loon LC. 1998. Induced resistance in plants and the role of pathogenesis related proteins. *European J. Plant Pathol* 103: 753–765.

Vasudeva RS and Samraj J. 1948. A leaf curl disease of tomato. *Phytopathol* 38: 364–69.

Vidhyasekaran P. 1988. Physiology of disease resistance in plants. Vol É. Boca Raton F L: CRC Press.

Vidhyasekaran P Sethuraman K Rajappan K and Vasumathi K. 1997. Powder formulations of *Pseudomonas fluorescens* to control pigeonpea wilt. *Biol. Control* 8: 166–171.

Viswanathan R and Samiyappan R. 2002. Role of oxidative enzymes in the plant growth promoting rhizobacteria (PGPR) mediated induced systemic resistance in sugarcane against *Colletotrichum falcatum. J. Plant Disease Prot* 109 (1): 88–100.

Walter MH. 1992. Regulation of lignification in defense. In: *Genes Involved in Plant Defences* Springer–Verlag,New York.327–352 pp.

Xue L Charest PM Jabaji–Hare SH. 1998. Systemic induction of peroxidase β– 13 glucanase chitinases and resistance in bean plant by binucleate *Rhizoctonia* species. *Phytopathol* 88: 359–365.

2015, **Recent Trends in Microbiology, Mycology and Plant Pathology**   *Pages* **339–348**
Editor: **Dr. H.C. Lakshman**
Published by: **DAYA PUBLISHING HOUSE, NEW DELHI**

Chapter 22

Management of Diseases of Horticultural Crops through Organics

Shripad Kulkarni and V.I. Benagi

Department of Plant Pathology,
UAS, Dharwad, Karnataka, India

Shift of Indian agriculture from a state of food deficiency to food sufficiency through introduction of fast growing, high yielding hybrid varieties, usage of high dose of chemical fertilizers, pesticides and weedicides. No doubt green revolution was yielded rich dividends but at the same time gifted serious pest problems and degradation of environment followed by deleterious effects on environment.

World over it has been estimated that more 67,000 different pest species attack crop. To protect them, many chemical pesticides are used which have unsafe environment impact, and hence there is a pressure for decreased reliance on such agents and greater regulatory control of their use. Besides many of the pathogens developed many fold of resistance to fungicide (Krishna Chandra *et al.,* 2005).

Most of Horticultural crops are eaten fresh or used for health care: hence any contamination in the form of pesticide/chemical residue may lead to health hazards; hence, organic horticulture offers a better possibility of producing healthy food. Plant diseases caused by different groups of organisms belonging Fungi, bacteria, viruses, rickettsia, spiroplasma, nematodes and few others have remained important in causing significant losses in different crops indicating the urgent need of their integrated management. The continuous and indiscriminate use of chemical pesticides has posed several serious problems such as pesticide residue, development of resistant strains, environmental pollution and adverse effect on beneficial micro-organisms and created a greater concern over global food safety and security.

Organic farming relies on crop protection, crop residues, animal manures, legumes, green manures, off farm organic wastes, cultural practices,mineral bearing rocks and aspects of biological pest control to maintain soil productivity and to supply plant nutrients and to control Diseases,insects, weeds and other pests.

1. Cultural Methods

By practicing the following methods we can regulate/modify the pest and disease incidence effectively.

a) Selection of Adopted and Resistant Varieties

By choosing varieties which are well adopted to the local environmental conditions (such as temperature, nutrient supply, pests and disease resistance) by which crop is allowed to grow healthier and stronger against attack of pests and pathogens.

Defense Mechanisms Influencing Diseases

i) Phenols:Host enzymes like plyphenol oxidase and peroxidase oxidize phenolics to quinines and the quinines are more fungitoxic than phenolics

ii) Phytoalexins:Phytoalexins are mostly isoflavonoids, terpenoids and poly acetylene compounds and synthesized *de novo* on infection by the pathogens. Phenylalanine and acetic acid may be involved in the biosynthesis of Phytoalexins and Phenylalanine ammonialyse (PAL) has been considered to be the key enzyme.

iii) Lignin:Phenylalanine and cinnamic acid are the important precursors. Lignin may act as physical barrier to the pathogens

iv) Callose: It is the substance found in the sieve tubes and may prevent the leakage of sieve tube sap or water in the cell walls. Penetration of incompatible pathogens into the host tissues results in the production of papillae and the papillae may mostly contain callose.

v) Sugars: Sugars are precursors of synthesis of phenolics, Phytoalexins, lignin and callose

vi) Amino acids: Aminoacids are the corner stones for synthesis of proteins and some of them are essential for the synthesis of phenolics, Phytoalexins and lignin

b) Cropping System

In a particular agricultural ecosystem plays major role. By adopting a suitable cropping system in right time, we can avoid most of the harmful pests and diseases. Some of the practical practices are as follows.

i) Activity in the soil and enhance the presence of beneficial organism. For instance plants colonized by arbuscular mycorrhizal may increase pests also. So careful selection of proper green manure is essential.

ii) Crop rotation: Cauliflower – paddy – cauliflower rotation is highly effective in controlling Stalk rot of Cauliflower disease caused by *Sclerotinia sclerotiorum* and it reduces infection by >60 per cent and > 161 per cent increase in seed yield has been observed.

iii) By adopting mixed cropping system pest and disease incidence can be minimized since pest has less host plants to feed on.

iv) By following crop rotation practices we can increase the soil fertility and reduce the chances of soil born diseases.

v) By cultivating green manure cover crops like Horse gram, Cow pea, Sun hemp, Sesbania, Dhaincha, Glyrisidia we can increase the biological activity in the soil and enhance the presence of beneficial organism. For instance plants colonized by arbuscular mycorrhizal fungi may increase pests also. So careful selection of proper green manure is essential.

c) Selection of Clean Seed and Planting Materials

Seeds and planting materials are the primary sources of diseses and hence selection and use of diseases and hence selection and use of disease free seeds after inspection for pathogens and weeds is very much essential. Further, it is advised to get seeds and planting materials from the reliable safe sources only. Use of healthy seed and it's hot water treatment at 50°C for 25 minutes or for sodium phosphate 90 gram/Lit. for 20 minute is Effective for tomato mosaic which is most dangerous disease of Tomato.

d) Selection of Optimum Planting/Sowing Time and Spacing

Most of the pests or diseases attack the crops only in a certain life stage. By adopting sufficient spacing between plants we can reduce the spread of a disease as well as allows good sunlight to the plants which facilitates less moisture on the leaves leading to hinder and of pathogen development and infection. In the same way more sunlight allows plants to do more photosynthesis. This practice not only avoids disease and pests in cropping system but also increase the crop productivity.

e) Balance Organic Nutrition

Gradual and steady growth makes plants less vulnerable to infection. So this steady growth could be achieved by applying organic fertilizers timely and moderately because, excess and indiscriminate use of fertilizers often results in damaging the roots. This damage facilitate to secondary infection. To overcome this problem we can adopt integrated nutrient management system with organic manures like FYM, compost, nutrients slowly when the plant needs. Further, by using liquid Bio – Fertilizers like Potash mobilizers namely *Frateuria aurentia* along with organic manures provides balanced potassium and contributes to the prevention of fungi and bacterial infections.

f) Addition of more and more Organic Matter

Organic content of the soil is directly related to density and activities of micro-organisms in the soil there by pathogenic and soil borne fungal population can be reduced. Besides this, organic matter provides.

☆ All the nutrients that are required by the plants.

☆ Corrects C:N ratio in the soil.

☆ Good physical chemical and biological support to soil

☆ More water holding capacity to the soil.

☆ Cover from evaporation losses of the moisture from the soil.

Ultimately, organic matter supplies substances which strengthen plants with their own protection mechanisms.

g) Soil Amendments

The decomposition of organic matter helps in alteration of the physical, chemical and biological conditions of the soil and the altered conditions reduce the inoculum potential of a soil borne pathogens. In addition, the practice improves soil structures, which promotes root growth of the host. Various biochemical substances like antibiotics and phenols are released during decomposition, which in turn induce resistance in the root system. Soil amendments like sunflower, rape seed cakes, mustard cake, gypsum and been straw can be used.

Rhizoctonia solani causal organism of damping off in nursery stage and wire stem, bottom rot and head rot after transplanting plants in crucifiers. Soil amended with neem cake 1 kg/m^2 and solarized by covering with white polythene sheet for two weeks and treatment of seeds with *Trichoderma harzianum* proved to be effective in the management off damping off of vegetables crops. Soil amended with cellulose powder was also found effective in reducing the disease incidence, which may be due to increase in C:N ratio in soil. This increase results in the decrease of fungal population but actinomycetes and bacterial populations increased.

Organic amendment is used here to mean organic material incorporation into the soil that comes from external sources such as processing residues or industrial waste products. Organic material added as fresh crop residue and grown in the field in rotation – break, cover, trap, antagonistic or green manure crops- are discussed below. Incorporation into the soil of large amounts of any organic material will reduce nematode densities. Oil cakes, coffee husks, paper wastes, crustacean skeletons, sawdust and chicken manure, amongst others, have been used with some success. Nematode control may be due to any one or more of the following mechanisms:

☆ Toxic and non-toxic compounds present in the organic material;

☆ Toxic metabolites produced during microbial degradation; or

☆ Enhancement of the soil antagonistic potential.

Chitin amendments have received much interest in the past as an organic amendment in that they stimulate the antagonistic potential in soil towards nematodes. Organic amendments have also been combined with various biocontrol agents with reports of enhanced levels of control. The use of organic amendments is often limited by availability and, in some cases, by the large quantities needed. In addition to their effects on nematode density, organic amendments also improve soil structure and water-holding capacity, reduce diseases and limit weed growth, all of which ultimately lead to a stronger plant and improved tolerance to nematode attack.

h) Biofumigation

This term normally refers to suppression of soil-borne pests and pathogens by biocidal compounds, principally isothiocyanates, released in soil; when glucosinolates in cruciferous crop residues are hydrolysed. Soil amended with fresh or dried cruciferous residues at 38 C day and 27 C night temeperatures reduced *Meloidogyne incognita* galling by 95-100 per cent after 7 days' incubation in controlled environment tests. It should noted here, however, that many cruciferous plants are good hosts of some important species of *Meloidogyne.*

The term biofumigation is now used more freely whenever volatile substances are produced through microbial degradation of organic amendments that result in significant toxic activity toward a nematode or disease. The release of toxic compounds already present in antagonistic plants used as amendments, *e.g.* neem, marigold and castor, or the production of toxic compounds due to microbial fermentation of nutrient-rich organic amendments, *e.g.* velvet bean, sunnhemp or elephant grass, lead to significant levels of nematode control.

Biofumigation under these circumstances is greatest when there is an optimum combination of organic matter, high soil temperature and adequate moisture to promote

Figure 22.1: Soil Solarization Method.

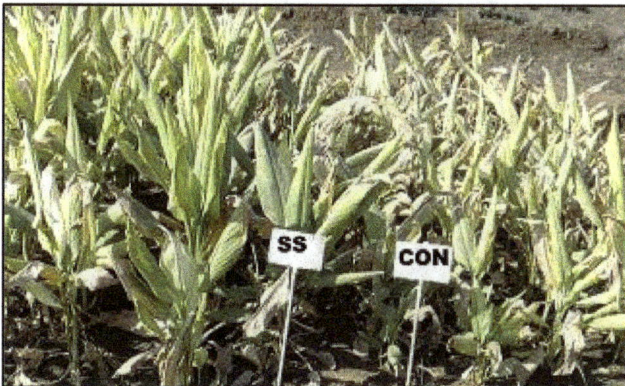

Figure 22.2: Healthy Turmeric Crop in Soil Solarized Plot.

Figure 22.3: Rajapuri (Highly Susceptible) Banana Variety against Sigatoka Disease.

Figure 22.4: Sakkarebale (Highly Resistant) Banana Variety against Sigatoka Disease.

microbial activity leading to toxin production. In tropical and subtropical production systems, plastic mulch and drip irrigation improve effectiveness of biofumigation. Transporting organic amendments to the field or incorporating cover crops that produce large amounts of biomass into the soil, together with plastic mulch and drip irrigation, should significantly increase the level of control attained.

Biofumigation using fresh marigold as an amendment is used effectively in root knot management in protected cultivation in Morocco. *Tagetes* is grown in the raised beds prior to the planting of susceptible horticultural crops. The crop is then incorporated into the soil after 2-3 months. The beds are fitted with drip irrigation and covered with plastic mulch. The soil in the bed is then biofumigated under conditions of the high temperature and optimum soil moisture.

Control due to any form of biofumigation is probably the result of multifaceted mechanisms including:

1. Non-host or trap cropping depending on the host status of the plant used.
2. Lethal temperature due to solarization.
3. Nematicidal action of toxic by-products produced during organic matter degradation.
4. Stimulation of antagonists in the soil after biofumigation.

i) Water Management

By practicing good water management water logging in the field and stress on plants can be avoided otherwise pathogens take chance and infect the crop. Further, sprinkling water on foliage shall be avoided as it increase diseases by giving chance to pathogenic fungal spores to germinate.

j) Use of Proper Sanitation Measures

"Pull and Burn" is the best method to control disease and removal of infected plant parts (leaves, fruits) from the ground to prevent the disease from spreading. Eliminate residues of infected plants after harvesting.

k) Soil Solarization

Soil solarization for 4 weeks during summer months coupled with application of neem cake @ 400 g/ha proved effective against damping-off (76.9 per cent) in nursery and resulted in significant, higher number of healthy transplants.

2. Mechanical Methods

a) By removing and burning Clipping of lower leaves upto 20 cm and weading to reduce the alternaria blight in tomato.

b) This is the best method, before reaching the loss beyond economic level. To save crop from fungal and bacterialdiseases pull and burn method is most effective.

3. Botanicals

Plants during their long evolution, have synthesized a diverse array of chemicals to prevent the colonization by pathogens. They produce secondary metobolites like terpenoids alkaloids, flavonoids, phenolic compounds. These secondary metabolites are having disease suppressing properties. In India several plants such as nicotinoids, natural pyrethrins, rotenoids, neem products have been used in the past for suppression of diseases. Among the botanical pesticides neem occupies very important place in the pest and disease control. Different parts of neem tree can affect more than 200 insects and diseases, some of them are effective against nematodes, fungi, bacteria and viruses.

4. Biological Control

Management of soil borne pathogens is difficult because of non-availability of desired level of resistance against major soil borne diseases caused by species of *Fusarium, Sclerotium, Macrophomina, Pythium, Phytophthora* and few others forced to search or resorting to new approaches to manage the diseases. Therefore, biocontrol agents or antagonists are means of plant disease control has gained importance in the recent years. The biocontrol agents multiply in soil and remain near root zone of the plants and offer protection even at later stages of crop growth.

Mechanisms of Biocontrol

- ☆ Competition
- ☆ Mycoparasitism
- ☆ Antibiosis and Enzymes
- ☆ Cross protection
- ☆ Induced systemic resistance (ISR)

Methods of Application

1. Broad cast application
2. Furrow application
3. Root zone application (Formulation mixed in soil @ 1 kg/plant)
4. Seed treatment (4g/kg seeds)
5. Wound application (Applied to wounds in peach, plum etc)
6. Spraying

Advantages of Biopesticides

1. Avoids environmental pollution (soil, air and water)
2. Avoids adverse an effect on beneficial organisms *i.e.* maintains healthy biological control balance
3. Less expensive thou pesticides and avoids problems of resistance
4. Biopesticides are self maintaining in simple application and fungicide needs repeated application

5. Biopesticides are very effective for soil borne pathogen where fungicidal approach is not feasible.

6. Biopesticides eco-friendly, durable and long lasting.

7. Very high control potential by integrating fungicide resistant antagonist.

8. Biopesticides help in induced system resistance among the crop species.

Table 22.1: Successful Control of Plant Diseases using *Trichoderma* Species

Chick pea wilt	Carbendazim seed treatment (ST) + *T. harzianum* or neem cake + *T. harzianum*
Redgram wilt	ICP8863 + ST with *T. harzianum*
Foot rot of paper	Neem cake + *T. harzianum* (soil solarization and application of Metalaxyl MZ (two spray)
Cotton root rot	*T. harzianum* + Benomyl
Sclerotium wilt of sugarbeet	Captan + *T. harzianum*
Damping off of tobacco, tomato, egg plant	Metalaxyl MZ + *T. harzianum*
Sclerotium rolfsii in tomato, beans, groundnut, sugar beet and chick pea	Furrow application of 130-160 kg/ha with 4 g/kg of seed treatment of (*T. harzianum T. viride, T. hamatum*)
Bean root rot *Macrophomina phaseolina*	Furrow application of *T. harzianum* (130-160 kg/ha)
Fusarium oxysporum in bean, chrysanthemum, cotton, melon and redgram	Soil application of *T. harzianum* at 130-160 kg/ha
Different wood rotting fungi in wood trees Grey mold, *Botrytis cinerea* in grapes	Dusting *T. viride* on wounds.Aerial spray with *T. viride* (106–108 cfu/ml)
Stalk rot of rabi sorghum	*T. harzianum* 90-125 kg/ha seed)

Mycoparasites

Mycoparasites Ampelomyces quisqualis for management of anthracnose, downy mildew and powdery mildew of grape. It is commercially available as AQ 10 (USA), Bio – Dewcon (India) 1000ml has to be mixed with 500 ml Neem Oil in 250 Liters of water.This also useful in managing powdery mildew of cucurbits caused by *Sphaerotheca fuliginea*.

5. Use of bleaching Powder

12 kg/ha found to reduce bacterial blight by 80 per cent when applied in furrows at the time of planting.

References

Krishna Chandra, Greep.S. and Srivatha. R.S. H., 2005, Bio control agents and Bio pesticides, published by Regional center of organic farming, Hebbal, Bangalore.

Gupta.S.K. and Thind T.S. 2006, Disease problems in vegetables production. Scientific publishers (India) P.O. Box 91, Jodhpur.

Ray.A.B., Sarma, B.K and Singh.U.P. 2004, Medicinal properties of plants-Antifungal, Antibacterial and Antiviral Activities. International Book Distributing Co-Lucknow- 226004 (U.P).

Singh R.S. 2005, Plant Diseases (Eight Edition) Oxford and IBH publishing Co. Pvt. Ltd. New Delhi.

2015, Recent Trends in Microbiology, Mycology and Plant Pathology   *Pages* 349–357
Editor: **Dr. H.C. Lakshman**
Published by: **DAYA PUBLISHING HOUSE, NEW DELHI**

Chapter 23

Interaction between Host and Pathogens in Development of Diseases

*B.S. Agadi[1] and H.C. Lakshman[2]**

[1]Department of Botany, K.L.E.'s P.C. Jobin Science College,
Vidyagiri, Hubli – 580 0031, Karnataka, India
[2]P.G. Department of Studies in Botany,
Karnatak University, Dharwad – 580 003, Karnataka, India
*E-mail: dr.hclakshman@gmail.com

Introduction

Disease develops as a result of host-pathogen interactions. The relationship between a host and pathogen/parasite is not merely a physical contact. Since these both come in contact with each other till the final stage of symptom development, a number of physiological processes are involved between the two. The mere contact of a pathogen with the host surface does not necessarily mean that is would penetrate. We have already seen how secretions of plant surface both in rhizosphere by root and in phyllosphere by shoot may inhibit the germination of spores of pathogens. There are of course several other host factors-structural and biochemical that prevent the spore germination and further penetration and invasion of the pathogen. Some pathogens fail to penetrate and some fail to cause infection even after penetration. Here again host's factors begin to operate.

Pathogens differ in respect of the precise way in which they enter the host and in nature of a attributes by which they attack their hosts. The whole sequence of pathogenesis or process of infection in disease development could be conventionally grouped into the following three stages, although these stages are not completely distinct. The stages are (i) Pre-penetration, (ii) Penetration, and (iii) Post-penetration.

Pre-penetration

This phase includes the growth of the pathogen before is actual entry of penetration into the host. During this phase, two main events will takes place, (i) spore germination (seeds/hatching of eggs), and (ii) growth of resulting structures (germ tubes in fungi)

Spore Germination

This occurs in various ways. The spore may be resting spore, produced asexually (chlamydospore) or sexually (teliospore, oospores, resting sporangia etc.), as well as the propagative asexual spore (thin-walled, short –lived, less nutrients, very in colour, shape physical factors (temperature, moisture etc.) is also largely influenced by the biological factors. The biological factors are the nature of the chemical substances secreted by plant surface (since these substances originate from plant, we call them biological factors) and presence of other micro-organisms on/near the plant surface. Plant exudates may exert a stimulatory or inhibitory effect on germination. The non-parasitic micro-organisms in the rhizosphere as well as phyllosphere also influences spore germination. Rhizosphere is the zone of soil around living roots characterized with intense microbial activity. Root exudates contain sugars and amino acids which serve as nutrients for rhizosphere micro-organisms. Chemicals secreted in rhizosphere by roots may stimulate or inhibit germination of spores of pathogens. For instance, oospores of Pythium mamillatum around turnip root, scletotia of *Sclerotium rolfsii* around alfalfa roots and spores of *Plasmodiopora brassicae*, *Urocystis tritici* and *Fusarium* spp. are stimulated to germinate by root exudates of their hosts.

Phyllosphere (phylloplane) is the leaf surface and immediate adjacent area, colonized by a variety of non-parasitic micro-organisms. Many of these phyllosphere microbes are antagonistic and inhibit spore germination of important leaf pathogens. Leaf exudates may also contain inhibitory substances. The leaf itself may contain some inhibitory substances. Orange peel onion, potato reduces spore germination. Waxy leaf surfaces of apple and other plants contain inhibitory substances that inhibit germination of spores of *Podosphaera leucotricha* and *Botrytis fabae*. Malic acid secreted by leaves of resistant varieties of gram inhibits germination of spores of *Mycosphaerella radiei*. Leaf exudates and phyllosphere microflora appear to have relatively less effect on pathogens of above graound plant parts but these are relatively more pronounced and important in root diseases.

Growth of Germ Tubes

The above said chemical stimulates or inhibits the growth of germ tube also. They are more pronounced in rhizosphere than in the phyllosphere. Because the leaf cuticle reduces exudation and rain removes shoot exudates. The formation of germ tube is affected by environmental factors and plant susceptibility. In different cases, they enter though wounds, natural openings, of directly through intact surface of host, often forming appressoria. The growth of germ tubes and subsequent appressorium formation are influenced besides other factors, by chemicals of host origin. Appressorial development is largely a thigmotropic response, but there is evidence that chemicals of host origin are also involved. Directional effects also occur near root and germinating seeds.

Thus spore germination and growth of germ tubes and formation of appressorium are complex and influenced not only by environment but also by the nature of plant surface, by substances secreted by it and by other micro-organisms present in the rhizosphere and phyllosphere.

Penetration

This is the actual entry of parasite in the host. Pathogens differ in ways of penetration. Penetration of pathogens may occur through various means such as through natural openings like stomata, through injuries caused by various reasons, through the physical and chemical forces. For a pathogen to infect a plant it must be able to make its way into and through the plant, obtain nutrients from the plant, and neutralise the defense mechanisms of the plant. Pathogens do these mostly through secretions of chemical substances that affect certain comp0onents or metabolic pathways of their hosts. Penetration and invasion, however, may be supplemented by, or in some cases be, entirely the result of mechanical force exerted by some pathogens on plant cell walls. We have already referred to such mechanical forces of different pathogens under penetration. These are appressoria of fungi, stylets of nematodes etc. Even in these cases perhaps cuticle and cell wall softening enzymes are involved. In most diseases, the activities of pathogens are largely chemical. Therefore, the effects caused by pathogens on plants are almost entirely the result of biochemical reactions taking place between substances secreted by the pathogens and those present in, or produced by the plant.

- ☆ **Post Penetration:** After successful penetration of pathogen into the host body the following events takes place.

- ☆ **Subcuticular development:** The hyphe grow mostly between the cuticle and outer wall of epidermal cell. Example: *Venturia inaequalis, Diplocarpon rosae*.

- ☆ **Parenchymatous tissues:** Many pathogens colonise the parenchyma of cortex and mesophyll. Examples: *Taphrina, Pythium debaryanum, Phytophthora*, Peronosporaceae, Albuginaceae, powdery mildews, rusts.

- ☆ **Vascular tissues:** Vascular tissues are colonized by Hymenomycetes, some ascomycetes, deuteromycetes and bacteria. Both parenchyma and vessels of xylem are colonized. They cause wilt and rot diseases.

- ☆ **Endoboitic development:** Fungi as *Olpidium, Plasmodiophora* viruses, some bacteria grow exclusively inside the host cells.

- ☆ **Systemic development:** They are highly specialized parasites. They have haustoria for nutrients and cause slight damage to tissue until later in life cycle. Some rusts, downy mildews, white rusts, and smuts are good examples.

Role of Enzymes in Disease Development

Pathogens have both constitutive as well as adaptive enzymes. Different kinds of enzymes that may play roles in disease development are as follows.

Cutinases

They catalyse the breakdown and dissolution of cutin, the main component of cuticle. Cuticular was outside or within the cuticle of many aerial plant parts is known to be penetrated by mechanical force alone and there is no evidence of its enzymatic breakdown. However, cutin, the main component of cuticle can be degraded enzymatically. Cutin is mixed with cuticular was and also merges into the outer walls of epidermal cells admixed with pectin and cellulose. Cutin is insoluble polyester of mostly unsaturated derivatives of C16 and C18 hydroxy fatty acids. Many fungi and a bacterium, streptomyces scabies are known to produce cutinases that can degrade cutin. Cutinases are esterases that break ester linkages between cutin molecules and release monomers of the component fatty acid derivatives from the insoluble cutin polymer.

Pectinases

They degrade pectic substances, the main component of the middle lamella of plant cells. Pectic Substances make also a large portion of primary cell wall, in which they form an amorphous gel filling the spaces between the cellulose microfibrils. Pectic substances are polysaccharides containing a very high percentage of galacturonic acid molecules. Several enzymes- the pectinases degrade pectic substances. The two major groups are: (i) Pectinesterases (PE) or Pectin methylesterases (PME), which remove small branches off the pectin chains and (ii) Polygalacturonases (PG) or Pectic glycosidases and Lyases, the chain-splitting enzymes, which cleave the pectic thain to release shorter chain portions containing one or few molecules of galacturonic acid. These enzymes are shown to be involved in many diseases and are widespread in many fungi and bacteria causing rots, dampling off, spots, blights and vascular wilts.

Cellulases

These are produced by several phytopathogenic fungi, bacteria and nematodes. In living tissues they play important role in softening and/or disintegration of cell-wall material, thus facilitating penetration and spread of pathogen. Cellulases are infact several enzymes which degrade cellulose by a series of enzymatic reactions. Cellulases are mainly studied in realation with deterioration of wood and textiles.

Breakdown of cellulose into the monomers, glucose is brought about by cellulose complex and other enzymes. One cellulose (C_1) attacks native cellulose by cleaving cross-linkages between chains. Second cellulose (C_2) also attacks native cellulose and breaks it into chorter chains which are then attacked by a third group of cellulases (C_x) which degrade them to the disaccharide cellobiose. Finally, cellobiose is degraded by B-glucosidase into glucose.

Hemicellulases

They are complex mixtures of polysaccharide polymers and major component of primary cell wall and to some extext middle lamella and the secondary wall. Several hemicelluases are produced by fungi. These include xylanases, galactanases, glucanases, arabinases mannoses etc.

Ligninases

Lignin is an amorphous three dimensional polymer. The most common basic tructural unit is phenypropanoid with one or more of its carbons having a-OH, - OCH_3 or =O group.

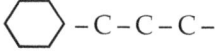

$$\langle \bigcirc \rangle - C - C - C -$$

Breakdown the most resistant component, lignin of cell walls. They are known among hymenomycetes only (Basidiomycetes), which alone are capable of decomposing wood. These are thus involved in wood decay. About 25 per cent the hymenomycetes are brown-rot fungi, which cause degradation of lignin but cannot utilize it. Most of the lignin is defraded and utilized by white-rot fungi, that secrete one or more ligniases. That enables them to utilize lignin.

Other Enzymes

These actually degrade the substances contained in the plant cell. The above mentioned enzymes degrade substances of the cell wall. There are also produced some enzymes that degrade proteins, starch, lipids present in the plant cell. Several bacteria, fungi and nematodes produce proteolytic enzymes, amylases, lipases, phosholipidases etc.

Role of Toxins in Disease Development

The term toxin is generally used for substances, usually but not invariably of pathogen origin which are injurious to plants and directly or indirectly play a role in diseases development Thus harmful enzymes are also toxins. There are different terminologies:

Vivotoxin is a substance produced in the infected plant by the pathogen and or its host which functions in production of disease.

Pathotoxin is a general term used for phytotoxic substances produced by live organism. **Endotoxin** is the intracellular toxin formed in bacterial cells and liberated only after their death. Exotoxin is an extracellular toxin which diffuses from live bacterial cells.

Toxins may act as poisons on host cell protoplast; they may alter permeability of cell memberanes; they may inhibit the activity of enzymes and may also act as antimetabolites (do not allow synthesis of a metabolite in plant cell). Some toxins act as general protoplasmic poisons and affect many species of plants of different families. There are general toxins. Some other toxins are toxic to only a few plant species or varieties and completely harmless to others. These are known as host-specific toxins. Fungi and bacteria may produce toxins in infected plants as well as in culture medium. Toxins are extremely poisonous substances and a effective in very low concentrations. Some are unstable or react quickly and are tightly bound to specific sites within the plant cell.

Host-specific Toxins

A host-specific toxin is a substance produced by a pathogen that, at physiological concentrations, is toxic only to the hosts of that pathogen and shows little or no toxicity against non-host plants. Most host-specific toxins must be present for the producing pathogen to be able to cause disease. So far, such toxins are shown to be produced only by some fungi belonging to the genera, *Helminthosporium*, *Altemaria*, *Periconia*, *Phyllosticts*, *Corynespora* and *Hypoxylon*, although some bacterial polysaccharides of *Pseudomonas* and *Xanthomonas* have also been reported to be host-specific.

Victorin or HV Toxin

This is produced by the fungus *Helminthosporium victoriae*, which appeared in 1945 on the introduced and widely spread variety of oat-Victoria and its derivatives. The pathogen infects basal parts of plant and then produces the toxin, which is carried to the leaves, causing leaf blight and destroying entire plant. The toxin produces histochemical and biochemical changes in plant, which include changes in cell wall structure, loss of electrolytes from cells, increased respiration, decreased growth and protein synthesis.

T-tpxin (*Helminthosporium maydis* race T toxin)

This is produced by race T of *Helminthosporium maydis*, cause of southern corn leaf blight. The toxin is a mixture of linear, long (35 to 45 carbon) polyketols. It acts specifically an mitochondria causing early loss of matrix density, rendering them non-functions.

AK-toxin

It is produced by a distince pathotype of *Alternaria alternate*, previously referred to *A. kikuchiana*, the cause of black leaf spot of Japanese pears (*Pyrus serotina*). The toxin causes the cells to instantaneously lose K^+ and phosphate.

AM-toxin

This is produced by the apple pathotype of *Alternaria alternate*, previously referred to *A. mail*, cause of Alternaria blotch of apple. The toxin is a cyclic depsipeptide. The site and mechanism of action are similar to Ak-toxin, but this also causes rapid loss of chlorophyll.

Other Host-Specific Toxins

Some other host specific toxins produced by fungi are listed below.

Pathogen	Toxin	Host plant
Helminthosporium carbonum Race 1	HC-toxin	Corn
H. sacchari	HS-toxin	Sugarcane
Alternaria citri(lemon race)	ACL- toxin	Lemon
Alternaria alternate. *lycoperisic*	AL-toxin	Tomato
Periconia circinata	PC-toxin	Sorghum
Phyllosticta maydis	PM-toxin	Corn
Corynespora cassiccla	CC- toxin	Tomato

Role of Growth-regulators in disease development

Auxins gibberellins, cytokinins ethylene and abscisic acid are known to be involved in symptom development of respective disease. Among auxins, increased levels of the IAA are reported in plants infected by *Phytophthora infestans* (potato), *Ustilage maydis* (corn), *Gymnosporangium juniperivirginianae* (cedar), and *fusarium oxysporum f. cubense* (banana). The auxins are mainly studied in relation with bacterial diseases, particularly crown gall of rosaceous plants caused by *Agrobacterium tumefaciens* and wilt of solonaceous plants by *Pseudonos solonacerum*. Gibberllins are produced by the fungus, *Gibberella fujikuroi*, cause of foolish seedling disease of reported in club rood galls, crown galls, rust galls etc.

Ethylene, naturally produced by plants exerts a variety of effects on plants including chlorosis and leaf abscission. This is also produced by several plant pathogenic fungi and bacteria. In the banana fruit infected with *Pseudomonas solanacerum*, ethylene content increases proportionally with the (premature) yellowing of the fruit. No ethylene could be detected in healthy fruit.

Abscisic acid, besides plants is also shown to be produced by some phytopathogenic fungi. It brings about several growth – inhibiting and hormonal effects in diseased plants. Implications of this growth inhibitor are shown in tobacco mosaic and cucumber mosai viral diseases; Southern bacterial wilt of tobacco and *Verticillium* (fungal) wilt of tomato. Diseased plant had higher levels of abscisic acid.

Cork Layers

Cork layers abscission layers and tyloses are formed or certain plants to resist further spread of the pathogen. Cork layers formed around the point of infection check further invasion. These layers do not allow the pathogen to penetrate further, block toxic substances of the pathogen and check the supply of nutrients and water from the healthy to the infected cells. Some examples of such diseases in which cork layers serve, as barriers are *Rhizopus* rot of sweet potato and common scab of potatoes (*Streptomyces scabies*).

Abscission Layers

By forming abscission layers, the plant discards the infected area which may be sloughed off and the rest of the leaf area is then protected from infection or from toxic substances released by the pathogen. Abscission layers are found in many fungal, bacterial and viral diseases *e.g.*, peach leaves infected by *Cladosporium carpophilum*, peach leaves infected by *Xanthomonas pruni* and necrotic ring spot virus infecting sour cherry.

Tyloses

The overgrowths of protoplast of living parenchymatous cells that protrude in xylem vessels care called tyloses. In verieties of sweet potato resistant to wilt fungus (*F. oxysporum* f. *batatas*), blocking further spread of the fungus. Tyloses interfering with transport can be harmful to the plant itself. If tyloses form after the pathogen has advanced, they may prove harmful to the plant. The speed of formation, the amount and the loction of tyloses are the important factors, which determine the type of effect on plants.

Gum

In apple varieties resistant to *Streum purpureum* (silver leaf disease) and *Physalospora cydoniae* (black rot), gum forms in parenchyma, intracellular spaces and xylem cells in advance of pathogen. This forms a strong barrier enclosing the pathogen. In rice varieties resistant to *Pyricularia oryzae* (rice blast disease), enlargement of lesion in leaf is limited because of gum deposition around the lesions.

Epidermal or Subepidermal Cells

In some plant diseases, the outer walls of epidermal or subepidermal cells swell and limit spread of the pathogen. Swelling of cell wall is known in cucumber varieties resistant to *Cladosporium cucumerinum* (scab disease) and in other host-parasite interactions. In certain plant diseases, a sheath formed by the host covers the hyphae penetrating the cell wall; this may slow down the progress of pathogen.

References

Bennet, C. W. 1973. A consideration of some of the factors important in growth of the science of plant pathology. *Ann. Rev. Phtopathol*. 11: 1–10.

Cook, R.J. and Baker, K.F. 1983. *The nature and practice of biological control of plant pathogens*. St. Paul, M.W. Academic press sciences, London, New York, pp. 538.

Dastur, J.F. 1963. A rambling talk on mycology and mycologists in India. *Indian Phytopathol*. 16: 321–332.

Hofte, H. and Whitely, H.R. 1989. *Microbial Review*. 53: 242–255.

Johri, J.K. 2007. Biopesticides–its need and future. In: *Rhizopshere Biotechnology*. A.K. Roy (Ed) Scientific publishers, Jodhpur, India, pp. 147–156.

Kreig, A. and Langenbroch, G.A. 1981. *Microbial control of pests and plant diseases*. H.D. Burgesed (Ed). Academic press, London. Pp: 838.

Lakshman, H.C. and Shwetha, J.S. 2011. Significance of GM crops in modern agriculture. In: *Microbial technology and Ecology*. Eds. Deepak Vyas *et al.,* Daya Publishers, New Delhi, pp. 484.

Large, E.C. 1962. *The advances of fungi*. Dover New York, USA, pp. 432.

Mahadevan, A. 1982. *Biochemical aspects of plant disease resistance*, Part I. Today and Ttomorrow Publishers, New Delhi, pp. 348.

Mishra, A., Bohra, A. and Mishra, A. 2005. *Plant pathology: disease and management*. Agrios publishers, Jodhpur, India, pp. 768.

Paulitz, T.M. and Belanges, R.R. 2001. Biological control in green house systems. *Ann. Rev. Phytopathol*. 39: 103–133.

Rangaswami, G. and Mahadevan, A. 1999. *Diseases of Crop Plants in India*. 4th edition, Prentice Hall of India Pvt. Ltd. New Delhi, pp. 568.

Sharma, P.D. 1999. *Microbiology and plant pathology*. Rastogi Publications, Meerut, India, pp. 878.

Sharma, R.C. and Sharma, J.N. 2005. *Integrated Plant Disease Management*. Scientific Publishers, Jodhpur, India, pp. 364.

Singh, S.S., Gupta, P. and Gupta, A.K. 2000. *Handbook of Agricultural Sciences*. Kalyani Publishers, New Delhi, pp. 826.

Tenhouton, J.G. 1972. Plant pathology, changing agricultural methods and human society. *Ann. Rev. Phytopathol*. 12: 1–13.

2015, Recent Trends in Microbiology, Mycology and Plant Pathology   Pages 359–372
Editor: Dr. H.C. Lakshman
Published by: DAYA PUBLISHING HOUSE, NEW DELHI

Chapter 24

Influence of Culture Media and Environmental Factors on Mycelial Growth and Conidial Yield of *Fusarium proliferatum* a Potential Pathogen of *Echinochloa crusgalli* (Major Weed in Rice)

G. Jyothi[1]*, K.R.N. Reddy[2], K.R.K. Reddy[2], Shefali Mishra[2] and A.R. Podile[3]

[1]*Department of Biotechnology, Jawaharlal Technological University, Hyderabad – 500 085, Andhra Pradesh, India*
[2]*Department of Plant Pathology, Sri Biotech Laboratories India Ltd., Hyderabad – 500 034, Andhra Pradesh, India*
[3]*Department of Plant Sciences, University of Hyderabad, Hyderabad – 500 046, Andhra Pradesh, India*
*E-mail: jyothibio84@gmail.com

Introduction

Barnyard grass, *Echinochloa crusgalli* (L.) Beauv. ranked as the world's third worst weed, is one of the most serious weeds in rice (*Oryza sativa* L.) (De Datta, 1981). Intense competition from *E. crusgalli* can reduce tillering in rice by up to 50 per cent. Weeds also serve as reservoir for plant pathogens that may cause significant economic loss in crop production. Besides, these weeds also cause environmental impact associated with its management such as non target injury of living organisms, contamination of ground and surface water etc (Turnera *et al.,* 2007). In this context, biological weed management practice seems to be a selective process against targeted weeds without damaging non-target living beings and environment. Biological control is the deliberated use of living organisms to control a pathogen or weed (Tamuli and Boruah 2000; Hallett 2005; Chutia *et al.,* 2006). Biological weed control practices have been developed for the sustainable use of biodiversity for economic benefit towards mankind. The idea of using plant pathogens for management of weeds was reported before the turn of century, but it is only in the last three decades that has received increasing interest (Charudattan 2001; Zhang *et al.,* 1996: Te Beest, 1992,1996).Various practices such as applying farming methods, planting resistant cultivars and application of herbicides have been suggested for controlling weeds (Motlagh, 2011). Currently the possibility of using indigenous plant pathogens as biological control agents to control these *Echinochloa* spp. in rice is being investigated (Rezvani *et al.,* 2002).

Biological control of weeds is an alternative approach, utilizing living organisms to control or reduce the population of undesirable weed species (Huang *et al.,* 2001). The classical approach with exotic plant pathogens to control weeds was developed in the beginning of the 1970s (Alan, 1991). An alternative approach to bioherbicide development is based on the idea that an endemic (*i.e.,* native) pathogen might control its weed hosts through a massive dose of inoculum at susceptible stages of weed growth (Yandoc-Ables *et al.,* 2006). Inoculative biological control and bioherbicides approach are the two steps in the microbial management of invasive weeds that have been applied successfully (Johnson *et al.,* 2002,Boyetchko and Rosskoff, 2006). Much research on the development of new mycoherbicide has been conducted during two past decades worldwide.

Some fungal species have been reported as candidates for biological control of barnyard grass *E. crusgalli* (L.) P. Beauv.(Safari motlagh,2010) *Exserohilum monoceras* (Drechsler) Leonard and Suggs, *E. rostratum* (Drechsler) Leonard and Suggs, *Curvularia lunata* (Wakker) Boedijin, *C. aeria* (Wakker) Boedijin, *Colletotrichum graminicola*, *Pyricularia grisea* (Ces.) Wils and *Ustilago trichophora* (Link) Körn. Tsukamoto *et al.* (1997) have been shown to have some promise as microbial agents for biocontrol of barnyard grass. *Exserohilum monoceras* has been evaluated as a potential bio herbicide for the control of *Echinochloa* species (Zhang and Watson 1996). In addition, the potential of *Cochliobolus lunatus* which induces necrosis on barnyard grass resulting in death of young seedlings has been reported as a biological control agent against this weed. However, no mycoherbicide has been registered at commercial level for biocontrol of barnyard grass.

From both practical and economic perspectives, biocontrol fungi that sporulate in liquid culture are favoured over those that require additional steps to induce sporulation. This factor alone has proved to be advantageous for the commercial development of a fungus as a mycoherbicide (Bowers, 1986). Nutrient balance can play an important part in sporulation of fungi in submerged culture. Studies with *Colletotrichum truncatum* have shown how carbon concentration and carbon to nitrogen (C: N) ratio influence propagule production (Jackson and Bothast 1990, Kim *et al.,*

2005). Moreover, a defined amino acid composition of the N source improved production of conidia. In addition, spore fitness in terms of germination and appressoria formation rate and subsequent disease production (Schisler *et al.,* 1991) was influenced by C: N ratios. Since infection, disease development and subsequently weed control efficacy of pathogens are usually dependent on the inoculum quantity is essential to determine optimal cultural conditions for conidial yield that should be overcome or bypassed for effective field control of *Echinochloa* spp. by the fungus *F. proliferatum.* Carbon and nitrogen sources have been recognized as two important parameters that contribute to the efficacy of a bio herbicide (Cook *et al.,* 1965). Various studies of the effects of different carbon and nitrogen sources on the efficacy of commercial and other potential bio herbicides have been reported and provided. Also, the role of nutritional factors on sporulation and virulence of two *F. oxysporum* isolates were studied as an effective factor in controlling *Orobanche aegyptiaca* (Ghotbi *et al.,* 2008). In another study fungal pathogens such as *F. oxysporum* and *Colletotrichum coccodes,* which are considered as good mycoherbicides were modified with soft genes that encode pectinase, cellulase and expansin, that is enzymes which facilitate the penetration and growth of fungi inside weed tissues (Hershkovitz *et al.,* 2007). In this study, *F. proliferatum* was isolated and to evaluate as biological control agent against weed by the application of massive dosage of inoculum, experiment was conducted to study the effect of culture media and environmental factors on the conidial yield of the fungus.

Materials and Methods

Fusarium proliferatum Strain and Maintenance

The *Fusarium proliferatum* strain previously isolated from infected leaves of *Echinochloa crusgalli* were used for this study (Jyothi *et al.,* 2010). This *Fusarium proliferatum* strain was maintained on PDA slants at 4°C for further use.

Optimization of Fermentation Process

The growth and conidial yield of fungal pathogen *F. proliferatum* was optimized by using different cultural parameters like Media, pH, Temperature, Carbon and Nitrogen sources.

Effect of Culture Media

To screen out the most suitable media for the mycelia growth of the pathogenic fungi, following eight culture media were tried. Each culture medium was prepared in 1 liter of water and autoclaved at 120°C at 15psi for 20min. These was cooled to 45°C and poured in 9cm Petri-dishes for solidification, the plates were inoculated with 5mm mycelial discs from the 7 day old culture of *F.proliferatum* grown on PDA and incubated at 25±2°C. Potato dextrose agar (PDA) medium (peeled and boiled potato starch-200g, Dextrose-30g, Agar-20g), Cornmeal-agar (CMA) (Cornmeal-20g, dextrose-20g, Agar-20g), malt-extract agar (MEA) (Maltextract-30g, Peptone-5g, Agar 20g), Yeast dextrose agar (YDA) (Yeastextract-15g, Dextrose-20g, Agar-20g), Sabourads maltose agar (SMA) (Peptone-10g, maltose-40g, Agar-20g), oat meal agar (OMA) (Oat meal-60g, Agar-20g), Richards-agar (RAM) (Potassium-nitrate-10g, Potassium-hydrogen orthophosphate-5g, Magnesium-sulphate-2.5g, Ferric-chloride-Trace,

Sucrose-30g, Agar-20g), Czapekdox-agar (CDA) (sodium-nitrate-3g, Potassium-chloride-0.5g, Magnesium-sulphate-0.25g, Di-potassiumhydrogenphosphate-1.0g, Ferrous-sulphate-0.01g, Sucrose-30g, Agar-20g).

Effect of different Carbon and Nitrogen Sources

Nutritional Study

The utilization of carbon and nitrogen nutrition was studied by replacing the sucrose and potassium nitrate in the basal medium with various nitrogen and carbon compounds. In these experimental studies Richards's medium was used as the basal medium for studying carbon and nitrogen. Fifty milliliters of the medium dispensed in 250ml conical flasks were sterilized and used for inoculation with the fungus. All the flasks were inoculated with 5mm of mycelia disc obtained from 7 day old cultures of *F.proliferatum* and incubated at 28±2°C for seven days. The cfu/ml was estimated using the Haemocytometer.

Nitrogen Sources

Four different nitrogen sources like yeast extract, peptone, sodium nitrate and ammonium nitrate was added to the Richard's broth in the place of potassium nitrate prior to autoclaving and the quantity of nitrogen sources added was determined on the basis of their molecular weights as described by Lily and Barnet (1951). Sucrose was added as the carbon source in all the treatments. Richard's broth without adding potassium nitrate was used as the control. All flasks were inoculated with 5mm of mycelia disc of seven day old fungal culture under aseptic condition and incubated for seven days. After screening of suitable nitrogen source, it was tried with different concentrations to screen out the amount of nitrogen to be added for one Liter media for commercialization process. The different nitrogen concentrations tried were ranges 0.5 per cent to 3 per cent.

Carbon Sources

Carbon compounds tested in the study were sucrose, dextrose, maltose and starch. Richard's medium without adding sucrose was used as control. Potassium nitrate was used as a source of nitrogen for all treatments. The quantity of carbon compound added was determined on the basis of their molecular weights as described by Lily and Barnet (1951).Different concentrations of carbon source that was screened were tried that ranges 1 per cent to 5 per cent.

Effect of Temperature

The fungal pathogen was subjected to different temperature conditions to study the best suited temperature level for the growth and conidial yield of the fungus. Richard's agar medium was used in this experiment to study the growth and conidial yield in solid medium. Twenty milliliters of Richard's agar medium was poured into each Petri plate under aseptic condition and inoculated with five mm of mycelia agar disc from seven day old culture plates and incubated at 15, 20, 25, 30, 35,40°C. The colony diameter and conidial yield were recorded seven days after incubation.

Effect of different pH Levels

Effect of pH on the growth of fungus was tested in vitro using liquid cultures containing different pH levels. Richard's broth medium was used to study the effect of pH of medium on the growth and sporulation of fungus. Fifty milliliters of liquid medium was poured into a 250ml conical flask under aseptic conditions. The reaction of the medium was adjusted to the different pH levels like 4, 5, 6, 7, 8 and 9 respectively by adding 0.1N NaOH or 0.1N HCl. Flasks were sterilized at 121°C at 15psi for 20min. Each flask was inoculated with 5mm of mycelia disc from 7-8 days old agar plates. Inoculated flasks were incubated at 25+2°C for seven days.

Bioassay Studies

A single batch of seeds of each of *E. crusgalli* and paddy collected from field populations was used in all experiments. Seeds were incubated in Petri dishes on moistened filter paper at room temperature for 48 h. Five germinated seeds (coleoptile and radicle visible) were planted in 10-cm-diameter plastic pots filled with water-saturated soil (Sand:clay:manure at 1:2:1). Seeded pots were placed on benches in the greenhouse, and a 2 to 3cm water level was maintained throughout the experimental period. Greenhouse conditions were 25 ± 2°C day/night temperature, a 12-h photoperiod, and average light intensity of 350µE s–1 m–2. Each treatment consisted of inoculating four pots of each of *Echinochloa* and rice (each pot containing five seedlings at the 2-4 leaf stage) with a conidial suspension at a rate of 10^5, 10^6, 10^7 conidia/m^2 containing 0.05 per cent Tween 20 as a wetting agent using hand sprayer. It should be mentioned that before inoculation, all pots were sprayed with distilled water. To create a relative humidity higher than 90 per cent, treated plants were immediately covered with plastic bags for 48 h (Ghorbani *et al.,* 2000). Evaluation was done at different intervals after inoculation based on lesion type and size in reaction to inoculation: 0 = lesions absent, 1 = small, unexpanded lesions, 2 = slightly to moderately expanded lesions, 3 = large lesions (Zhang *et al.,* 1996). Therefore, standard evaluation system and Horsfall- Barratt system were applied for *Echinochloa* spp. (Zhang *et al.,* 1996; Bertrand and Gottwald, 1997)

$$\text{Disease rating} = \frac{(N_1 x1) + (N_2 x2) + \dots \dots (N_t xt)}{(N_1 + N_2 + \dots \dots N_t)}$$

Where, N is number of leaves in each of rate, t is number of treatments.

Stastical Analysis

Data obtained from all the experiments were analyzed by analysis of variance ANOVA using SPSS, version 17.0. Least significance difference (LSD) at 5 per cent level of significance (P = 0.05) was used to compare the mean values of different treatments in the experiment.

Results and Discussion

Effect of Culture Media

The results of the experiment revealed that, Richard's agar and potato dextrose agar media were the best for the radial growth of *F. proliferatum* f. sp. *Echinochloa* isolates with a mean maximum growth of 84.5 and 81.6 mm, followed by Czapek's dox, malt extract agar and yeast extract agar with mean maximum growth of 70.25, 68.0, 71.69 mm respectively (Table 24.1). These results were in confirmation with

Ingole (1995), Gangadhara *et al.* (2010) who reported that PDA and Richard's agar supported best mycelia growth of *F. udum* and *F. oxysporum.* Jamaria (1972) also reported maximum growth and sporulation of *F. oxysporum* f. sp. *niveum.* on potato dextrose agar, Richard's agar and Czapek's dox agar. Anjaneya Reddy (2002) observed maximum growth of *F. udum* on Richard's agar and potato dextrose agar. Different synthetic and non synthetic media has profound effect on the cultural and morphological characteristics of fungus (Shaikh 1974).

Table 24.1: Effect of different Culture Media on Growth of *F. proliferatum*

Culture Media	Mycelial Growth (mm)
Potato dextrose agar (PDA)	81.60
Cornmeal agar (CMA)	50.25
Malt extract agar (MEA)	68.00
Yeast dextrose agar (YDA)	71.69
Sabourad maltose agar (SMA)	61.20
Oat meal agar (OMA)	57.80
Richards gar (RMA)	84.50
Czapexdox agar (CDA)	70.25
CD (P≤0.05)	8.95
CV (per cent)	1.21

Effect of different Carbon and Nitrogen Sources

The results of this experiment indicated that all the carbon sources were suitable for the growth of the *F. proliferatum.* However sucrose was found to be the best carbon source for the growth of fungus followed by dextrose and maltose with mean maximum spore count of 3.53×10^7 respectively, after seven days of inoculation (Tables 24.2 and 24.3). These findings are in conformity with the reports of (Gangadhara *et al.,* 2010) that sucrose was the suitable carbon source for the growth of *F. proliferatum.* Similar observations were made by (Khanzada *et al.,* 2003) for the mycelial study of *Macrophomina phaseolina.* Hussain *et al.* (2003) obtained the maximum growth when glucose and peptone were used as carbon and nitrogen sources respectively. The carbon use by every fungus is depend on the availability of sugar in the substrate (Carlile and Watkinson, 1994), and also determined by the structure and the sugar configuration besides the availability of fungal enzymes (Lilly and Barnett, 1951). Sucrose is a disaccharide which consists of glucose and fructose (Lilly and Barnett, 1951). Disaccharide use by cell metabolism is determined by the sugar residual of carbon source, the closer their configuration to glucose, the easier they are being used by fungi (Lilly and Barnett, 1951).

As it is evident that, out of four different forms of nitrogen sources utilized, yeast extract was found to be the best nitrogen source for the growth and conidial yield of the fungus. Sodium nitrate also had stimulatory effect on the growth of the fungus. However the least growth was observed with the ammonium nitrate followed by peptone powder. Sajidfarooq *et al.* (2005) obtained the maximum growth of *F.oxysporum* f.sp.ciceri when Glucose and peptone were used as carbon and nitrogen sources, respectively. Organic nitrogen in form of amino acids also contain element of carbon which may also be used by the fungi. This may explain the better result achieved on organic nitrogen sources compared to inorganic nitrogen (Carlile and Watkinson, 1994). The addition of carbon sources from amino acids support the growth of the fungi, therefore more biomass production can be achieved (Griffin, 1981; Lilly and barnett, 1951).

Table 24.2: Effect of different Carbon and Nitrogen Sources on Growth of
***F. proliferatum* in Richards Broth**

Carbon Source	Conidial Yield (10⁷ cfu/ml)	Nitrogen Source	Conidial Yield (10⁷ cfu/ml)
Sucrose	3.53	Yeast extract	4.03
Dextrose	3.03	Peptone powder	1.20
Maltose	2.63	Sodium nitrate	3.40
Starch	1.53	Ammonium nitrate	2.36
CD (P≤0.05)	1.26	CD (P≤0.05)	2.34
CV (per cent)	0.90	CV (per cent)	1.34

Table 24.3: Effect of different Concentrations of Sucrose and Yeast Extract on
Growth of *F. proliferatum* in Richards Broth

Sucrose Conc. (per cent)	Conidial Yield cfu/ml (10⁷)	Yeast Extract conc. (per cent)	Conidial Yield cfu/ml (10⁷)
1 per cent	1.2	0.5 per cent	2.3
2 per cent	7.5	1.0 per cent	6.8
3 per cent	12.0	1.5 per cent	13.0
4 per cent	4.2	2.0 per cent	9.4
5 per cent	0.76	2.5 per cent	3.2
		3.0 per cent	0.65
CD (P≤0.05)	0.37	CD (P≤0.05)	0.31
CV (per cent)	4.00	CV (per cent)	2.37

To determine the definite amount of carbon and nitrogen sources various concentrations were tried of which Sucrose at 3 per cent concentration yielded the highest conidial yield of 12.0×10^7 followed by 2 per cent with 7.5×10^7.least growth was observed with high amount of carbon source and poor growth at high concentrations. Yeast extract at 1.5 per cent concentration yielded the highest conidial yield of 13.0×10^7, whereas least growth at lower conc. of nitrogen source. Pathogens to be used as a successful mycoherbicide has to produce abundant inoculum in artificial media and conidial viability during storage conditions are essential. Carbon and nitrogen sources ratio play an important role in the growth and development of fungus (Jackson and Bothast, 1990).

Effect of Temperature

Differences in the growth rate of the fungus *F.proliferatum* f.sp. *Echinochloa crusgalli* were recorded at different temperature levels *viz.*, 15 °C, 20 °C, 25 °C, 30 °C, 35 °C and 40 °C. The maximum growth of the fungus was observed in the temperature range of 25 °C. Slight growth was observed between 15-20 °C and there was a drastically decrease of growth with the temperature range of 30-35 °C and no growth was observed at 40 °C (Table 24.4). Sajidfarooq *et al.* (2005) reported similar findings regarding the requirement of temperature for this fungus. These studies are in conformation with Gangadhara *et al.* (2010) who reported that the growth of *F.oxysporum* was maximum at 25 °C. In vitro studies conducted by Chi and Hansen (1964) indicated that *F. solani* isolates grew well at higher temperature of 28 °C. Gupta *et al.* (1986) reported similar findings regarding temperature requirements to this fungus. Soil temperature relationship indicated that suitable temperature for the development of chickpea wilt is 25-30 °C (Chauhan, 1965)

Table 24.4: Effect of different Temperature and pH Levels on Growth of
***F. proliferatum* in Richards Broth**

Temperature	Conidial Yield (10^7 cfu/ml)	pH levels	Conidial Yield (10^7 cfu/ml)
15°C	1.20	4.0	0.40
20°C	3.20	5.0	3.20
25°C	3.66	6.0	3.63
30°C	2.40	7.0	2.86
35°C	1.40	8.0	1.20
40°C	0.05	9.0	0.01
CD (P≤0.05)	1.20	CD (P≤0.05)	1.15
CV (per cent)	0.87	CV (per cent)	0.46

Effect of different pH Levels

Differences in the growth rate of *F. proliferatum* were recorded at different pH levels. Growth of the fungus was observed at all the pH levels, but maximum growth was attained at pH range 5.0-6.0. No growth was observed at the pH below 4.0 and least growth was observed above 7.0-8.0 and completely declined at pH 9.0 (Table 24.4). The studies conducted by Jamaria (1972) on *F. oxysporum* f.sp. *nivium* indicated that, as the pH decreases or increases from the optimum, the rate of amount of growth gradually decreases. This fungus can tolerate a wide range of pH 5.0-6.5 (Shaikh, 1974).

Bioassays

Bioassay results revealed that weed seedlings applied with 10^7 conidial/ml showed 100 per cent disease mortality than 10^6, 10^5 conidia/ml. No disease symptoms were observed on rice seedlings treated with conidial suspension (Table 24.5). The disease symptoms were observed on weed seedlings after 48hrs of inoculation, light yellowing of leaf was started and later necrosis followed by complete wilting of leaves 14 DAT (Figure 24.1). In this research, many *Fusarium* species were isolated from rice paddy weeds; however, a limited number of isolates had high disease ratings in weeds. Results of researches conducted by Charudattan *et al.* (2005) were consistent with this fact that of 71 *Fusarium* isolates obtained from eight weeds in

Figure 24.1: Application of *Fusarium proliferatum* Spore Suspension onto Rice, Weed Seedlings under Greenhouse Conditions.

Control (Barnyard grass)

Diseased weed seedlings of 14DAT with 10^7 conidia/ml

Contd...

Figure 24.1–*Contd...*

Control (Paddy)

Treated Rice Seedlings without Infection after 14 DAT.

wheat and soybean fields, only six fungi had the capability of causing high disease rating in weeds. This finding could be relevant to the domain of fungi hosts which affect weeds (Charudattan *et al.,* 2005). Research done by Safari Motlagh *et al.* (2011b) revealed that *F. equiseti* was effective in controlling *the S. trifolia and Echinochloa* spp. at a conidial concentration of 10^6 conidial/ml on the height and fresh weight of weed seedlings. Here, the weed was more susceptible to the fungus during its early growth stages in a way that the high disease rating of this fungus in these growth stages of this weed resulted in the positive effect of the fungus on the studied traits.

Table 24.5: Bioefficacy of *F. proliferatum* Conidial Suspension on Target Weed and Host Crop

Conidial Conc.	Seedlings	Days after Inoculation and Disease Score		
		5	7	14
10^5	Weed	0.7	1.5	2.0
	Rice	0.0	0.0	0.0
10^6	Weed	1.2	2.5	3.0
	Rice	0.0	0.0	0.0
10^7	Weed	2.5	3.0	4.5
	Rice	0.0	0.0	0.0

Conclusion

This study concludes that *F. proliferatum* isolated from *E. crusgalli* can be used as biological agent for controlling *E. crusgalli* weed in rice. This is the first report on identification of *F. proliferatum* from rice weed. Further studies on characterization of toxins produced by *F. proliferatum*, development of formulations, and field evaluation are under progress.

Acknowledgements

We gratefully acknowledge the DBT for financial support under small business innovative research initiative (SBIRI) programme.

References

Alan KW. The classical approach with plant pathogens. In: *Microbial Control of Weeds* (ed. D.O. TeBeest) Chapman and Hall, London. pp. 3–23 (1991).

Alam MS, Begum MF, Sarkar MA and Islam MR. 2001. Effect of temperature, light and media on growth, sporulation, formation of pigments and pycnidia of *Botryodiplodia theobromae*. *Pak. J. Biol. Sci.* (4): 1224–1227.

Anjaneyareddy B. 2002.Variability of *Fusarium udum* and evaluation of Pigeon pea (*Cajanus cajan* (L). Mills) genotypes. M.Sc. (Agri.) Thesis. *Univ. Agril. Sci.* Bangalore. pp.115.

Bowers RC. 1986. Commercialization of Collego– An industrialist's view. *Weed Sci.* *34*: 24–5.

Boyetchk S and Rosskopf EN. 2006.Strategies for Developing Bioherbicides for Sustainable Weed Management,In: Handbook for Sustainable Weed Management, Singh HP, DR Batish and RK Kohli(Eds).Haworth press,Inc, New York,USA. pp. 393–420.

Bertrand PF and Gottwald TR. 1997. Evaluation fungicides for pecan disease control. In: Hickey KD(ed) Methods for Evaluating Pesticides for Control of Plant Pathogens. 2nd ed. APS Press, pp. 149–164.

Charudattan R. 2001. Biological control of weeds by means of plant pathogens: significance for integrated weed management in modern agroecology. *Bio Control*. 46: 229–260.

Charudattan R, Shabana YA, Devalerio JT and Rosskopf EN. 2005. Broad–spectrum bioherbicide to control several species of pigweeds and methods of use. U. S. Patent, 5393728.

Chauhan SK. 1965. The interaction of certain soil conditions in relation to the occurrence of Fusarium wilt of gram. *Indian J. Agric.* Sci 35: 52.

Chutia M, Mahanta JJ, Saikia, R, Boruah AK.S and Sarma TC. 2006. Effect of leaf blight disease on yield of oil and its constituents of Java Citronella and *in vitro* control of the pathogen using essential oils. *World J. Agric.Sci.* 2: 319–321.

Chi CC and Hansen EW. 1964. Relation of temperature, pH and nutrition to growth and sporulation of *Fusarium* spp. from red clover. *Phytopathology* 54: 1053–1058.

Cook RJ and Schroth MN.1965. Carbon and nitrogen compounds and germination of chlamydospores of *Fusarium solani* f. *phaseoli. Phytopathology* 55: 254– 256.

Carlisle MJ and Watkinson SC.1994. The Fungi. Academic Press, New York. 482 pp.

De Datta SK.1981. Principles and Practices of Rice Production. John Wiley and Sons, New York.618pp.

Ghotbi M, Amini Dehaghi M, Montazeri M, Ghotbi MA and Kambouzia J. 2008. Comparison of two liquid Media in increasing virulence and desiccation tolerance of two isolates of Fusarium oxysporum for biocontrol of Broomrape (Orobanche spp.) In Proceeding at 5th International Weed Sciense congress, Vancouver, BC University, Canada, 23–27July, p. 56.

Gangadhara Naik B, Nagaraja R, basavaraja MK and Krishna Naik R.2010. Variability studies of *Fusarium oxysporum* f.sp. *Vanillae* isolates *I.J.S.N.* Vol.1 (1): 12–16.

Gupta O, Khare MN and kotasthane SR.1986. Variability among six isolates of *Fusarium oxysporum* f.sp.*ciceri* causing wilt of chick pea. *Indian Phytopathol.* 39: 279.

Griffin DH. 1981. Fungal Physiology, John Wiley and Sons, New York, 102–145, 195–201, 209–239.

Ghorbani R, Seel W, Litterick A and Leifert C. 2000. Evaluation of *Alternaria alternata* for biological control of *Amaranthus retroflexus.Weed Sci.* 48: 474–480.

Hallett SG.2005. Where are the bioherbicides? *Weed sci.*53: 404–415.

Hussain A, Iqbal SM, Ayub N and Haqqani AM.2003. Physiological studies of *Sclerotium rolfsii. Pakistan J. Plant Pathol.* 2: 102–106.

Huang SW, Watson AK, Duan GF and Yu LQ. 2001. Preliminary evaluation of potential pathogenic fungi as bioherbicides of barnyardgrass (*Echinochloa crusgalli*) in China. *International Rice Research Institute Notes*.26(2): 36–37.

Hershkovitz MS, Larroch C, Al–Ahmad H, Amsellem Z and Gressel J.2007. Enhancing mycoherbicide activity. Novel and Sustainable Weed Management in Arid and Semi–Arid–Ecosystems, Rehovot, Israel.

Ingole MN. 1995. Estimation of losses, variability among isolates and management of pigeon pea wilt caused by *Fusarium udum Butler.* M.Sc. (Ag.) Thesis, Dr. PDKV, Akola, pp.146.

Jackson MA and Bothast RJ.1990. Carbon concentration and carbon to nitrogen ratio influence submerged culture conidiation by the potential bioherbicides *Colletotrichum truncatum* NRRL 13737. *Appl. Env. Microbiol.* 56: 3435–3438.

Jamaria SL.1972. Nutritional requirement of *Fusarium oxysporum* f.sp. *niveum. Indian Phytopath.* 25: 29–32.

Johnson RL, Blossey B. 2002 U.S. Dept.Agr.ForestService, Morgantown, WV, USA, Biological control of invasive plants in the Eastern United States, FHTET–2002–04, pp. 79–90.

Jyothi G, Vijayavani S, Reddy KRK and Sreenivas V. 2010. Pathogenicity of *Fusarium oxysporum* and *Curvularia lunata* as a mycoherbicide for the control of *Echinochloa crusgalli* (Barnyard grass). *J Biopest*.3(3): 559 –562.

Kim YK, Xiao CL and Rogers JD. 2005. Influence of culture media and environmental factors on mycelial growth and pycnidial production of *Sphaeropsis pyriputrescens. Mycologia.* 97: 25–32.

Khanzada SA, Iqbal SM and Haqqani AM. 2003. Physiological studies on *Macrophomina phaseolina. Mycopathol* 1: 31–34.

Lilly VG and Barnet HL. 1951. Physiology of fungi. McGraw Hill, book companies Inc.New york.pp.464.

Montazeri M, Mojaradi M and Rahimian–Mashhadi H. 2006. Influence of adjuvant on spore germinaton, desiccation tolerance and virulence of *Fusarium anthophilum* on barnyard grass (*Echinochloa crusgalli*). *Pak. J. Weed Sci. Res.* 2(1–2): 89–97.

Montazeri M and Greaves MP. 2002. Effects of culture age, washing and storage conditions on desiccation tolerance of *Colletotrichum truncatum* conidia. *Biocontr. Sci. and Technol.* 12: 95–105.

Rezvani A, Izadyar M and Faghih A. 2002. A guide to pests, diseases and rice weeds. The ministry of agricultural crusade publication. *J. Org.Agric. Res. Train.* 5: 1–5.

Shaikh MH.1974. Studies on wilt of gram (*Cicer arietinum* L.) caused by *Fusarium oxysporum* f.sp. *Ciceri* in Marathwada region.M.Sc. Thesis, Marathwada Krishi Vidyapeeth, Parbhani, India.

Sajidfarooq SH, Muhammad Iqbal H and Abdul rauf. 2005. Physiological studies of *Fusarium oxysporum* f.sp. *ciceri. Int. J. Agri. Biol.* 7: 275–277.

Safari Motlagh MR.2010. Isolation and characterization of some important fungi from *Echinochloa* spp. the potential agents to control rice weeds. *Austr. J. Crop Sci.* 4(6): 457–460.

Safari Motlagh and Armin Javadzadeh. 2011. Study of the reaction of major weeds and some rice cultivars to *Fusarium equiseti. J. of Medicinal plants* Vol.5 (24).pp.5796–5802.

Schisler DA, Jackson MA and Bothast RJ. 1991. Influence of nutrition during conidiation of *Colletotrichum truncatum* on conidial germination and efficacy in inciting disease in *Sesbania exaltata. Phytopathol.* 81: 458–461.

Tamuli P and Boruah P. 2000. Biological control of *Rhizactonia solani* Kuhn. In: aromatic Cymbopogonus by *Trichoderma* sp. *Plant Arch.* 2: 77–80.

TeBeest DO, Yang XB and Cisar CR. 1992. The status of biological control of weeds with fungal pathogens. *Ann. Rev. Phytopath.* 30: 637–657.

Turnera RJ, Davies G, Moore H, Grundy AC and Mead A. 2007. Organic weed management: A review of the current UK farmer prespective. *Crop Prot.* 26: 377–382.

Tsukamoto H, Gohbara M, Tsuda M and Fujimori T. 1997. Evaluation of fungal pathogens for biological control of paddy weeds, *Echinochloa* sps. by drop inoculation method. *Annual Phytopathol Soc.* 63: 366–372.

TeBeest DO. 1996. Biological control of weeds with plant pathogens and microbial pesticides. *Adv. Agron.* 56: 105–113.

Yandoc –Ables CB, Rosskopf EN and Charudattan R. 2006. Plant pathogens at work: Progress and possibilities for weed biocontrol. Part 1: Classical vs. bioherbicidal approach. APSnet Feature Story Aug.

Zhang WM, Moody K and Watson AK. 1996. Responses of *Echinochloa* species and Rice (*Oryza sativa*) to indigenous pathogenic fungi. *Plant Dis.* 80: 1053–1058.

2015, Recent Trends in Microbiology, Mycology and Plant Pathology   *Pages* 373–388
Editor: **Dr. H.C. Lakshman**
Published by: **DAYA PUBLISHING HOUSE, NEW DELHI**

Chapter 25

Importance of Plant Disease Forecasting and Disease Management

*Romana M. Mirdhe and H.C. Lakshman**

*Microbiology Laboratory, P.G. Department of Studies in Botany,
Karnatak University, Dharwad – 580 003
E-mail: dr.hclakshman@gmail.com

Introduction

Plant disease forecasting is a management system used to predict the occurrence or change in severity of plant diseases. At the field scale, these systems are used by growers to make economic decisions about disease treatments for control. Plant disease forecasting systems (synonym: plant disease warning systems) have been developed to help growers make economic decisions about disease management. The forecasting of plant disease has become considerably important in the present day situation in India as well as in other agriculturally important countries of the world due to large scale adoption of new agricultural strategies for increasing food production.

Forecasting systems are based on assumptions about the pathogen's interactions with the host and environment, the disease triangle. The objective is to accurately predict when the three factors–host, environment, and pathogen–all interact in such a fashion that disease can occur and cause economic losses. The ultimate aim of plant science pathology is to have a complete understanding of plant diseases- host, pathogen and environmental interactions- so that it becomes possible to control diseases economically. The practical implications of forecasting have been incorporated in definition: " Forecasting involves all activity in ascertaining and notifying the owners of a community that conditions are sufficiently favourable for

Human

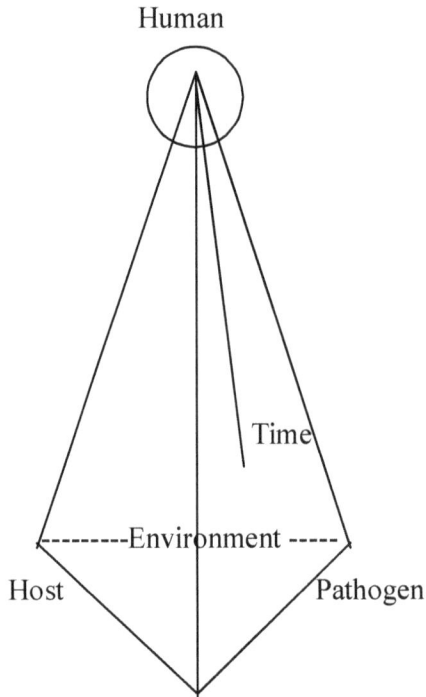

Figure 25.1: A Disease Pyramid Showing the Interrelationships of the different Components of Disease.

certain diseases that application of control measures will result in economic gain, or on the other hand, and just as important that the amount of disease expected is unlikely to be enough to justify the expenditure of time, energy and money for control". The economic value of forecasting is obvious enough. Some consideration should be given, however, to its position as a part of the science of plant pathology. It is this inherent constructive feature that makes the development of forecasting one of the most interesting and rewarding phases of plant pathological investigation.

Development of Forecasting System

The prediction of plant diseases has emerged as a well established component of epidemiology that is rapidly being incorporated into disease management. The mathematics of the disease progress has matured to a point of becoming a powerful and respected component in the management and prediction of epidemics. However, many models for prediction of plant diseases are theoretical and have not proven useful for disease management. The assumption is that a disease prediction model should make projections of the main events in the development of diseases, which most models do not (Seem, 2001). This would be especially valuable for disease management if models would eliminate unnecessary pesticide applications and reduce production cost.

To develop a forecasting system a complete knowledge of on the factors affecting disease development should be available. Conceptual models are the first step in model-building and aim at sensibly arranging available information or processing through critical analysis. Traditionally, plant disease models have used leaf wetness duration (LWD) combined with temperature, to predict infection and colonization, and then identify the risks of an epidemic. These types of models have been used with observed climate records to track the favorable periods, indicating tactics or strategies of control (Jabrzemski and Sutherland, 2006). In some cases, inoculum levels can be incorporated into disease models (Peres *et al.,* 2002), but with many diseases, inoculum is in excess and not useful in disease prediction (Bhatia *et al.,* 2003; Biggs and Turechek, 2010). Weather is an important factor that influences disease development.

Computer Based Models

The availability of computers has allowed pathologists to write programmes that allow stimulation epidemics of several diseases. Revolutions in web-based technologies are bringing major changes in the development and use of decision-support systems by producers and specialists in the management of plant diseases (Fernandes *et al.,* 2007).

☆ TOM- CAST (Pitalbo, 1992)- Modern *et al.* (1978) developed a computer based forecasting system for tomato early blight.

☆ PLASMO (Rosa *et al.,* 1993)- This is a forecaster developed in Northern Italy to control downy mildew of grapes.

☆ BLIGHTCAST (Krause *et al.,* 1975)- for late blight of potato.

In a computer stimulation of an epidemic, the computer is given data describing the various sub components of the epidemic and control practices. The computer then provides continous information regarding not only the spread and severity of the disease over time, but with regard to the final crop and economic losses likely to be caused by the disease under control conditions of disease epidemics.

Computer stimulation of epidemics is fairly used in plant disease forecasting systems. Such systems allow farmers to take appropriate control measures against a disease as soon as the conditions that are likely to lead to disease development.

Conceptual Models

Conceptual models are the first step in model building and aim at sensibly arranging available information or processing thoughts for critical analysis. The conceptual model show the consequence of operations and decision making level. Weather is an important factor that influences disease development where host and pathogen coincide, weather is the only variable that influences epidemic development.

☆ Everding's method (1926)- was the first to base his forecasting system for late blight of potato. His forecast was based on the following rules;

1. Minimum temperature (night temperature must not be lower than 10°C)

2. Dew (leaves must remain wet for atleast 4hours the day)

3. Cloudiness (the day following the night dew must be cloudy)

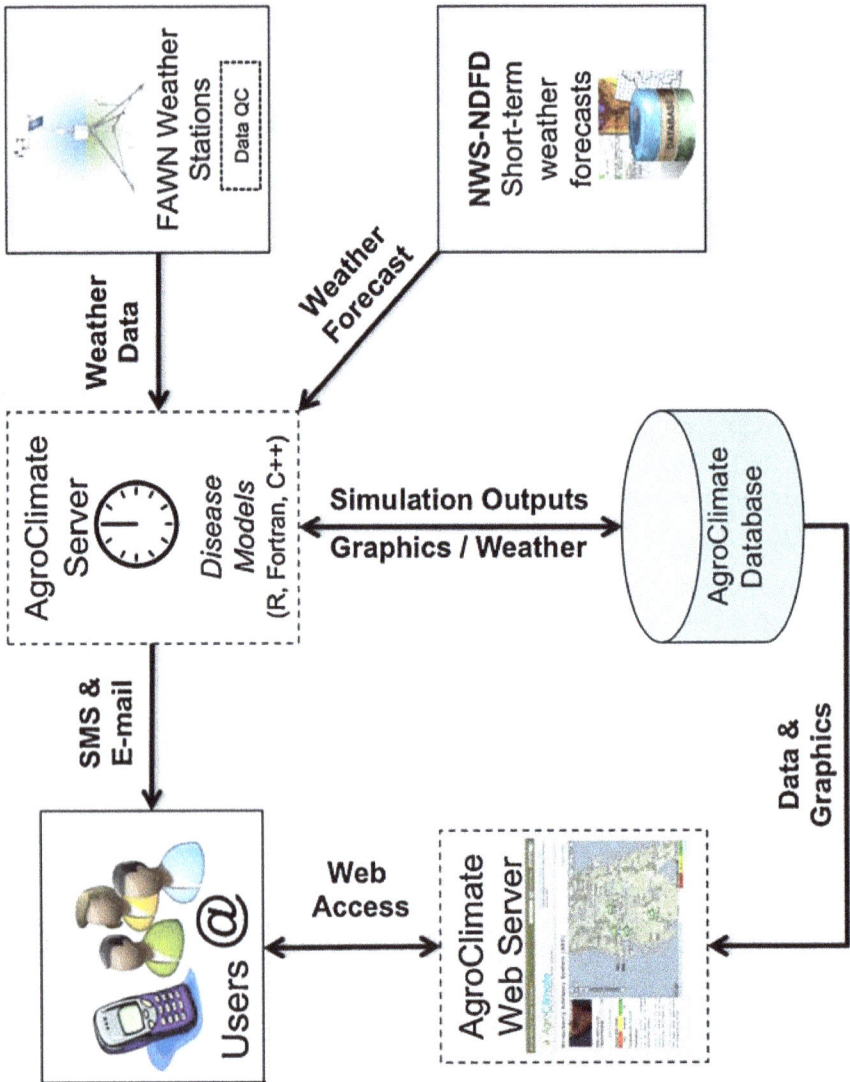

Figure 25.2: A Generalized Computer Based Model (*Source:* Pavan *et al.,* 2010).

4. Rainfall (During such periods described above there must be atleast 0.1mm of rainfall)

☆ Reaumont and Stainlands method (1934, 1938)modified the rules for England

1. Minimum temperature 10°C

2. Relative humidity of 75 per cent or above for 2 consecutive days

The blight could be expected approximately 10 days afterward. These periods are called Beaumont's periods are the basis for forecasting system in England and Wales.

☆ Hyre's method (1954)- Developed a criteria for the north eastern region of the U.S.A which were essentially the same as Cook's expecting that a 10-day moving graph method was followed and which according to him was more accurate than the 7-day moving graph method.

☆ Wallin and Hoyman's method- the forecasting system was very much successful in Eastern United states for prediction of blight in advance and development did not seen to be applicable in the Midwest (U.S.A).

☆ Hyre and Bonde's method (1955)- this is a modification of Cook's (1949) and Hyre's (1954) method of 7-day moving graph in which a 10- day rainfall and a moving 10-day average mean temperature are considered suitable. In this case also the critical rainfall line is calculated on the basis of total crop season as above.

Methods Used in Plant-Disease Forecasting

Monitoring the weather is the most important consideration in disease forecasting; because of the overriding effect that weather has on disease development. While broad scale weather data has been used for disease forecasting, it is well known that microclimate within the crop has a more impact on disease. Devices have been developed to monitor microclimate factors such as duration of leaf wetness and temperature, and with time, they will be affordable and accurate enough for widespread use on individual farms.

In developing a plant disease, one must take into account several factors of the particular pathogen, host and environment. The methods used in disease forecasting mostly fall into 5 main groups based on:

1. Weather conditions during the intercrop months particularly those affecting the survival of the inoculam.

2. Weather conditions during the crop.

3. The amount of disease in the young crop.

4. The number of propagules of the pathogen in the air, soil or planting material.

5. Changes in the physiology.

Biological Characteristics of a Fungal Pathosystem

Once the objective of a model is clearly established, the second step is usually to determine the variables that are needed for developing the model. These variables generally represent key features in the development of epidemics, including initial inoculum, progeny/parent ratio and the length of latent period (Van der Plank, 1982) as well as major external factors. Recently, de Vallavieille-Pope *et al.* (2000) assessed the use of various epidemiological parameters in modelling.

Fungal Pathogens

A. Inoculum

Fungal inoculum is of prime concern; its source, density and type will greatly inuence the design of the forecasting scheme. Without inoculum, there is no epidemic. A fungal population consists of individuals at various stages of their life-cycle. The population can be described by the proportion or absolute quantity of individuals at each stage, *i.e.* age structure of thepopulation (Shaw, 1998).

B. Inoculum Dispersal

Inoculum dispersal fulls essentially three functions: (1)population survival, (2) colonisation of new habitats and (3) reproduction (Ingold, 1971). Dispersal can occur either by mycelium or spores. Spore dispersal comprises three phases; liberation, which can be passive or active (Dix and Webster, 1995), transport, and deposition. The dispersal scale depends on inoculum properties as well as the transport vector, and may range from a few metres through rainsplash (Madden, 1992), to 100–10 000 m for airborne spores, such as those of powdery mildew of cereals (Andrivon and Limpert, 1992).

C. Latent and Infectious Periods

In epidemiology, the latent period is the interval between the onset of spore germination and the appearance of the next spore generation. The rate of epidemic development is largely influenced by the length of latent period, which determines the number of potential infection cycles that can be completed during a growing season (de Vallavieille-Pope *et al.*, 2000). The shorter the latent period, the more reproduction cycles the fungus can have per season. In contrast to polycyclic diseases, monocyclic diseases have only one reproductive cycle throughout a single season.

D. Pathogen Dynamics Regulation

Brasier (1990) showed that the transfer of virus pathogens of fungi would occur more efficiently when the fungal population is at a high density. Another natural regulation comes from pathogen population competition, *e.g.* the competition between the eyespot pathogens *Tapesia yallundae*, *T. acuformis* and the sharp eyespot pathogen *Rhizoctonia cerealis* (Bateman *et al.*, 1995). This aspect of microbial community interaction (symbiosis or competition) is, in general, poorly understood for fungal pathogens; it is now becoming gradually more important with the present moves to more integrated disease management strategies.

Environmental Factors

Wind and rain are essential for pathogen dispersal; rain provides free water on host surfaces for most pathogens to infect and sporulate and sun provides favourable temperatures for disease development (Lacey, 1996). The duration of each event as well as its timing is also important. For example, for apple scab, night rain results only in a much smaller proportion of mature ascospore being discharged compared to daytime rain (MacHardy, 1996). Prediction of actual wetness duration is preferable to prediction of occurrence because many pathogens cause more damage as the duration of wetness increases (Hosford *et al.,* 1987; Francl and Panigrahi, 1997). Artificial neural network models have been developed to predict wetness on wheat flag leaves from both dew and rain (Francl *et al.,* 1995). The relationships between disease development and environmental factors are the key component and often the only component of disease forecasting systems. Both past and future weather forecasts can be used in these systems for predicting epidemic development. Forecasting systems provide an indication or quantification of disease development, especially when the disease is likely to exceed an economic-injury threshold and thus warrants a treatment.

Conditions for Successful Forecasting

1. When the number of parameters necessary to accurately infection increases the probability of making accurate forecast decreases. Less number of parameters should be used.

2. The infection rate of the pathogen should not exhibit variability each year. Then only the rate can be used to predict future epidemic development.

3. Forecasting system should be developed only for the disease which is not controlled by genetic resistance and which can completely destroy a major food crop. It should also be amenable for chemical control.

Plant Disease Management

Plant diseases have caused severe losses to humans in several ways. The goal of plant disease management is to reduce the economic and aesthetic damage caused by plant diseases. Traditionally, this has been called plant disease control, but current social and environmental values deem "control" as being absolute and the term too rigid. More multifaceted approaches to disease management, and integrated disease management, have resulted from this shift in attitude, however. Single, often severe, measures, such as pesticide applications, soil fumigation or burning are no longer in common use. Further, disease management procedures are frequently determined by disease forecasting or disease modelling rather than on either a calendar or prescription basis. Disease management might be viewed as proactive whereas disease control is reactive, although it is often difficult to distinguish between the two concepts, especially in the application of specific measures.

1. Cultural Methods (Stevans, 1960)

a. *Avoidance of pathogen-* Many plant diseases can be prevented by proper selection of land or field, choice of time of sowing and selection of varieties, seed and planting stock, and by modification of cultural practices.

b. *Selection of field-* The selection of suitable area or field for cultivation is very important from the point of view of better yields as well as protection of the crop from the ravages of pathogens.

c. *Choice of time of sowing-* Pathogens are able to infect susceptible plants only under certain environmental conditions. Therefore the crop must be sown inn the period when the pathogens are less susceptible.

d. *Disease escaping varieties-* Certain values of crops escape damage because of their growth characters. These varieties escape the onslaught of the pathogens and resist the attack due to their inherent characteristics.

e. *Selection of seed and planting stock-* Since many plant propagate by vegetative parts, the selection of disease free seeds and indirectly controlling disease.

All the host plants herefore infected by or suspected to harbour the pathogen are to be removed and burnt. This result in elimination of the pathogen and preventing greater loss from the spread of the pathogen to more plants. Host eradication should be done routinely in nurseries, green houses and fields. Some pathogens of annual crops overwinter or oversummer mainly in the perennial wild plants. Eradication of such hosts helps in eliminating the inoculam. Similarly eradication of alternate hosta helps control of disease as *Puccinia graminis tritici, Cronartium ribicola.* This eradication is used in (i) eradication of overwintering/oversummering hosts, (ii) eradication of alternate hosts, (iii) eradication of diseased plants etc. Removal and destruction of plants as they become diseased is also called roguiging. This is useful in viral diseases.

Sanitation includes all activities aimed at eliminating or reducing the amount of inoculam present in a plant, field, or warehouse and at preventing the spread of the pathogen to other healthy plants and plant products. Sanitation means destruction of diseased plant material whether in the form of plant parts in the field or crop residues after harvest. Crop rotation is an old practice where sources of primary inoculam structures are eradicated. Soil pathogens whose host range is narrow can be controlled by this method. Cultural practices designed to improve the plant vigour often help increase its resistance to pathogens. Thus proper fertilization drainage of fields, irrigation, proper spacing of plants, proper method and deapth of sowing, date of sowing, avoiding the injury to plant improve the plant growth.

2. Exclusion of Inoculam

These methods aim at preventing at preventing new pathogens and diseases from reaching an uninfected area and avoiding contact between the pathogen and the crop of field.

a. *Seed treatment-* Seeds, tubers, grafts, bulbs and other propagating material can be given heat, gas or chemical treatment to keep then free of pathogens.

b. *Inspection and certification-* The crops grown exclusively for seeds are inspected periodically for the presence of disease that are disseminated by seeds, and necessary precautions are taken to remove the diseased plants.

c. *Quarantine regulation-* A quarantine can be defined as a legal restriction on the movement of agricultural commodities for the purpose of exclusion, prevention or delay in the spread of plant pests and diseases in uninfected areas. The agricultural pests and disease act of different states prevent interstate spread of pests and pathogens. The plant quarantine stations are at work at major sea airport. Some of the worst plant disease epidemics that have occurred throughout the world. *e.g.* downy mildew of grape in Europe, the bacterial citrus canker, the chestnut blight, the dutch elm disease in the U.S.A are all introduced from other countries. Pathogens are likely to be introduced in or on live plants, seeds, tubers and other planting materials, in imported grain, fruit, vegetables and other food stuffs, in plant material, in industry (as cotton fibres etc), in plant material used in packing, or in soil or from disease free areas. Even if the material is already inspected by the quarantine services of the exporting country, it should be rechecked at the point of import. Special precautions are needed when cultures of pathogenic fungi, bacteria, etc. are exchanged for scientific purposes. According to gram (1960) " Professors are among the best quarantine breakers".Several voluntary inspection and certification systems are also in effect in various areas in which appreciable amounts of nursery stock, potato seed tubers etc. are produced. Growers interested in sale of these materials submit for voluntary inspection and/or indexing of their crop in the field and in the storage by a regulatory agency. Some woody plants are attacked by non-vectored viruses and the crop plants attacked by vectored viruses, mycoplasmas, fungi, bacteria and nematodes can be protected from disease by this nematode. It must be ensured that the propagating materials (seeds,tubers, grafts, root stocks, corms, cuttings, rhizome etc) are free from pathogens.

Eradication of Pathogens

These measures primarily aim at breaking the "infection chain" by removing the foci of infection and starving the pathogen.

a. *Rouging-* This practice entails careful removal the "foci" of infection and preventing wide dissemination of the pathogen.

b. *Eradication of alternate or collateral hosts-* Many plant pathogens complete their life cycle on two hosts; one of them is generally a wild plant.

c. *Crop rotation-* Very often, continuous cultivation of the same crop or related or related crops lead to perpetuation of a pathogenic soil- borne fungus and the gradual increase in intensity of the disease.

d. *Sanitation-* Like plant sanitation field sanitation is necessary for control of many plant diseases.

e. *Heat and chemical treatments-* The most important application of hot air treatments is to eliminate vegetatively propagated material. Treating the seeds with fungicides, especially organomercurials, is a common practice to remove the superficially-borne inoculam.

Biological Control

In a healthy, balanced ecosystem, biological control by natural predators is constantly occurring. The more diverse a cropping system becomes, the greater the spectrum of insect species and micro-organisms within it. This leads to the development of more natural predators within the ecosystem. Ladybugs, ambush bugs, hoverfly larvae, lacewings, spiders, birds, frogs, toads and a host of other insects are predators of aphids, bertha armyworm larvae, sunflower beetles, beet webworms, and both grasshopper eggs and adults. The destructive wheat midge may also be partially controlled by a parasitic wasp, but crop damage may still occur. Various types of fungi are insect parasites and can either kill their insect hosts or reduce their ability to reproduce. Very few biological controls are available to reduce the effects of plant diseases, as most commercial products do not perform well if the disease is already established in the crop. Mycoparasitism is a form of bio-control where one fungus parasitizes another.

Biological pest control provides an important route to environmentally harmless plant protection (Diercks. 1983). This route reduces risks, e. g. of becoming aware too late of unfortunate side effects of plant protection agents. Thus there are examples ofside effects offungicides which, applied regularly, wipeout the ground beetle population of a field within 12 years (Basedow, 1991). Intense use of active substances can be harm beneficial organisms and sometimes even promote pests. In apple orchardsit was quickly found that a benzimidazole fungicide kills earthworms in the soil and hence indirectly encourages infestation with apple scab in the following year (Niklas and Kennel, 1981). Unfortunately, market for the so called "biological" agents is much disorganised. They are generally sold as "plantstrengthening agent" without further testing of their efficacy. Adverse side effects against useful micro-organisms, which also occur here, are neglected or remain undiscovered. The legislator should play a guiding role here so that the introduction of biological control systems, which in practice encounters many obstacles and requires a great deal of persuasion (Albert and Meinert, 1991), can also be successful. An effective system which has demonstrated its advantageous influence on the balance of nature should be appropriately promoted by the legislator (Langenbruch and Huber, 1990) by, for example, adapting the registration to the requirements in biological pest control and perhaps even simplifying them.

There are situations where biological methods can be used to eradicate or reduce the pathogen inoculam. These are as follows:

1. *Suppressive soils*- Several soil borne pathogens, such as *Fusarium oyxsporium, Phtyoptora cinnamomi* and *Heterodera avenae* thrive cause diseases in some soils, called conducive soils, whereas they develop much less and cause much milder diseases in other soils known as suppressive diseases. A range of hyperparasitic and antagonistic microbes have been found to increase in suppressive soils, most of which are fungi, such as species trichoderma, pencillum, or bacteria of genera *pseudomonas*. Adding suppressive soils containing these microbes to areas of conducive soil reduces the disease. Continuous cultivation of same crop on conducive

soil reduces disease naturally. In some diseases, crop rotation helps in control of disease through encouragement of such. Hyperparasites attack mycelium and resting spores – oospores, sclerotia, of the target pathogen.

2. *Trap plants*- Rye, corn or other tall plants, planted around fields of bean, peppers or squash act as trap plants for aphid –borne viruses of beans, peppers etc. Viruses are non-persistent and many aphids feeding on trap plants loose the viruses infecting beans, peppers, squash etc.

Selection of Biocontrol Agents for Plant Disease Control

There is little difference, in principle, between screening chemicals and testing biological agents for disease control. Both selection processes require an appropriate assay system to detect activity against the target pathogen, or pathogens.

Screening for biocontrol agents should, therefore focous on specific natural sources chosen to optimize the changes of isolating strains with the correct biological properties. Random screening of various micro-organisms for suppressive effects on plant disease has often been used to select the most promising strains for further development. The success of screening should be aimed on rational approach based on ecology. For effective results, a biocontrol agent (BCA) should be able to colonize a particular habitat, or occupy a specific niche, in introducing a randomly selected microbial antagonist, it would be better to introduce one known to be adapted to the habitat concerned. Hence the best place to look for BCAs is in the specific environment in which they are to be used. For example, if the target is a pathogen, which infects plant roots, the logical place to look for an antagonist is in the rhizosphere. Any micro-organism isolated from this habitat is likely to have biological and physiological attributes enabling successful multiplication in the root zone, and therefore rhizosphere component. Similar concepts apply to leaf surfaces or ther substrates such as straw where a particular stage in the pathogen life cycle might be disrupted.

Production and Formulation of Biocontrol Agents

BCAs are originated from living organism and usually needs to be metabolically active and effective. The commercialization of a BCA is the major difficult steps including its production, packing and delivery in sufficient quantities to the agent in a viable and stable form. The other problems are as follows:

☆ Variation is a normal feature of micro-organism, but each batch needs to have similar activity. This can be particular problem in scaling–up production, as microbial populations can change their properties during growth in fermenters. The BCA then has to be harvested and distributed in a formulation, which ensure viability.

☆ BCA need to have a reasonable shelf life so that it can be stored for a period without significant loss of activity.

☆ It needs to be applied to the crop, or into soil, in a way which ensures that the antagonist grows and persists in the environment for sufficient time to exert control.

Examples of Successful Biocontrol Agents

BCAs have been used or are being used on a commercial sale to control plant pathogens. *Bacillus thuringiensis* – about 100 species of bacteria infect insects. Nearly all the entomapathogenic bacteria are from the class schizomycetes. Although species of both sporulating and non-sporulating bacteria are considered potential candidates for the development into bacterial insecticide, more success has been achieved with spore formers, especially species of the genus *Bacillus* family bacillaceae.

Mode of Action of Biocontrol Agents

The action of BCAs involves protection of an infection court by prior treatment with a microbial antagonist. For effective suppression of the pathogen establishment of a metabolically active threshold population of the BCA at the infection site is required. BCAs act against pathogens in soil or on plant debris and affect the survival of resting structures or propagules, rather than preventing infection. The exact mechanism of antagonism is still to be established.

Suppression by BCAs may be due to occupation of a particular niche by the BCA. It leads to physical exclusion of the pathogen, or to competition for essential nutrients. At one time it was believed to control of heterobasidion by peniphora was due to this type of effect. Now it is suspected that there is more direct interactions between the hypahe of the two fungi called hyphal interference.

Environment Interactions

☆ Competition for macronutrients

☆ Siderophores

☆ Antibiotics and other nutrients

☆ Alterations in pH

Antibiotics have also been shown to be important in the biological control of root infecting fungi including take-all. Siderophores are low molecular weight compounds. They have high affinity for iron and aid transport into cells. These chemicals are efficient scavengers of iron and thus may mop up all of the available supply in the immediate environment. Many pathogens require iron as an essential mineral nutrient for growth, and in some cases iron is required for virulence. Hence, production of a siderphore by BCA may reduce the growth of a pathogen, of its ability to attack the host. Florescent pseudomonas produce several siderphores such as the pigmented compound pyoverdin, and the most convincing evidence that these contribute to disease control again comes from studies on non-producing mutants, which are less effective than wild type strains.

Induction of host resistance- BCAs can act by supression of disease through the induction of host defence mechanisms. It has been known for many years that inoculation of plants with a virulent strains of pathogens, or agents causing necrosis, can trigger both local and systematic resistance. More recently, it has been shown that rhizobacteria applied to seeds or roots can induce a systematic resistance response expressed against pathogens infecting with aerial tissues. For example, treatment with *Pseudomonas* spp. Increases resistance to vascular wilt fungus *Fusarium oxysporum*

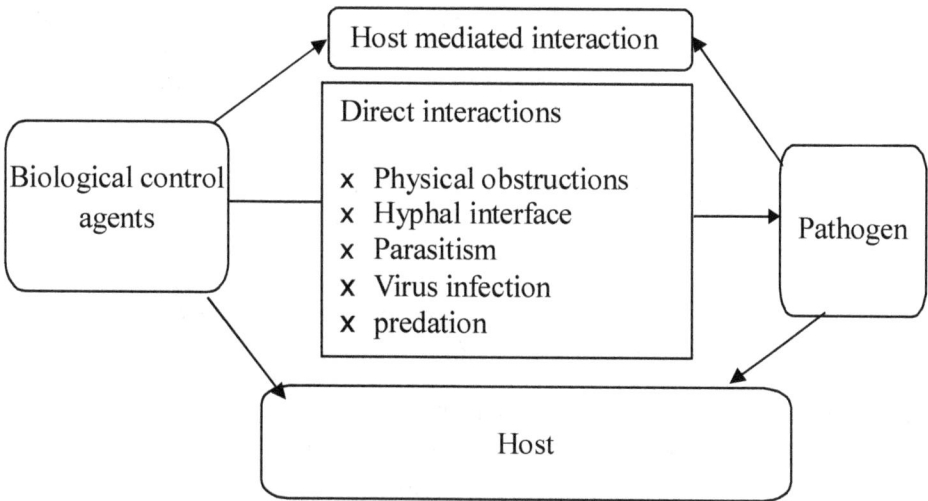

Figure 25.3: Modes of Action of Microbial Biocontrol Agent.

in carnation, leaf pathogens in bean. This resistance appears to be associated with the induction of pathogenesis related (PR) proteins as seen in a typical systemic acquired resistance (SAR) response. Presumably the colonizing BCA produces signal molecules, which activates the SAR pathway. However, the specific mechanism of induction is still not clear.

Contrians for Production of Biocontrol Agents

Control of plant diseases using BCAs has been a disappointed.

Bacterial Antagonists

Several bacteria of different genera could protect plants from their respective pathogens. Some examples are given below

a. *Seed bacterization-* Treatment of seeds of cereals, sweet corn, with water suspensions, slurries or powders containing bacteria, *Bacillus subtilius* strain A13 or *streptomyces* sp protected the plants from root pathogens. *Pseudomonas flourescens* group bacteria reduced soft rot of several plants when these were applied to seeds, seed pieces and roots.

b. *Post harvest diseases-* Stone fruits like peach, apricot, plum etc. treated after harvest with suspension of the bacterium, *Bacillus subtilius* remained free of brown rot caused by *Monilinia fructicola.*

c. *Biocontrol of crown gall-* Crown gall of several fruits caused by *Agrobacterium tumefaciens* is successfully controlled by treating seeds, seedlings and cuttings with a suspension of strain K84 of related but non-pathogenic bacterium, *Agrobacterium radiobacter*. This produces an antibiotic, agrocin 84.

d. *Biocontrol of bacteria-*mediated frost injury- Some of the phylloplane bacteria, such as *Pseudomonas syringe* and *Erwinia herbicola* act as ice- nucleation-active (INA) catalysts which cause ice-formation inside the cells of their host plants at as high temperature as -1°C, causing frost injury can be prevented.

e. *Biocontrol of diseases in the phylloplane-* Several saprophytic gram- negative bacteria of the genera, *Erwina, Pseudomonas* and *Xanthomonas* and some gram-positive genera, such as *Bacillus*, and *Corynebacterium* are very common in the phylloplane of plants. In several cases spraying leaves with of these bacteria protected them from infection by pathogenic bacteria like *Pseudomonas syringae, Erwina amylovora*.

Integrated Disease Management

In order to manage a disease or pest, let us recall the disease pyramid with its three principal components – the host, pathogen and the environment. None of these can operate alone to cause a disease. Management should tackle all the three at same time instead of only one of them. In control, usually this is one of these three which is tackled at a time. For management approach, instead there should be integration of all the methods directed in favour of host, against the pathogen, and for modification of environment. This is what is now known as integrated disease or pest management (IPM). Management of the host involves the practices directed to improve plant vigour and induce resistance through cultural practices and breeding for disease resistance. Management of the pathogen involves the practices directed to eradicate or reduce the inoculam or preventing it through legislation. Direct protection of plant surface by chemicals also discourage the pathogen. Management of environment involves water management, soil management and crop management.

An integrated control program can be developed against all diseases affecting the crop, *e.g.* apple, citrus, banana, potato etc. Sometimes an integrated control program is designed against a particular destructive and common disease e.g apple scab, late blight of potato etc. For an orchard crop as apple, peach, one must take into account the following (i) disease free nursery stock is to be taken, (ii) for possibility of presence of nematodes, stock must be fumigated, (iii) the location where trees will be planted must not be infested with *Phytopthora, Armillaria.* If any it is to be fumigated before planting the rootstocks (iv) proper drainage of the area must be done, and improved, (v) young trees should not be planted between or next to old trees that may be carrying canker fungi, bacteria, (vi) once the trees begin to bear fruit, they should be fertilized, irrigated, pruned and sprayed for the most common insects and diseases.

In an integrated control program of an annual crop *e.g.* potatoes one must again start with (i) healthy stock, free from diseases (viruses, late blight fungus, ringrot bacteria and other fungi), (ii) this is to be planted in a field free of old tubers (that may harbour some of the above pathogens) (iii) it is better to rotate with legume, corn or other crop, (iv) the soil is to be fumigated and field to be properly drained (v) a few weeks after young plants, appear, they are protected from blight (*Alternaria*) and late blight (*Phytopthora*) through foliage sprays throughout the season.

Conclusion

Operation of the services is simple: close watch is kept over all the conditions affecting infection and spread of the parasite, and growers are warned in time to take appropriate action. Benefits from the warning services are increased efficiency and reduced cost of control because the forecasts allow growers to time their operations properly or to omit them when not needed. In general, the predictions are based on known facts about abundance and source of inoculum; environmental requirements of the parasite for infection, development and reproduction; manner of disease spread in relation to growth stages of the host; plus weather forecasts to show probable trend of disease development. In some cases formulae and charts have been devised for speedy translation of observed factors into disease prognostication. For diseases transmitted by insects, the source and abundance, manner and direction of spread, and feeding preferences of the vectors are considered, as well as environmental conditions affecting them in all their phases. In the biological control of plant diseases, the coincidence of pathogen and useful micro-organism constitutes a particular problem. The harmful micro-organism generally has an advantage over the useful one. The useful micro-organism therefore has to be applied in large amounts, and the growth conditions should be optimised for it. In addition to these difficulties, living micro-organisms can give rise to allergies and often eliminate toxic metabolic products, Hence, the importance of biological control of plant diseases is always restricted to a few special cases where the biology of the pathogen and of the useful micro-organism have been very carefully investigated beforehand.

References

Andrivon D and Limpert E. 1992. Origin and proportions of the components of composite populations of *Erysiphe graminis* f.sp. *hordei. Journal of Phytopathology* 135: 6–19.

Bhatia A Roberts PD Timmer LW.2003. Evaluation of the alter–rater model for timing of fungicide applications for control of alternaria brown spot of citrus. *Plant Disease* 87 (9): 1089–1093.

Biggs AR Turechek WW.2010. Fire blight of apples and pears: epidemiological concepts comprising the maryblight forecasting program. *Plant Health Progress* http://www.plantmanagementnetwork.org/sub/php/research/2010/fire.

de Vallavieille–Pope C Giosue S Munk L Newton AC Niks RE Ostergard H Pons–Kuhnemann J Rossi V and Sache I.2000. Assessment of epidemiological parameters and their use in epidemiological and forecasting models of cereal airborne diseases. *Agronomie* 20: 715–727.

Dix NJ and Webster J. 1995. *Fungal Ecology*. Chapman and Hall, London, New York, USA.

Francl LJ Panigrahi S Pahdi T Gillespie TJ and Barr A.1995. Neural network models that predict leaf wetness. *Phytopathology* 85: 1128.

Hosford RM Jr Larez CR and Hammond JJ.1987. Interactions of wet periods and temperature on *Pyrenophora tritici–repentis* infection and development in wheat of different resistance *Phytopathology* 77: 1021–1027.

Ingold CT.1971. Fungus Spores: Their Liberation and Dispersal. Clarendon Press, Oxford.

Jabrzemski R Sutherland A. 2006. An innovative approach to weather–based decision–support for agricultural models. In: 22nd International Conference on Interactive Information Processing Systems for Meteorology, Oceanography, and Hydrology. American Meteorological Society, Washington, DC, USA.

Kruase RA and Massie LB. 1975. Predictive systems: Modern approaches to disease control. *Ann. Rev.Phytopath* 13: 31–47.

Lacey J.1996. Spore dispersal – its role in ecology and disease: The British contribution to fungal aerobiology. *Mycological Research* 100: 641–660.

MacHardy WE.1996. Apple Scab: Biology, Epidemiology, and Management. *American Phytopathological Society*, St. Paul, MN, USA.

Madden LV. 1992. Rainfall and dispersal of fungal spores. *Advances in Plant Pathology* 8: 29–79.

Niklas J and Kennel W. 1981. The role of the earthworm. Lumhr;Cl/s terrestns(L.)in removing sources of phytopathogenic fungi in orchards. *Gartcnbauwissenschaft* 46: 138–142.

Oicrcks R. 1983. Alternativen im Landbau. Stuttgart. Eugen Umer. Fanirher. M., Kast, W.K., and GroBmann. F. (1991): Alternative Spritzfolgen im Test. Weinwirtschaft Anbau 5,12–15.

Pitbaldo RE. 1992. The development and implementation of TOM–CAST, Ontario ministry of agriculture and food, Toranto, Canada.

Rosa M R Genesia B Gozzini G Maracchi and S Orlandini. 1993. PLASMO: a computer programme grapvine downy mildew development forecasting. *Comput. Electron. Agric.* 12: 311.

Seem R.2001. Plant disease forecasting in the era of information technology. In: Plant Disease Forecast: Information Technology in Plant Pathology, Kyongju, Republic of Korea.

Sharma P D. 1999. Microbiology and plant pathology. Rastogi Publications. 64–81 pp.

Shaw MW.1998. Pathogen population dynamics. In: Jones DG (ed) The Epidemiology of Plant Diseases (pp 161–180) Kluwer Academic Publishers, Dordrecht, the Netherlands.

Van der Plank JE.1982. Host–Pathogen Interactions in Plant Disease. Academic Press, New York, USA..

Index

A

Actinomycetes, 9, 67, 230

Actinophages, 17

Aflatoxin, 190

Anchorage, 4

Antagonist, 246, 254, 259, 279, 437, 382

Antibiosis, 250

Apprasorium, 261, 270, 350, 351

Arbuscular mycorrhizal fungi, 12, 18, 27, 28, 40, 47, 48, 52, 95-102, 143-145, 250, 276, 277, 341

B

Bacetriophages, 17

Bioagents, 245

Bioassays, 368

Biofumigation, 343, 345

Bioremediation, 61-66, 68, 79

C

Cephaloporins, 221, 222

Chitosan, 325, 328, 329, 331, 333, 334

Citrinin, 191

Commensalism, 268

Cyanobacteria, 12, 13, 14, 16, 105, 112

Cyanophages, 17

E

Echinochloa, 359-361, 363, 366-369

Epilepsy, 155, 156

Eumycota, 177, 178

F

Forecasting, 373-376, 378

Frankia, 10

Fumonisis, 192

H

Heterotrophic, 9

Hydrocarbons, 65-67

Hyperparasite, 250, 268

L

Latex, 286, 287, 288, 291

M

Meloidogynae, 261, 343

Mushroom, 121-123, 124-127, 153-155, 160, 164-168

Mycoparasitism, 269, 281

Mycoplasma, 300, 313, 314, 320-322

Mycotoxins, 188, 189

Mycoviruses, 253

N

Neutralism, 268

O

Ochratoxins, 191

Oomycota, 177

Organic farming, 340

P

Parasitic, 299, 315

Pectinases, 177-209, 352

Pharmaceuticals, 213, 217-219, 230

Phyllosphere, 250

Phyllosphere, 350

Phytotoxic, 326

Polyamines 43

Proliferation, 4

R

Rhizobium, 20

Rhizopshere, 5,6,27, 31, 35, 95, 250, 254

Rhizoremediation, 5

S

Salicylic acid, 43

Siderophore, 259, 326, 384

Single Cell Protein, 115-121, 127-132, 342

Sustainability, 4

T

Therapeutics, 220

Trichoderma, 83-91, 96-102, 251, 253, 256, 257, 259, 260, 266, 267, 270, 275, 276, 286, 290, 291, 292